Construction Hazard and Safety Handbook

Other Butterworth books on Safety

A Safe Place of Work, D.W.B. James

Industrial Hazard and Safety Handbook, Revised impression, Ralph W. King and John Magid

Safety at Work (Second edition), John Ridley

Redgrave's Health and Safety in Factories—Second edition, His Honour Judge Ian Fife and E. Anthony Machin

Health and Safety at Work, His Honour Judge Ian Fife and E. Anthony Machin

Occupational Health Practice, Richard Schilling

Lifting Tackle Manual, D.E. Dickie and E. Short

Crane Handbook, D.E. Dickie

Handbook of Reactive Chemical Hazards (Second edition), L. Bretherick

Electrical Safety Engineering (Second edition), W. Fordham Cooper

Construction Hazard and Safety Handbook

R W King, BSc, CEng, FIChemE
R Hudson, MIOSH

Butterworths
London   Boston   Durban   Singapore   Sydney   Toronto   Wellington

All rights reserved. No part of this publication may be reproduced or transmitted in any form or by any means including photocopying and recording, without the written permission of the copyright holder, application for which should be addressed to the publishers. Such written permission must also be obtained before any part of this publication is stored in a retrieval system of any nature.

This book is sold subject to the Standard Conditions of Sale of Net Books and may not be re-sold in the UK below the net price given by the Publishers in their current price list.

First published 1985

© Butterworth & Co (Publishers) Ltd, 1985

British Library Cataloguing in Publication Data
King, R.
  Construction hazard and safety handbook.
  1. Building—Safety measures.
  I. Title   II. Hudson, R.
  624′.028′9   TH438
  ISBN 0–408–01347–8

Library of Congress Cataloging in Publication Data
King, Ralph W. (Ralph William), 1918–
  Construction hazard and safety handbook.
  Bibliography: p.
  Includes index.
  1. Building—Safety measures.   I. Hudson, R. (Roland)
  II. Title.
  TH443.K375   1985   624′.028′9   85-3762
  ISBN 0–408–01347–8

Typeset by Scribe Design, Gillingham, Kent
Printed in Great Britain by Page Bros Ltd, Norwich, Norfolk

Foreword

V.C. Jordan, JP, LLB, MIOSH, Barrister
Formerly HM Deputy Chief Inspector of Factories and Chairman of the Construction Industry Advisory Committee

Year in and year out, in both the industrialized and the developing world, the construction industry produces a toll of fatal and serious accidents far greater than that of any other manufacturing industry. Where the traditional heavy industries have declined the contrast has become even more marked.

Construction accidents by their nature usually occur singly and to those employed within the industry so that they attract but a fraction of the public concern which is felt about major industrial catastrophes. Yet in some post-war decades as many have died in construction accidents in Britain alone as perished at Bhopal.

Informed pressure for improved occupational health and safety will, it is hoped, continue to find expression in improved national and international legislation and standards. However, with only some 100 inspectors of factories in Britain concerned directly with the construction industry it is to the industry itself rather than Government that we must look for real and lasting improvement.

That is why this book is both useful and timely. It is useful in that it contains a fund of knowledge and sound advice on the hazards of construction and the achievement of a safe and healthy site. It is timely because as the industry changes its character with an increasing number of individuals and small firms of sub-contractors operating outside the scope of the professional safety and training services of the best of the larger firms, the need for such knowledge to be compressed within the covers of an easily accessible reference work becomes more pressing. Effective self-regulation depends upon adequate knowledge.

In devoting as much attention to occupational health as to safety the authors strike an unusual note. Because of the difficulties of monitoring health in a largely transient workforce it has been something of a Cinderella in a Cinderella industry. However, expanding public awareness of such issues as the carcinogenic properties of commonplace substances such as asbestos and the hearing loss associated with the high levels of noise produced by some construction plant and equipment are transforming the scene.

The introduction of new chemical processes and techniques demands an increasingly professional approach to the control of the working environment, but everyone who works in construction has some interest in the subject matter of the book. There is need for management to provide the spur and incentive for better performance, to appoint and support the safety and health professionals who will interpret the needs of the firm and offer sound advice, and for site management with the co-operation of the trade union safety representatives to see that the advice is put into practice.

Two hundred years ago Dr Johnson had the following to say about the construction industry:

'Were you to tell men who live without houses, how we pile brick upon brick, and rafter upon rafter, and that after a house is raised to a certain height a man tumbles off a scaffold and breaks his neck, they would laugh heartily at our folly in building, but it does not follow that men are better without houses.'

May this book prove an antidote for folly.

Preface

The seeds for this book were sown in 1982 when one of us, Ralph King, was asked by the International Labour Office, Geneva, to draft a report for the tenth session of its Building, Civil Engineering and Public Works Committee on *'The Improvement of Working Conditions and of the Working Environment in the Construction Industry'*. Part of this report is reproduced as Chapter 4.3 'The future'. Feeling that there was a real need for a comprehensive book dealing with the varied hazards faced by workers in different trades and occupations in the construction industry, Ralph started work on the book early in 1983. He was joined in this task later in the year by Roland Hudson who has many years of varied experience as a safety professional in the construction industry. Both authors have had previous books published by Butterworths. *'Industrial Hazard and Safety Handbook'* by King and Magid in 1979 and *'Mobile Crane Manual'* by Dickie and Hudson in 1985.

Rather than taking the usual approach of concentrating on safe working methods and procedures, we have focused our attention on the hazards to safety and health of workers in the wide range of construction trades and occupations. We have devoted special attention to the numerous and varied health hazards facing most construction workers which in the past have been mainly treated on a piecemeal basis in specialised books and journals. Although the treatment of illnesses caused by these hazards lies with doctors, the causes and their alleviation, e.g. solvent vapours, harmful dusts, excessive noise and exposure to bad weather lie largely within the control of the industry itself.

In dealing with mobile plant, cranes and lifting appliances, we have stressed the role of training and of characteristic hazards such as noise and vibration. We have not attempted to give detailed guidance on the mechanical systems involved or their safety. These are dealt with in considerable detail in three other recent Butterworth publications, *'Lifting Tackle Manual'*, *'Crane Handbook'* and *'Mobile Crane Manual'*.

Since Appendix B to this book was written, the following Guidance Notes on demolition and steel erection have been issued by the Health and Safety Executive.

Demolition GS 29
Part 1. Preparation and Planning
Part 2. Legislation
Part 3. Techniques
Part 4. Health Hazards

Safe Erection of Structures GS 28
Part 1. Initial Planning and Design
Part 2. Site Management and Procedures

Parts 3 and 4 will have been issued by the time this book is published.

We regret that it was too late to refer to these Guidance Notes in the relevant sections of this book, although most of the main points in them are found in these pages.

We are indebted to many firms, organizations and individuals for illustrations, information and constructive criticism for which we hope we have made appropriate acknowledgement in the text, but to any whom we have failed to acknowledge, in the complex task of assembling a text such as this, we offer our apologies. We also express our thanks to our typists, wives and families who have put up with all the inconvenience and stress of working and living with this book and its authors for many months.

Ralph King
Roland Hudson

Contents

Introduction 1

Part 1   Scope, activities and safety aspects of the industry 7

Chapter 1.1   The scope of construction and its hazards 9
    1.1.1 Building 10
        1.1.1.1 High-rise buildings 11
        1.1.1.2 Prefabricated buildings 12
        1.1.1.3 Typical houses in industrialized countries 13
        1.1.1.4 Third World shanty towns 13
        1.1.1.5 Primitive rural dwellings 16
    1.1.2 Civil engineering 17
    1.1.3 Process and power plant construction 17
    1.1.4 Offshore oil and gas installations 18
    1.1.5 Demolition 18

Chapter 1.2   Trades and activities in construction 20
    1.2.1 Trades and numbers employed 20
    1.2.2 Comprehensive list of occupations 24
    1.2.3 More recent data 27

Chapter 1.3   Safety and health statistics 30
    1.3.1 Comparative international fatal injury statistics in construction 34
    1.3.2 Accidental injury statistics in construction in the UK 36
    1.3.3 Statistics of occupational diseases in construction in the UK 39
    1.3.4 Risk comparison with other activities 41

Chapter 1.4 General trades and their hazards 43
    1.4.1 Bricklayers and allied crafts 43
        1.4.1.1 Bricklayers 43
        1.4.1.2 Stonemasons 44
    1.4.2 Carpenters, joiners and allied crafts 45
        1.4.2.1 Carpenters and joiners 45
        1.4.2.2 Formworkers (shuttering carpenters) 45
        1.4.2.3 Shopfitters 46
        1.4.2.4 Woodworking machinists 46
    1.4.3 Painters and decorators 47
        1.4.3.1 Painters 49
        1.4.3.2 Decorators 49
    1.4.4 Plumbers and allied craftsmen 49
        1.4.4.1 Plumbers 49
        1.4.4.2 Heating, ventilating and air-conditioning engineers 49
        1.4.4.3 Gas services engineers 50
        1.4.4.4 Welders and gas cutters 50
    1.4.5 Electricians 51
    1.4.6 Plant operators and related trades 52
        1.4.6.1 Pile driving operators 52
        1.4.6.2 Slingers 53
    1.4.7 Concreters and plasterers 53
        1.4.7.1 Concreters 53
        1.4.7.2 Plasterers 53
    1.4.8 Glaziers and cladders 54
        1.4.8.1 Glaziers 54
        1.4.8.2 Cladders 54
    1.4.9 Wall and floor tilers 54
    1.4.10 Lift engineers 55
    1.4.11 Other categories 55
        1.4.11.1 Storekeepers 55
        1.4.11.2 Labourers 55
        1.4.11.3 General workers 55

Chapter 1.5 High risk occupations 57
    1.5.1 Steel erection 57
        1.5.1.1 Falling 59
        1.5.1.2 Instability of partially erected structures 60
        1.5.1.3 Falling material 60
        1.5.1.4 General precautions and recommendations 60

1.5.2 Scaffolding (including falsework)   62
   1.5.2.1 Wood and bamboo scaffolds   62
   1.5.2.2 Tubular metal scaffolding and falsework   66
   1.5.2.3 Lightweight and modular metal scaffolding (proprietary systems)   73
   1.5.2.4 Mobile access platforms   73
   1.5.2.5 The Joint Advisory Council Report on scaffolding   74
   1.5.2.6 Support scaffolds. Falsework   74
   1.5.2.7 Steeplejacks scaffolds   75
1.5.3 Roof work   75
   1.5.3.1 Roof edge falls   76
   1.5.3.2 Falls through fragile roofing material   76
   1.5.3.3 Falls from internal roof structures   77
   1.5.3.4 The special problems of roof maintenance   77
   1.5.3.5 Precautions   78
   1.5.3.6 Work in adverse weather   83
   1.5.3.7 Electric cables   83
   1.5.3.8 Access ladders   83
1.5.4 Demolition   83
   1.5.4.1 Statistics and the Joint Advisory Council Review   84
   1.5.4.2 Need and features of safe systems of work   85
   1.5.4.3 Precontract survey   86
   1.5.4.4 The Contract and its requirements   87
   1.5.4.5 Safety requirements during demolition   87
   1.5.4.6 Demolition by knocking down and taking to pieces   89
   1.5.4.7 Demolition by pulling and pushing down   90
   1.5.4.8 Demolition by explosives   90
   1.5.4.9 Demolition of structures with special risks   91
   1.5.4.10 Trends in safer demolition   92
1.5.5 Excavation   93
   1.5.5.1 Site investigation   95
   1.5.5.2 Site records, including underground hazards   95
   1.5.5.3 Underground cables and services   95
   1.5.5.4 Effect of excavation on adjacent buildings   97
   1.5.5.5 Flooding   97
   1.5.5.6 Side protection   97
   1.5.5.7 Fencing   98
   1.5.5.8 Excavated material   98
   1.5.5.9 Unidentified and suspicious objects   99
   1.5.5.10 Use of explosives and noisy plant in excavation   99

1.5.5.11 Access to, exit and escape from excavations　99
　　　1.5.5.12 Risks of toxic and flammable gases in excavations　99
　　　1.5.5.13 Toxic and radioactive materials in soil　100
　　　1.5.5.14 Biological hazards in soil　100
　　　1.5.5.15 Risks to ground workers from cranes, mobile plant and other site operations　100
　　　1.5.5.16 Risks to timbermen engaged in supporting the sides of excavations　100
　　　1.5.5.17 Statutory inspections　100
　　1.5.6 Tunnelling and caisson work　101
　　　1.5.6.1 Principal hazards　103
　　　1.5.6.2 Some historical examples　104
　　　1.5.6.3 Development of sub-aqueous tunnelling techniques　105
　　　1.5.6.4 Accident prevention in tunnelling　106
　　　1.5.6.5 Work in compressed air atmospheres　108
　　1.5.7 Diving　110
　　　1.5.7.1 Diving contractors　112
　　　1.5.7.2 Diving supervisors　113
　　　1.5.7.3 Divers　113
　　　1.5.7.4 Diving team　113
　　　1.5.7.5 Diving rules　114
　　　1.5.7.6 Qualifications of divers　115
　　　1.5.7.7 Certificate of medical fitness to dive　115
　　　1.5.7.8 Plant and equipment　115
　　　1.5.7.9 Maintenance examination and testing of plant and equipment　116
　　　1.5.7.10 A tragic example　116
　　1.5.8 Steeplejacks　117

Part 2　Health hazards of construction workers　121
　　Introduction　123
Chapter 2.1　How healthy are construction workers　125
　　2.1.1 Occupational mortality of construction workers in the UK　127
　　2.1.2 Health of construction workers in other countries　134
　　　2.1.2.1 Notes of papers covered by CIS abstracts　135
　　　2.1.2.2 Notes from EEC survey　142

Chapter 2.2 Occupational diseases and their causes   144
    2.2.1 The amended ILO list   145
    2.2.2 Occupational diseases in the UK   145
        2.2.2.1 The official list of occupational diseases in the UK   148
        2.2.2.2 Pneumoconiosis and byssinosis and occupational disorders   161

Chapter 2.3 Health hazards—general   163
    2.3.1 Occupational health professionals   164
        2.3.1.1 The occupational health physician   165
        2.3.1.2 Occupational (or industrial) hygienists   168
        2.3.1.3 Nurses, first aiders and other voluntary workers   171
        2.3.1.4 Other occupational health professions   173
        2.3.1.5 Information sources and training in occupational health   175
    2.3.2 Modes of entry and attack. Target organs, allergies   176
        2.3.2.1 The inhale route and respiratory hazards   176
        2.3.2.2 The dermal route and skin hazards   179
        2.3.2.3 The oral route   180
    2.3.3 Other target organs of the body   180
        2.3.3.1 The liver   181
        2.3.3.2 The kidneys   181
        2.3.3.3 The blood   181
        2.3.3.4 The nervous system   182
        2.3.3.5 The eyes   182
        2.3.3.6 Other organs   182
    2.3.4 Allergies   183

Chapter 2.4 Chemical health hazards   186
    2.4.1 Occupational exposure limits for airborne substances   188
    2.4.2 Gases   189
    2.4.3 Organic solvents   202
        2.4.3.1 Hydrocarbons   207
        2.4.3.2 Alcohols   208
        2.4.3.3 Ketones   208
        2.4.3.4 Esters   209
        2.4.3.5 Ethers   209
        2.4.3.6 Chlorinated hydrocarbons   209

      2.4.3.7 Nitro-paraffins   210
      2.4.3.8 Conclusions and precautions for solvents   210
  2.4.4 Water soluble compounds used in solution   211
      2.4.4.1 Inorganic acids   211
      2.4.4.2 Alkalis   214
      2.4.4.3 Salts   214
  2.4.5 Metal fumes and metal pigment dusts from welding, brazing, cutting and paint spraying   216
  2.4.6 Other non-solvent organic compounds (excluding those used as resin components)   218
  2.4.7 Components of resins produced on site   221
  2.4.8 Other airborne particles   225
      2.4.8.1 Airborne and respirable dusts   225
      2.4.8.2 Nuisance dusts   227
      2.4.8.3 'Semi-nuisance dusts'   228
      2.4.8.4 Mineral dusts containing silica   228
      2.4.8.5 Asbestos   228
  2.4.9 Precautions   229
      2.4.9.1 Storage   230
      2.4.9.2 Working conditions   230
      2.4.9.3 Disposal of surplus and unused chemicals   230
      2.4.9.4 Personal protective clothing and equipment, general   231
      2.4.9.5 Respiratory protection   232
      2.4.9.6 Hand protection   233
      2.4.9.7 Eye protection   233
      2.4.9.8 Use of rubber boots and PVC overalls   234

Chapter 2.5   Man-made physical health hazards   236
  2.5.1 Noise   236
      2.5.1.1 Sound and its velocity   237
      2.5.1.2 Sound frequency   238
      2.5.1.3 Sound power, intensity, pressure and level   239
      2.5.1.4 Hearing and the dB(A) scale   242
      2.5.1.5 Hearing loss and audiograms   246
      2.5.1.6 Recommended limits of noise exposure at work   248
      2.5.1.7 Noise levels on construction sites   252
      2.5.1.8 Noise reduction on construction sites   253
      2.5.1.9 Noise level monitoring and personal hearing protectors   257

        2.5.2 Mechanical vibration   259
            2.5.2.1 Physical characteristics of vibration   260
            2.5.2.2 Whole body vibration   262
            2.5.2.3 Hand and arm vibration   263
            2.5.2.4 Vibration standards, measurement and legal control   264
        2.5.3 Electromagnetic radiation   267
            2.5.3.1 UV radiation   268
            2.5.3.2 Visible light (except lasers)   269
            2.5.3.3 Infra-red radiation   269
            2.5.3.4 Lasers   270
            2.5.3.5 Ionizing radiation   274
        2.5.4 Electricity   276
            2.5.4.1 The importance of voltage in AC supply   278
            2.5.4.2 Treatment for electric shock   279
            2.5.4.3 Burns   280
            2.5.4.4 Accidental electrical injuries in construction   280
            2.5.4.5 Further points concerning shock protection   281
            2.5.4.6 Electric shock protection for welders   281

Chapter 2.6   Other health hazards   284
        2.6.1 Biological hazards   284
            2.6.1.1 Animal hazards (excluding insects and visible parasites)   284
            2.6.1.2 Insect and parasitic hazards   285
            2.6.1.3 Micro-organisms   287
            2.6.1.4 Poisonous plants   289
        2.6.2 Weather   289
            2.6.2.1 Temperature   290
            2.6.2.2 Windspeed   293
            2.6.2.3 Rain, hail, snow and frost   294
            2.6.2.4 Weather forecasts and meteorological data   294
        2.6.3 Work in cramped and unnatural positions   295
            2.6.3.1 The slipped disc and its causes   296
            2.6.3.2 Beat conditions   299
            2.6.3.3 Tenosynovitis   300
        2.6.4 Living conditions and welfare facilities   300
            2.6.4.1 Construction camps   301
            2.6.4.2 Food (with special reference to canteens)   302
            2.6.4.3 Drinking water   302
            2.6.4.4 Sanitary conveniences   303

    2.6.4.5 Facilities for clothing 303
    2.6.4.6 Temporary accommodation 305
  2.6.5 Work related mental stresses 305
    2.6.5.1 Warning signs of overstress 306
    2.6.5.2 Classification of jobs with high hazard potential to others 306
    2.6.5.3 Selection tests for jobs with high hazard potential 307
    2.6.5.4 Training and trainee testing and certificates for jobs with high hazard potential 308
    2.6.5.5 Discussion of workers problems 308
    2.6.5.6 Motivation 309
    2.6.5.7 Alcoholism and drug addiction 309

Chapter 2.7 Hazard monitoring 311
  2.7.1 Airborne chemical hazards 312
    2.7.1.1 Official methods for the detection of toxic substances in air 313
    2.7.1.2 HSE method of airborne lead monitoring 314
    2.7.1.3 HSE method of airborne asbestos dust monitoring 317
    2.7.1.4 Physical methods of monitoring chemical hazards 318
    2.7.1.5 Apparatus and methods involving a known air volume 319
    2.7.1.6 Personal monitoring kits relying on diffusion 321
    2.7.1.7 Firms offering portable air monitoring kits and equipment 321
  2.7.2 Monitoring physical hazards 321
    2.7.2.1 Noise and hearing measurement, training requirements 322
    2.7.2.2 Sound level meters 323
    2.7.2.3 Personal noise dosemeters 325
    2.7.2.4 Audiometry 326
    2.7.2.5 Vibration monitoring 326
  2.7.3 Monitoring other health hazards 327
    2.7.3.1 Biological hazards 327
    2.7.3.2 Weather hazards 328

Chapter 2.8 Personal protective clothing and equipment 333
  2.8.1 Philosophy and problems 333

  2.8.2 Legal requirements for protective clothing and equipment in the UK construction industry   335
  2.8.3 Clothing and equipment available and guidelines to their selection, general   341
  2.8.4 Eye protection   341
    2.8.4.1 Spectacles   345
    2.8.4.2 Goggles   345
    2.8.4.3 Other types of eye protection   346
  2.8.5 Protection of the respiratory system   346
    2.8.5.1 Breathing apparatus   348
    2.8.5.2 Respirator   351
  2.8.6 Face protection   352
  2.8.7 Head protection   353
  2.8.8 Ear protection   354
    2.8.8.1 Earplugs   354
    2.8.8.2 Ear muffs   355
    2.8.8.3 Noise helmets   355
  2.8.9 Hand protection   355
    2.8.9.1 Gloves   356
    2.8.9.2 Barrier and other hand creams   358
  2.8.10 Foot protection   358
  2.8.11 Body protection   359
    2.8.11.1 Normal working clothes for building workers   360
    2.8.11.2 Foul weather clothing   361
    2.8.11.3 High visibility clothing   362
    2.8.11.4 Chemical protection   363
  2.8.12 Fall protection   364
    2.8.12.1 Safety nets   365
    2.8.12.2 Safety belts, harnesses and lanyards   365
    2.8.12.3 Fall arrest devices   367
    2.8.12.4 Fixed anchorages for industrial safety belts and harnesses and fall arrest devices   367

Chapter 2.9   Portable ladders   369
  2.9.1 General   369
  2.9.2 Regulations and standards   370
  2.9.3 Purchasing   372
  2.9.4 Records, inspection, maintenance and storage   372
  2.9.5 Carrying ladders   373
  2.9.6 Erection and lowering of ladders   374
    2.9.6.1 Short plain rung ladders or standing ladders   374

2.9.6.2 Long plain rung standing ladders and extension ladders   374
2.9.6.3 Stepladders and trestles   375
2.9.6.4 Roof ladders   375
2.9.7 Securing plain and extension ladders   376
2.9.8 Use of ladders   376
    2.9.8.1 Plain and extension ladders   376
    2.9.8.2 Use of steps and trestles   376
    2.9.8.3 Use of roof ladders   376
2.9.9 Other precautions   377

Part 3 Fire, explosion and allied hazards   379
Introduction   381

Chapter 3.1 Compressed and liquefied fuel gases supplied in containers   382
3.1.1 General features of the gases   382
3.1.2 The LPG gases   383
    3.1.2.1 Use of LPG in construction   385
    3.1.2.2 The hazards of LPG in portable cylinders   387
    3.1.2.3 Transport, unloading and storage of LPG cylinders at site   388
3.1.3 Acetylene, hydrogen and oxygen   396
    3.1.3.1 Acetylene and hydrogen   396
    3.1.3.2 Oxygen   397

Chapter 3.2 Other fire and explosion hazards   399
3.2.1 Fire and its effects on personnel   400
3.2.2 Common fire causes on construction sites   401
    3.2.2.1 Rubbish   402
    3.2.2.2 Children playing on site and arson by other trespassers   402
    3.2.2.3 Cigarettes and matches   403
    3.2.2.4 Heaters and drying clothes   403
    3.2.2.5 Bitumen and tar boilers   403
    3.2.2.6 Cutting and welding equipment   403
    3.2.2.7 Blow lamps   404
    3.2.2.8 Electrical faults   404
    3.2.2.9 Chimneys, flues and plant and vehicle exhausts   405

          3.2.2.10 Site huts and caravans   405
          3.2.2.11 Storage and use of flammable liquids   405
     3.2.3 Fire organization and first aid fire fighting   407
          3.2.3.1 Organization   407
          3.2.3.2 Fire detection and warning   408
          3.2.3.3 Escape   408
          3.2.3.4 Rescue   409
          3.2.3.5 Emergency firefighting and appliances   409
     3.2.4 Special fire and explosion problems   411
          3.2.4.1 Use of explosives in demolition, excavation and tunnelling   411
          3.2.4.2 Use of cartridge-operated fixing tools   413
          3.2.4.3 Fire and explosion hazards in pressurized workings   414
          3.2.4.4 Reactive and explosive chemicals   415

Part 4   **Special problems of construction and the future   417**
          Introduction   419

Chapter 4.1   Special safety and health problems of the construction industry   420
          4.1.1 Temporary duration of work sites   420
          4.1.2 Seasonal employment   421
          4.1.3 Extensive use of migrant labour   421
          4.1.4 Small size of construction firms   422
          4.1.5 Extensive use of sub-contractors   422
          4.1.6 Effects of weather   423
          4.1.7 Clandestine work   423
          4.1.8 High labour turnover   424
          4.1.9 Welfare problems   424
               4.1.9.1 Food   425
               4.1.9.2 Housing and shelter   426
               4.1.9.3 Transport to and from work   426
               4.1.9.4 Other welfare amenities and services   427
          4.1.10 Competitive tendering   427
          4.1.11 Special problems in developing countries   429

Chapter 4.2   Training   431
          4.2.1 General   431

4.2.2 Training for different vocational groups in construction   433
    4.2.2.1 Director and principals   433
    4.2.2.2 Site managers and supervisors   434
    4.2.2.3 Operatives and new entrants   435
    4.2.2.4 Planners and designers   437
    4.2.2.5 Safety officers and advisors   437
    4.2.2.6 Safety representatives   437
4.2.3 How effective is safety training in construction   438

Chapter 4.3 The future   440
4.3.1 General remarks   440
4.3.2 Action at the undertaking level   441
4.3.3 Action at the national level   442
    4.3.3.1 Government action   433
    4.3.3.2 Action by employees' organizations   443
    4.3.3.3 Action by workers' organizations   443
4.3.4 Action at the international level   444

APPENDICES

Appendix A   Sources of help   448

A.1 International organizations   448
    A.1.1 ILO   448
    A.1.2 WHO   449
    A.1.3 ISO and IEC   449
    A.1.4 Le Comité International de Prevention des Risques Professionnels du Batiment et des Travaux Publics   449

A.2 UK organizations   450
    A.2.1 Government and official but independent national organizations   450
        A.2.1.1 The Law   450
        A.2.1.2 The Health and Safety Commission and Executive   453
        A.2.1.3 The British Standards Institution   453
    A.2.2 Employees organizations   454
    A.2.3 Workers' organizations   455
    A.2.4 Joint bodies   456

  A.2.5 Universities and technical colleges 456
  A.2.6 Professional and voluntary organizations 457
  A.2.7 Other UK organizations 457

A.3 Other national organizations 458
  A.3.1 USA 458
  A.3.2 Canada 458
  A.3.3 Brazil 458
  A.3.4 Japan 458
  A.3.5 India 459
  A.3.6 Kenya 459
  A.3.7 France and EEC 459

Appendix B Recommendations of advisory committees for construction safety 460
  B.1 Safety in steel erection 460
  B.2 Safety in scaffolding 462
  B.3 Safety in demolition 464

List of tables and figures

Table Number

1.2.1 UK construction industry. Employment by occupation unit groups, 1971 census. 10% sample *20*
1.2.2 Construction craft operatives employed in the UK, 1978 *23*
1.2.3 Employment in construction firms in the UK in 1978 *24*
1.2.4 Construction occupations, professional, trade, etc. as found in the UK *25*
1.3.1 Fatal accident frequency rates in construction for selected countries over consecutive 8 year period. Number of fatal accidents per 10^6 man hours worked *34*
1.3.2 Fatal accident incidence rates in construction in selected countries. Deaths per 1000 employees *35*
1.3.3 Incidence rates of total and fatal accidental injuries in construction and manufacturing industries and total fatalities in the UK (1973) *37*
1.3.4 Distribution of fatal accidents in construction by causation (1974–76 and 1976–79) and total for 1980–82 *37*
1.3.5 Occupations of fatally injured workers (1977 to 1979) *38*
1.3.6 Distribution of fatal accidents by type of work, 1977 to 1979 *39*
1.3.7 Construction workers qualifying for injury benefit from prescribed industrial diseases *40*
1.3.8 Death certificates mentioning asbestos-related diseases 1968 to 1979 *40*
1.3.9 Life expectancy for a UK citizen if exposed only to risk specified *41*
1.5.1 Fatal accident incidence rates in steel erection compared with all construction and manufacturing in the UK *58*
1.5.2 Accidents as percentage of study group and fatalities to steel erectors *59*
1.5.3 Some common types of tubular metal scaffolds *68*
1.5.4 Some tubes used for parts of tubular metal scaffolds *69*
1.5.5 Reported accidents in demolition and all building operations *84*
1.5.6 Reported demolition injuries related to type of building involved (fatalities in brackets) *85*

1.5.7	Analysis of reports of fifty fatal accidents to men employed in demolition	*85*
1.5.8	The main hazards of excavation	*94*
1.5.9	Types and purpose of tunnels in civil engineering	*102*
2.1.1	Standard mortality ratios and relative age gradients for construction trades and occupations	*130*
2.1.2	Diseases among construction workers attributable to occupational factors	*140*
2.1.3	Site construction operations with highest frequency of occupational disease	*142*
2.2.1	List of occupational diseases as amended by the ILO/WHO meeting of experts January 1980	*146*
2.2.2	List of prescribed diseases with the occupations for which they are prescribed from 'The Social Security (Industrial Injuries) (Prescribed Diseases) Regulations 1980	*150*
2.3.1	List of hygiene consultants	*170*
2.4.1	Occupational exposure limits of airborne materials encountered in construction	*190*
2.4.2	Asphyxiant gases liable to be found in construction	*198*
2.4.3	Occupational exposure limits for vapours of organic solvents used for paints, etc.	*204*
2.4.4	Special boiling points—standard UK grades	*207*
2.4.5	Occupational exposure limits for other airborne dusts and fibres	*226*
2.5.1	Sound velocities in various media	*238*
2.5.2	Sound levels, acoustic intensity and sound pressures for typical environmental noises	*241*
2.5.3	Calculation of combined sound level of a small band from a point in mid audience	*242*
2.5.4	Maximum recommended exposure to noise at various levels per working day	*249*
2.5.5	Sound power levels and sound levels experienced by workers for various construction activities and plant	*250*
2.5.6	Noise levels (dB(A)) inside driver's cab of standard (1974) and modified site vehicles	*255*
2.5.7	Maximum sound power levels of hand held power picks and breakers	*256*
2.5.8	Natural frequencies of vibration of various parts of human body for vertical vibrations	*261*
2.5.9	Recommended design and minimum lighting levels on construction sites	*269*
2.5.10	Definitions and labelling requirements for laser products	*272*
2.5.11	Effects of electric current on a man's body	*277*
2.5.12	Calculated electrical characteristics of human body at 50 Hz in dry conditions	*279*
2.5.13	Approximate threshold shock voltages at 50 Hz ac	*279*
2.5.14	Types of electrical burns	*280*
2.5.15	Electrical accidents on UK construction sites over one year	*281*
2.6.1	Preferred environmental temperatures for various tasks	*290*
2.6.2	WBGT and recommended work/rest regimes in hot weather	*292*

2.7.1	Official methods for the detection of toxic substances in air	*314*
2.7.2	Some UK firms offering air monitoring kits and equipment	*322*
2.7.3	Weather adverse to building work	*329*
2.8.1	General requirements of personal protective clothing and equipment	*334*
2.8.2	Parts of body and hazards against which protection may be needed	*336*
2.8.3	British regulations and orders prescribing protective clothing and personal protective equipment which apply in the construction industry	*338*
2.8.4	Summary of specified processes in construction for which eye protection is required under Protection of Eyes Regulations, 1974	*340*
2.8.5	British Standards for protective clothing and equipment used by construction workers	*342*
2.8.6	Hand hazards and suggested protection for construction occupations	*357*
2.9.1	Accidental and fatal fall injuries from ladders and stepladders, 1978–1982	*370*
2.9.2	Ladder inspection checklist	*373*
3.1.1	Properties of liquefied petroleum gas as used in the UK	*384*
3.1.2	Minimum separation distances for open air LPG stores	*389*
3.1.3	Heights of stacks, widths of gangways and maximum amounts of LPG per stack	*389*
3.1.4	Properties of acetylene and hydrogen	*396*
3.2.1	Classification of combustible liquids by flashpoint (closed cup)	*406*
3.2.2	Flammability classes. Flash points and ignition temperatures of some organic solvents and other liquids used in construction	*406*
3.2.3	Types of portable fire extinguishers	*410*
A.1	Principal regulations which apply to safety and health in construction	*451*
A.2	Other Rules and Orders which apply for safety and health in construction	*452*

Figure Number

1.1.1	The scope of construction	*10*
1.1.2	Value of output by contractors and direct labour on construction work at 1980 prices	*14*
1.1.3	A typical Brazilian favela	*15*
1.1.4	Building in the Cameroons	*16*
1.4.1	Bricklayers are exposed to the hazards of falling, to the weather and other dangers	*44*
1.4.2	The formworker is exposed to most of the hazards of carpentry and joinery as well as contact with cement and release agents	*45*
1.4.3	In 1979 there were 1530 notifiable accidents involving painters and decorators in the UK of which 570 were due to falls from heights	*47*
1.4.4	The electrician must always be prepared for unexpected hazards	*51*

1.5.1	Most fatal accidents in steel erection result from falling and too few workers have as much thought for their safety as these men 200 metres above Berlin	*57*
1.5.2	The single ladder scaffold	*64*
1.5.3	Double ladder scaffold providing wider deck	*64*
1.5.4	Ledgers, splices and bases for pole scaffolds	*65*
1.5.5	A young roofing worker, who was not wearing a safety helmet, died as a result of injuries when scaffolding collapsed into a shopping centre in Faversham, UK	*67*
1.5.6	Typical independent tied and putlog scaffolds erected on substantial timber poles with secure handrails and toeboards	*71*
1.5.7	Reveal ties and through ties are technically acceptable but a nuisance to occupants or other tradesmen	*72*
1.5.8	A high proportion of roofing accidents occur during maintenance, particularly on roofs of industrial premises	*77*
1.5.9	Free-standing counterbalanced frames may be used on strong flat roofs	*79*
1.5.10	Scaffolding erected from the ground can be used on buildings where the roof is not more than five metres above ground level	*80*
1.5.11	Edge protection clamped to edge beam of framed building	*81*
1.5.12	Precautions should be taken to protect the general public and enclosed chutes should be used to bring down loose material	*89*
1.5.13	Italian miners in the Great St Bernard Tunnel	*105*
1.5.14	A diver inspecting one of the legs of Comorant 'A' production platform in the North Sea	*111*
2.1.1	How healthy are construction workers	*125*
2.1.2	Mortality by social class and cause of death; standardized mortality ratio for men and married women (by husbands occupation) ages 15–64	*129*
2.3.1	'…the dermatologist has to be a bit of an epidemiologist and a bit of a psychologist as well'	*173*
2.3.2	The human respiratory system	*177*
2.3.3	The approximate size range of a number of commonly encountered airborne particles	*177*
2.4.1	Structures of common constituents of organic solvents	*203*
2.5.1	Amplitude and frequency of sound waves	*241*
2.5.2	Structure of the human ear	*243*
2.5.3	Normal equal loudness contours for pure tones	*243*
2.5.4	Comparison of octave, ⅓ octave and narrow band spectra of the same sound	*245*
2.5.5	Audiometer (Bruel and Kjaer, 1800)	*246*
2.5.6	Audiograms in hearing loss due to age, noise and middle ear disease	*247*
2.5.7	A precision computing sound level meter and frequency analyser	*258*
2.5.8	Typical performance of ear protectors	*259*
2.5.9	Simple mars/spring system	*260*
2.5.10	Exponential decay due to damping	*261*

2.5.11	Reiher-Meister scale for vertical vibration on standing subjects	262
2.5.12	The organ of balance	262
2.5.13	Standard for exposure to whole body vibration	265
2.5.14	Standard for exposure to hand-arm vibration, with various appliances superimposed	266
2.5.15	The electromagnetic spectrum	267
2.6.1	Wet bulb of globe temperature index	291
2.6.2	Wind chill index	293
2.6.3	Anemometer and wind direction indicator	294
2.6.4	Front and end virus of the human spine	296
2.6.5	Pressure transmitted to dural tube and nerve root	297
2.6.6	St Thomas's Hospital posture chart	298
2.7.1	Orifice type sampling head used for airborne lead	315
2.7.2	NIOSH charcoal tube	321
2.7.3	Nomogram for calculating wind chill index	330
2.8.1	'...the use of protective items by workers varies considerably'	334
2.8.2	Compressed-air line breathing apparatus	349
2.8.3	Anchorage line greatly extends the zone in which a worker can operate	362
2.8.4	'In hotter climates...'	364
2.8.5	Safety and secure harness	366
3.1.1	Typical compound for storing LPG cylinders	390
3.2.1	The Fire Pyramid	400

Introduction

The unacceptable face of construction is veiled behind emotionless figures which record the dead, maimed and crippled human by-products of the industry. Quantification of this carnage on an international scale is impossible, for while data (in many cases of questionable validity) is available from a number of nations, from the vast majority there is none whatsoever. While conjecture could extrapolate figures to fill this void the value of such an exercise is dubious. Suffice it to say that in tangible terms the historical path is littered with the graves of those 'sacrificial lambs' on the altar of construction.

However, a morbid predilection with statistics *per se* has only the counterproductive effect of deflecting attention from the real essence of such figures which is in their rôle as a record of past performance. Treated in this way and prepared on a sufficiently broad data base it is possible to identify discrete causes and trends. In this manner statistics can be a productive instrument in determining future priorities and programmes of action. Unfortunately, if U.K. experience is a measure, it would appear that the construction industry is not destined to learn from its mistakes. Given that in this country 'falls of persons from heights' has accounted for over half the fatal accidents in construction during the last decade one is inevitably drawn to the conclusion that the industry will never come to terms with basic principles of accident prevention. More alarming though is that such circumstances exist in an industry which practices methods of working which have changed little over the decades and in some cases over centuries.

Construction is and has been an important part of Man's activities from his earliest days and irrevocably intertwined with all stages of his development. In some aspects it can be said to predate Man himself as most birds, mammals and insects practice some form of construction to provide a protected environment in which to raise their young and store victuals against lean times. So, with Man, when the cave was no longer a desirable residence he took his first tentative steps into the arena of construction to provide these basic necessities.

As Man became an increasingly gregarious animal his group cultures developed hierarchical social orders which made further demands on construction. More grandiose edifices were required to reflect the standing of the rulers within their society, both during life, for religious purposes and in death—the Taj Mahal, the temple of Karnak, the Pyramid of Cheops and the Ming Tombs being prime examples. The construction of those edifices and monuments to the glorification of an elite were carried out by workers whose lives were of little concern to the elite class. In essence the building of these prestigious structures was regarded in the same light as war—a noble task justifying the loss of many lives. So accidental fatalities caused little concern, and in some instances, such as the Ming Tombs, the workforce was deliberately 'sacrificed' by interment to preserve the secrets of their location and construction.

Many of these 'follies' of human gratification served also as a hallmark of the particular society and its culture. Not surprisingly, therefore, after inter society supremacy struggles it was usual for the victor not only to pillage and commit genocide but to undertake the wholesale destruction of the defeated adversary's buildings to ensure others could not be infected by that culture. The razing of Carthage by the victorious Romans is a case in point. In some instances this dismantling was confined to the superstructures with the existing foundations being used by the conquerors for their buildings. This proved highly practical in Cusco, Peru, where the Inca foundations were far more resistant to the frequent earthquakes of the area than the Spanish superstructures.

For all their enduring beauty and noble proportions, there was little scientific about the ancient architecture of Egypt, China and Greece. Columns for example were built with little consideration of their proportions in relation to the load being supported, as the forest of massive pillars which occupy so much of the space in the Temple of Karnak, in upper Egypt, testifies. The builders of these eras solved structural problems by methods probably handed down within the secrecy of their guilds[1]. Scientific study of the strength of materials and of structural mechanics did not really have any impact until the seventeenth century with the experiments of Leonardo da Vinci and subsequently several other famous men—Galileo, Wurtz, Joseph Moxon, Christopher Wren, Robert Hooke, Phillippe de la Hire and Newton[2].

With mankind's technological advancements and increasing social awareness, demands for more complicated infra-structures grew apace. So, besides providing basic housing and sheltered environment, the construction industry extended its field of activity into such areas as hospitals, roads, bridges, sewage works, factories, offices and all the other structures which form part of modern day to day living. These developments represent a considerable investment on the part of society devoted to providing greater protection against the hazards of nature, greater comfort of its inhabitants and generally improving the standard of living. In the past this has been used as a justification for the expenditure, and the incidental loss of life when measured against the benefit of the community as a whole was considered an insignificant overhead. Not surprisingly this attitude was symptomatic of the prevailing socio-economic climate; wages were determined by the minimum required for the bare subsistence of the labourer and his dependants, birth rates were high, and mortality from diseases was also high so that fatal accidents were of little consequence[3]. Thus Graunt, in recording the causes of 97,306 deaths in 130 London parishes in 1665 lists 'the plague' as the cause of 68,596 and 'killed by feveral accidents' as responsible for only 46 deaths[4].

This picture of construction and the conditions described are by no means a thing of the past and can still be witnessed today in corners of the world. In third world and developing countries especially, limited resources have to be devoted not only to the provision of a protected environment, which is seen as a basic human need of even the poorest who are seldom able to afford the standards of 'civilised society', but to the large scale development of a modern industrial infra-structure. Thus anomalous situations arise where a modern sophisticated official building industry exists side by side with primitive construction which flouts most rules and regulations. Visitors to Latin America can observe prime examples of this situation in the 'shanty towns' which spring up overnight on the outskirts of large urban conglomerations.

In industrialised countries advancements in social sciences over the past three centuries have prompted a greater awareness of the sanctity of life and the unacceptability of premature death due to industrial accidents. This movement has culminated in intervention by the state to regulate these activities and require minimum standards of performance with the objective of eliminating such incidents. It would be expected that in more enlightened societies this lead would be vigorously pursued but the statistics unfortunately do not support this contention. This is indeed surprising considering that many of the precautionary measures to reduce the carnage are neither costly or time consuming.

Such circumstances do not auger well for the prospects of reductions in fatalities particularly in third world and developing countries. International construction organisations freed from the constraints of exacting requirements appear to adopt the principle of the 'lowest common denominator' rather than try to inculcate the emerging nations with higher standards.

Perhaps the time is fast approaching when a 'Safety Revolution' would be in order, though what form it should take is far more difficult to establish. Self-regulation, thought of as a panacea, is unlikely to be a serious contender in view of the recent U.K. experience regarding the wearing of safety helmets on sites, which was an abject failure. Therefore motivation to effect change appears to rest in the hands of government bodies whose principal remedy lies in the further liberal use of legislation which impinges increasingly on every part of daily life. It is unfortunate that such a vehicle has to be used as legislative language tends to be unintelligible to the layman. Also, as no subject can be comprehensively covered in all its finer technical points in legal language, literature of varying degrees of authority in the form of standards, codes of practice and guidance notes proliferate. The picture is not complete however until an effective and efficient policing function coupled with realistic financial penalties for non-performance have been established. Without these the laws would be impotent.

Such policing is common in most industrialized and many developing countries and has achieved some success. Third world countries however encounter particular problems in introducing appropriate legislation, enforcement procedures and standards particularly for the new construction technologies. The International Labour Office, a special tripartite United Nations Agency which represents the interests of governments, workers and employers assists such countries in overcoming these problems. To this end the I.L.O. has published its own Code of Practice and guidance material on safety and health in building and civil engineering[5,6,7]. Its unenviable task is made no easier by the need to steer between the Scylla of promulgating lower standards for third world countries than those prevailing elsewhere and the Charybdis of high standards which for technical and economic reasons would be very difficult or impractical to enforce in third world countries at the present time. Examples of this dilemma can be seen in the ILO's current recommendations on the use of timber scaffolding and of asbestos in construction.

The inherent problems of construction are not however either simplistic or easily resolved when considered in isolation. Many factors have a bearing on, and are symptomatic of, the industry's poor safety record and of necessity must be entered into the equation if a lasting solution is to be found. These, not necessarily in order of importance include:

- the minimal outlay required to start and carry on a business and the concomitant ease with which it can be liquidated.
- the small size of many undertakings.
- the acquisition of work by competitive tendering.
- the extensive use of sub-contract and self-employed labour-only sub-contractors.
- the use of migrant labour.
- the temporary nature of the work.
- the seasonal nature of the employment.

Further discussion on these problems will be found in Chapter 4.1.

In safety and health circles the main preoccupation until recently has been with the prevention of accidental injuries, an area in which the construction record is far worse than that of manufacturing industries. The past few years have seen a gradual change of emphasis as it is recognised that occupational illness and disease among construction workers may be an even more serious problem. Improvements in the general health (and prosperity) of the populations of industrialised countries during the sixties and seventies caused more attention to be devoted to the various occupationally related black spots.

Regrettably statistical analysis of occupationally related diseases is generally considered in the same light as that of accidental injuries. This approach unfortunately does not give cognizance to the one great disparity between these sets of figures, that being the different time scale involved. Accidental injury figures are an immediate record of incidents where there is an immediate relationship between cause and effect. Occupational disease statistics can have no such immediate relationship between causal exposure and the onset of the disease or its development to a recognisable state. Failure to acknowledge this particular point can lead to the waste of much time and effort in the pursuit of increasingly stringent controls to eliminate a hazard that may have been already reduced by existing standards to such an extent as now to be of negligible consequence. This is of particular concern to health and safety practitioners as it diverts attention from the problems of today which will be tomorrow's history.

Health hazards are not confined solely to diseases or illnesses consequent upon the individual's exposure to dangerous substances or infection. Especially in construction the nature of the work itself involves other risks to health—exposure to a frequently hostile and changing climate which leads to colds, bronchitis, rheumatism or conversely heat stroke; hard physical work in cramped and unnatural body positions leads to bursitis; manual handling of heavy objects causes 'slipped discs' or other spinal disorders, to name but a few. Government and employers are under increasing pressure to eliminate these hazards, particularly where 'free' national health services are available to all citizens since the cost of treatment is considerable. In third world countries, however, which have limited resources and meagre social security and public health services, these occupationally related illnesses are less significant when compared with the widespread poverty, ill-health and unemployment and consequently take second place in such countries' social priorities.

The wide range of hazards to which construction workers the world over are exposed every day of their working lives is now becoming increasingly obvious. Fortunately many hazards are quite specific to one or two of the myriad of trades which constitute the construction industry while some affect a wide spectrum of the different skills. In both cases, the role of 'Safety' in its widest context, should concentrate on prevention not cure. This philosophy transcends the self-imposed limitations of national legislation enabling attention to be concentrated on the preparation of effective procedures to eliminate or control the hazards to which workers are exposed. For this reason the theme of this book is on hazards rather than health and safety. In this context the definitions of 'hazard', 'accident' and 'safety' in one of the author's earlier works make this clear[8].

Hazard   "is a condition with the '*potential*' of (causing) an accident or ill health."

Accident   "is an unplanned event which has a probability of causing personal injury or property damage."

Safe   "a thing is provisionally categorised as safe if its risks are deemed known and in the light of that knowledge judged to be acceptable."

Thus a condition of safety is only reached by removing the risks or *hazards* which could result in an *accident*. This is the task of the safety specialist.

References

1. WARE, I., *The Architecture of Andrea Pallodio* (1518–80)
2. MOXON, J., *Mechanick Exercises*, (2 series of monthly parts starting 1667)
3. MARX, K., *Capital*, 1887.
4. GRAUNT, J., *Natural and Political Observations, mentioned in a following Index and made upon the Bills of Mortality*, 5th ed 1676.
5. ILO CODE OF PRACTICE, *Safety and Health in Building and Civil Engineering Work*, International Labour Office, Geneva (1972).
6. ILO OCCUPATIONAL SAFETY AND HEALTH SERIES 42, *Building Work. A Compendium of Occupational Safety and Health Practice*, International Labour Office, Geneva (1979).
7. ILO OCCUPATIONAL SAFETY AND HEALTH SERIES 45, *Civil Engineering Work. A Compendium of Occupational Safety Practice*, International Labour Office, Geneva (1981).
8. KING, R.W. and MAGID, J., *Industrial Hazard and Safety Handbook*, 3rd imp., Butterworths, London (1982).

PART 1
SCOPE, ACTIVITIES AND SAFETY ASPECTS OF THE INDUSTRY

1.1
The scope of construction and its hazards

The term 'construction' is so comprehensive that anything constructed by man or living creatures could be included in it. Thus it is as well to be clear which construction activities this book covers. Let us start with some exceptions. Ships, cars, aeroplanes and trains are all constructed, but they are not covered here, nor are toys, furniture, caravans, garden tools, cranes, or other machinery. Houses, bridges, factories, churches, cinemas, roads, office buildings, tunnels, dams, offshore platforms and the building of power and process plant are the types of construction whose hazards this book tries to cover. Besides modern construction it discusses the traditional, as well as primitive construction methods of developing countries. The common features of those objects of construction covered here are that they are

1. permanently attached to the ground and usually built on foundations; tunnels and other underground structures are included;
2. for the most part constructed or assembled on site, although they may contain many factory-made components or sub-assemblies.

Construction or assembly 'on site' is perhaps the key to the activities covered by this book.

There are, of course, hazards concerned with the factory construction of sub-assemblies such as doors, window frames, lift cages, baths, roofing tiles, electrical fittings and an almost endless list of items. The hazards of these factory processes are not considered here, nor are the hazards involved in transporting such artefacts to the site. Thus a more fitting title for this book might have been 'Hazards of Site Construction'.

This book does not cover the hazards experienced when the constructed object is complete and in use. These hazards, whether to people or property, may arise from many causes—structural deficiency, fire, earthquake, etc.

Since demolition is closely linked to construction its hazards are included. The book thus deals with the hazards to life and property, particularly the construction workers and the equipment they use—scaffolding, cranes, excavators, etc.—both during construction and demolition.

SECTION 5		CONSTRUCTION
Group	Activity	**CONSTRUCTION**
500	**5000**	**GENERAL CONSTRUCTION AND DEMOLITION WORK** Establishments engaged in building and civil engineering work, not sufficiently specialised to be classified elsewhere in Division 5, and demolition work. Direct labour establishments of local authorities and government departments are included.
501	**5010**	**CONSTRUCTION AND REPAIR OF BUILDINGS** Establishments engaged in the construction, improvement and repair of both residential and non-residential buildings, including specialists engaged in sections of construction and repair work such as bricklaying, building maintenance and restoration, carpentry, roofing, scaffolding and the erection of steel and concrete structures for building.
502	**5020**	**CIVIL ENGINEERING** Construction of roads, car parks, railways, airport runways, bridges and tunnels. Hydraulic engineering e.g. dams, reservoirs, harbours, rivers and canals. Irrigation and land drainage systems. Laying of pipe-lines, sewers, gas and water mains and electricity cables. Construction of overhead lines, line supports and aerial towers. Construction of fixed concrete oil production platforms. Construction work at oil refineries, steelworks, electricity and gas installations and other large sites. Shaft drilling and mine sinking. Laying out of parks and sports grounds. Contractors responsible for the design, construction and commissioning of complete plants are classified to heading 3246. Manufacture of construction steelwork is classified to heading 3204. The treatment of installation work is described in the introduction to the SIC.
503	**5030**	**INSTALLATION OF FIXTURES AND FITTINGS** Establishments engaged in the installation of fixtures and fittings, including, such as gas fittings, plumbing, heating and ventilation plant, sound and heat insulation, electrical fixtures and fittings.
504	**5040**	**BUILDING COMPLETION WORK** Establishments specialising in building completion work such as painting and decorating, glazing, plastering, tiling, on-site joinery and carpentry, flooring (including parquet floor laying), installation of fireplaces, etc. Builders' joinery and carpentry manufacture is classified to heading 4630; shop and office fitting to heading 4672.

Figure 1.1.1 The scope of construction.

The main areas covered are generally known as 'building' and 'civil engineering' with a smattering of other engineering activities—mechanical, electrical, telecommunication, process industries and offshore oil platforms thrown in.

For U.K. readers, the scope of 'construction' is defined under Division 5 of 'Standard Industrial Classification'[1] reproduced here under *Figure 1.1.1*[2]. To this we would add "the on-site construction, assembly and installation of plant for the process industries", which is covered under heading 3246 of the Standard Industrial Classification (SIC).

1.1.1 Building

By 'building' we generally understand houses, offices, shops, hotels and restaurants, flats, schools, churches, and cathedrals, as well as factories, railway stations, hospitals, theatres, pagodas, public lavatories, town halls, museums, libraries, brothels and so on. Building, in short, covers most forms of construction built on terra firma with walls, floors, doors, windows and roofs where people live, shop, worship, debate, amuse themselves or simply shelter from the weather. Traditionally they smack of bricks, mortar, timber and tiles with an earlier ancestry of stone and slate and some are of very ancient vintage.

Most buildings today are designed by architects who are the lineal descendants of the ancient master builder, a mongrel breed of artist, engineer, businessman and sometimes courtier and politician. Architecture, said Sir Henry Wotton[2], must fulfill three conditions, "Commoditie, Firmeness and Delight".

Building contracts in the U.K. other than for government buildings, generally use the JCT Standard Form of Contract, formerly the RIBA Form of Contract. For British Government Contracts, a single form, Form GC/Works/1 is used for both building and civil engineering works, with the same general conditions.

The architect is referred to in this latter form as the 'Superintending Officer', although for an engineering contract this man will usually be a civil or mechanical engineer.[3]

1.1.1.1 High-rise building

Again there is no universally-accepted definition of what constitutes a high or a 'high rise' building. It just depends on what you are used to.

In quite recent Imperial times in Peking, all private dwellings were, in effect, single-storey bungalows since the emperor (or empress) did not want to be overlooked when carried through the streets in a palanquin. There any building of three storeys or more qualifies as 'high'. In New York, however, a 'high' building would need to have at least forty storeys.

There is no doubt that scaling tall buildings has for many the same challenge to their machismo as mountaineering or rock climbing. It even breeds a pride in taking risks, as the following passage illustrates:

> 'But Andrew Eken, the chief builder, is still alive. I saw him in his office against a background of innumerable windows and the measured zigging lines of his gigantic creations: a friendly, rather tired old man with thin steel-edged spectacles. And he spoke again of his favourite, his problem child, the Empire State. Within the next few days his men were to set up an aerial for television on the highest point of the building. As he spoke of it his grey face began to lighten. "The wind up there", he said, "blows and beats. We wanted to set up a protective scaffolding for the safety of the workmen, but the steel construction people laughed at us. They'll rope their boys on. If one of them blows down, he'll hang on a line like a fish. He may have a scare but he won't be hurt.'[4]

Such attitudes constitute a serious psychological obstacle to reducing construction hazards. It is difficult to deal with workers who have become addicted to the kicks they get out of the risks of their jobs. It is as difficult as it is to get people to give up smoking.

Whilst we think of high rise buildings as essentially modern we should not forget the spectacular heights reached by many much older buildings. The Almighty, however, did not altogether approve of such activities:

> 'And they said to one another, Go to, let us build us a city and a tower whose top may reach unto heaven ... and the Lord said, let us go down and there confound their language that they may not understand one another's speech... Therefore is the name of it called Babel'[5].

Perhaps Kipling was right when he wrote:

> 'How very little since things were made
> Things have altered in the Building Trade.'

The pyramid of Ceops completed before 2100 BC reached 147 metres. The Qutub Minar south of Delhi, built in AD 1200 reached 72 metres. The Eiffel Tower in Paris completed in 1889 reached 301 metres. The Empire State Building in New York, completed in 1931, reached 451 metres.

High rise buildings bring additional hazards into the lives of builders, maintenance men and painters. The higher wind speeds, the greater exposure to cold caused by them, the psychological fear of heights and the movement of tall buildings caused by wind and earth tremors are some of the main ones.

High rise construction puts stringent demands on designers, planners, contractors and workers themselves in foreseeing and eliminating hazards. But the greatest hazard of high rise construction in many developing countries lies in the use of traditional wooden or bamboo scaffolding. The hazards of scaffolding are discussed in Chapter 1.5.

The use of unskilled and inexperienced labour and labour-intensive methods also adds to the hazards of high rise construction in the Third World.

On a more hopeful note, most modern high rise buildings incorporate many prefabricated parts which are repeated on each storey. Such repetition makes it easier and cheaper to plan and use safe systems of work and safety equipment, thereby reducing many hazards.

1.1.1.2 Prefabricated buildings

Since Roman times most organised building in Europe has contained a high proportion of prefabricated elements—bricks, tiles, standard sawn and dressed timber, nails, pipes, lime mortar and the like. One has to go back to pre-industrial societies such as the Incas of Peru to find substantial stone buildings where the stones, individually cut on site, fitted very closely in an irregular jigsaw. In fact we could not build in that way today even if we tried, as the art has been lost.

In speaking of prefabricated building today we usually mean far more than the use of prefabricated bricks, concrete blocks, doors and window frames. Rather we mean the rapid assembly on site of large and complete modules weighing many tons, complete with doors, windows, floors, ceilings and all interior fittings. This sort of pre-fabrication reaches its ultimate in the living quarters built on offshore oil and gas platforms.

Prefabricated construction generally reduces the manual skills required while increasing the need for skills in planning, organisation and in the use and operation of plant and machinery.

The main hazards of prefabricated construction on site lie in trying to install large prefabricated components by a labour force familiar only with more traditional methods, and without the aid of appropriate modern mechanical handling plant and equipment. Problems arise when bricklayers used to handling standard bricks weighing two to three kg find themselves asked to lay much larger concrete blocks of more than twice the weight. Until they reach the building site these have been handled entirely by machinery.

Large prefabricated components have to be handled entirely by cranes and other machinery on site. Many accidents have occurred through attempting to use unsuitable cranes of inadequate capacity, as well as untrained operatives.

On the other hand, the use of prefabricated parts leads to standardisation and repetition. This facilitates the adoption of safe systems of work and special jigs and equipment for the safe handling of prefabricated parts.

1.1.1.3   Typical houses in industrialized countries

Houses usually require building permission from the local authority after the submission of plans complying with building standards and local regulations. They are usually of brick, timber and concrete block with prefabricated metal or wooden doors and windows.

Despite the long experience in this type of building, it is still beset by a bad accident record. This is especially true among the numerous small building firms which make up a large part of the industry. A high proportion of the accidents are due to falls from unsafe ladders and scaffolding and from unprotected sloping roofs. The 'one off' nature of much of the work and the high relative cost of providing adequate safety measures in relation to the value of the work done, leads to a strong temptation to cut corners. The more standardised the building becomes and the larger the contracting organisation, the easier it becomes to provide proper safety equipment, training and supervision of work.

At the time of writing the world as a whole and Britain in particular is still in the deepest industrial recession since the 1930s and construction is suffering particularly badly. At the same time, a marked change in the composition of construction output is taking place in the U.K. An increasing proportion is being devoted to repair, maintenance and improvement at the expense of new construction. This trend is shown graphically in *Figure 1.1.2.*[6]

This change in pattern of construction output is likely to be reflected by a change in pattern of hazard exposure.

1.1.1.4   Third World shanty towns

Whilst many residential dwellings in Third World countries are built to similar standards as those in industrialised countries, there is no doubt that a substantial proportion of third world inhabitants are housed in unofficial, usually home-made, constructions.

14 The scope of construction and its hazards

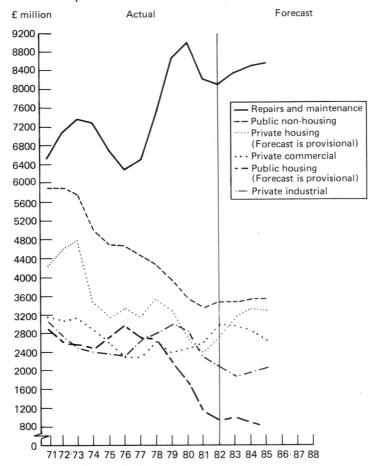

Figure 1.1.2 Value of output by contractors and direct labour on construction work at 1980 prices.

Because this type of building is unofficial, often illegal, one is unlikely to find official reports of the accidents of those involved in it. These dwellings are clustered in 'shanty towns' on the least desirable sites on the outskirts of large cities and the risks to the inhabitants living in these structures are probably greater than the risks of constructing them even though the majority are only single storey. Without such dwellings, however, their lives would be even less secure. The following passage gives a good description of such construction in Brazil.[7]

'The favelas of beautiful Rio de Janeiro are perhaps best known. Well over 650,000—perhaps a million—people are said to live in favelas. These favelas consist of a jumble of rude shacks made of wood, cardboard, scraps of metal and other bits of materials. They are crowded together, sometimes only a few feet apart on the steep hillsides which rise above the city... The shacks are often perched most

Figure 1.1.3 '. . . a jumble of rude shacks made of wood, cardboard, scraps of metal . . .' A typical Brazilian favela (Popperfoto).

precariously on steep slopes, supported by fragile wooden piles; sometimes in a heavy tropical rainstorm, several come tumbling down, killing the inhabitants. The alleyways between the shacks are of course unpaved and the houses follow no regular order. Sewers are totally lacking.

Favelas also lack a public water supply, except at a few points where the city may have set up a public fountain.

There is no public illumination. Oddly enough, however, many houses have electricity—a few weak bulbs, radios and even TVs. The electricity is obtained from enterprising sub-contractors who string wires from their own homes near the edge of the settlement to the shacks of their clients.

Such favelas come to be established in a characteristic manner throughout Brazil. They often begin with an invasion, a number of squatters suddenly occupy a piece of vacant property which for one reason or another is of low value but lies within the city. In Rio, for example, the hillsides had no water or sewers and in Bahia the valleys were hot and full of mosquitoes. Such invasions generally take place at night and are well organised. A landlord awakes one morning to find several shacks on his property....

As such favelas become permanent parts of great cities, a further step is taken. The residents begin to improve their property and to build more permanent dwellings. This is especially noticeable in Bahia where shacks are turned into adobe houses; next a painted facade is added, and then the municipality paves the street and allows electricity and other services to be brought into the area. Finally, the invasion that was a slum becomes a permanent residential district for the better-off members of the city's lower class.'

It is hard to comment usefully on accident prevention in these conditions. They must, nevertheless, be borne in mind as part of the background to the more regulated and controlled construction which most readers are used to. Indeed, such conditions are becoming increasingly common in our industrialised societies in the present economic climate.

1.1.1.5 Primitive rural dwellings

Primitive rural dwellings in third world countries vary from the tents of nomads to simple single-storey buildings with dried mud brick walls and earth floors. They have flat roofs (based on wooden poles and matting) in dry climates, sloping roofs in wet.

In some African countries with moderate to heavy rainfall the thatched adobe hut has a very short life. The thatch must be repaired several times a year and completely replaced every third year. The walls often disintegrate through rain erosion within six years. The hut is then abandoned and another one built.

Figure 1.1.4 Building in the Cameroons is a labour intensive activity and roofing methods are different as is the climate. Safety headgear, however useful, would look mildly ridiculous (Popperfoto).

The advent of corrugated galvanized iron roofing sheets which protect the tops of the walls from rain erosion provided a marked technical advance which unfortunately not all can afford in rural Africa.

The hazards of constructing these dwellings are again probably small compared to the hazards of living in them.

1.1.2 Civil engineering

Civil engineering is mainly concerned with the building of roads, bridges, tunnels, dams, harbours and other works, but it encroaches to a large extent on the work of the builder. Civil engineering contracts in the U.K. are usually based on the ICE (Institution of Civil Engineers) Conditions, apart from Government contracts where the same Form GC/Works/1 applies to both building and civil engineering. They employ many of the same professional skills (structural engineers, clerks of works, site agents) and tradesmen (scaffolders, concrete workers, ground workers, plant operators, carpenters), to mention but a few. However, several activities common in civil engineering, e.g. tunnel work, diving, welding and steel erecting, are less common in building.

Civil engineering enterprises are on average much larger than those in building. Several civil engineering contractors are very large companies; some are international household names. Within the building sector, especially private housing, the small firm is the norm. Many large construction contractors do, however, undertake both building and civil engineering work.

The smaller contractors generally have poorer safety records than the medium-sized and large ones (see Chapter 4.1). Civil engineering workers are more at risk than building workers to the possible collapse or flooding of the structure they are constructing. Bridges, tunnels and dams are all more vulnerable than houses in the construction stage since they are subject to much higher and sometimes unknown stresses. The post war box girder bridge disasters (Milford Haven, 5 lives; Yarra river, Australia, 35 lives and Lieser river, Austria, 10 lives (1975)) are examples.[8]

The railway navvies and tunnel builders in nineteenth century England suffered appallingly from the exposed conditions of their work and living quarters in improvised construction camps. The same applies today to the migrant workers on large construction projects in many third world countries.

These hazards are discussed in Chapter 4.1.

1.1.3 Process and power plant construction

Oil refineries, chemical plant, cement factories, mineral treating plant, boilers and power stations and a vast range of other process plant account for a sizeable proportion of permanent fixed construction. In some cases the plant is housed in buildings while in others it is erected in the open. The construction of buildings, roads, foundations and sewers falls under building and civil engineering, whilst most of the plant, equipment and machinery is constructed elsewhere and simply installed on site. Much of the plant installed is very heavy and requires large cranes and other lifting equipment.

A great deal of structural steel and pipework is involved in this type of work and has to be erected and installed, often at considerable heights. Once erected, pipes and vessels have to be lagged and clad, instruments and services have to be added, and painting is usually required. The lifting and positioning of pipes and other equipment makes it difficult to use scaffolding which may obstruct the lifts. Thus the construction crew are exposed to increased risks of falling or being hit by falling objects. Asbestos insulation, whilst now generally banned in most industrialised countries, is still in use in several third world countries. It is a serious health hazard to laggers and insulators.

The hazards of erecting plant and equipment inside a building are usually less than doing the same in the open. The building offers protection against the elements and often has cranes or other lifting gear built into it.

Large tanks and pressure storage vessels often have to be site-fabricated by welding. A great deal of planning is needed to ensure that the welders and other workers are properly protected from falling and other hazards.

1.1.4 Offshore oil and gas installations

Most offshore oil and gas work is potentially very hazardous. It includes the installation of:

- offshore platforms and the buildings, plant and equipment on them;
- the pipework, much of it submarine, linking the platforms to the shore; and
- the drilling and completing of gas and oil wells.

It requires great and thorough engineering skill in designing and planning, and experienced disciplined teams following carefully worked-out procedures to accomplish the work.

Divers have to work at ever-increasing depths. Reports of deaths and casualties among divers working on offshore platforms and their pipelines are not uncommon. Submarine pipework has to be installed and connected to a subterranean oil or gas reservoir under high pressure. This is a highly specialised, as well as a hazardous task.

1.1.5 Demolition[9]

Demolition is included here as a construction activity. It is often a necessary preliminary to construction and the skills required and the hazards associated with it are very similar to those found in construction. An intimate knowledge of construction and the order in which things were done is essential in demolition. Demolition generally proceeds in the reverse order to construction.

Demolition is usually and rightly regarded as a 'high risk activity'. Its hazards are discussed in more detail in Chapter 1.5.

Four types of contractor are involved in demolition work, general demolition, dismantling (where recovery of scrap metal is the main objective), explosives engineers and civil engineers.

General demolition workers in the U.K. are classified in three categories, general labourers, mattock men who work at intermediate heights and topmen who should be capable of working safely at any height.

Bricks laid with lime and mortar—found principally in old buildings—are much easier to dismantle and recover than bricks laid with cement mortar which is often stronger than the bricks themselves. Reinforced concrete is still more difficult to demolish. Pre-stressed concrete, either pre-tensioned or post-tensioned, is even more of a problem. The stresses in the material may be released with considerable and unexpected force when the anchorages or tendons are cut or the load on the stressed member is reduced.

The hazards commonly found in demolition include the following:

- falls
- being hit by falling objects
- eye injuries from flying particles
- inhalation of hazardous dusts, fibres and fumes
- explosions
- electric shock through contact with live wires and cables.

The hazards of demolition are by no means confined to the workers themselves. They extend to many others in the locality, as well as adjacent property. Underground pipes, sewers and electrical and telephone lines may be damaged by vibration and heavy falls of materials. The damage is often not immediately apparent. Further hazards may then arise (e.g. escape of fuel gas leading to explosions and escape of sewage leading to disease).

References

1. CENTRAL STATISTICAL OFFICE, *Standard Industrial Classification*, HMSO, London (revised 1980).
2. WOTTON, H., SIR, *The Elements of Architecture* (1624).
3. JOHNSTON, J.E., *The Clerk of Works in the Construction Industry*, Crosby Lockwood Staples, London (1975).
4. JUNGK, R., *Tomorrow is Already Here*, Hart-Davis, London (1954) p. 166.
5. *The Holy Bible*, Authorised King James Version, Genesis XI.
6. NATIONAL ECONOMIC DEVELOPMENT OFFICE, *Construction Forecasts 1983–1984–1985*, HMSO (1983).
7. WAGLEY, C., *An Introduction to Brazil*, revised edition, Columbia University Press, New York (1971).
8. SCOTT, G., *Building Disasters and Failures*, Construction Press, Lancaster (1976).
9. PLEDGER, D.M., *A complete guide to Demolition*, Construction Press, Lancaster (1977).

1.2 Trades and activities in construction

1.2.1 Trades and numbers employed

Labour in the construction industry is sub-divided into a variety of inter-dependent crafts and occupations. Each is subject to a characteristic spectrum of hazards from sudden accidents and illness caused by exposure to toxic substances, bad weather and physical exertion in cramped or unnatural positions.

The early specialization of building crafts prior to mechanization led to the development of several highly skilled crafts, though much unskilled labour was also employed. As mechanization, standardization and rationalization increased, the nature of much of the work changed. Many of the early manual skills became redundant. At the same time there was a greater need for mechanics, electricians, men to work closely with machinery—circular saws, mechanical diggers and grinders—and men who could follow written instructions and read drawings.

The process of automation and factory production of prefabricated parts has also proceeded fast. Fewer men now work as manual complements to machines (e.g. feeding timber to a circular saw), but instead there is a higher proportion of white-collar workers dealing with planning, ordering, inventories, standards, technical calculations, drawing and the like.

TABLE 1.2.1. U.K. construction industry. Employment by occupation unit groups, 1971 census. 10% sample

Group No.	Occupation	Number	Percent-age
005	Gardeners & groundsmen	201	0.12
009	Coal mine workers below ground	181	0.11
026	Linesmen, cable joiners	142	0.09
027	*Electricians*	9058	5.43
028	Electrical & electronic fitters	302	0.18
030	Electrical engineers (so described)	532	0.32
033	Sheet metal workers	146	0.09
034	Steel erectors, riggers	1577	0.94
035	Metal plate workers, riveters	243	0.15
036	Gas & electric welders, cutters, braziers	706	0.42

TABLE 1.2.1. Continued

group No.	Occupation	Number	Percentage
039	Machine tool operators	137	0.08
041	Motor mechanics, auto engineers	256	0.15
042	Maintenance fitters & engineers, millrights	1360	0.81
043	Fitters, n.e.c. machine erectors etc.	1827	1.09
045	*Plumbers, gas fitters, lead burners*	8697	5.2
046	Pipe fitters, heating engineers	3311	1.98
054	Other metal working	1393	0.83
055	*Carpenters & joiners*	15978	9.57
057	Sawyers & wood working machinists	308	0.14
059	Woodworkers, n.e.c.	134	0.08
091	Craftsmen, n.e.c.	276	0.17
093	*Bricklayers, tile setters*	11467	6.87 ⎫
094	Masons, stone cutters, slate workers	938	0.56 ⎭
095	Plasterers, cement finishers, terrazzo workers	3999	2.40
096	Builders (so described) Clerk of Works	6440	3.86
097	Bricklayers' etc. labourers n.e.c.	1083	0.65
098	*Construction workers n.e.c.*	15773	9.45
100	*Painters, decorators*, n.e.c.	15261	9.14
102	Boiler firemen	150	0.09
103	Crane & hoist operators, slingers	857	0.52
104	Operators of earth moving & other construction machinery	4400	2.64
113	*Labourers & other unskilled workers*	22093	13.24
122	Drivers of road goods vehicles	4406	2.64
127	Telephone operators	199	0.12
136	Warehousemen, storekeepers & assistants	1083	0.65
138	Office managers	387	0.23
139	Clerks, cashiers	6156	3.69
140	Office machine operators	487	0.29
141	Typists, shorthand writers, secretaries	3388	2.03
143	Proprietors & managers sales	105	0.06
144	Shop salesmen & assistants	242	0.14
148	Commercial travellers, manufacturers agents	311	0.17
150	Salesmen, services, valuers, auctioneers	155	0.09
153	Guards and related workers	334	0.20
161	Canteen assistants, counter-hands	252	0.15
166	Charwomen, office cleaners, window cleaners, chimney sweeps	543	0.33
176	Managers in building & contracting	6890	4.13
177	Managers in mining and production n.e.c.	196	0.12
179	Sales managers	880	0.53
180	Managers n.e.c.	650	0.39
195	Civil, structural and municipal engineers	2153	1.29
196	Mechanical engineers	610	0.37
197	Electrical engineers	156	0.09
200	Planning, production engineers	114	0.07
201	Engineers, n.e.c.	148	0.09
209	Accountants, professional	112	0.07
210	Company secretaries	480	0.29
211	Surveyors	1912	1.15
212	Architects, town planners	321	0.19
218	Draughtsmen	687	0.41
220	Technical & related n.e.c.	1238	0.74
	Not elsewhere described		1.91
	Gross total	166913	100.00

Occupations which contribute more than 5% to the workforce are in italic type.

Most prefabricated parts such as doors, windows, wall panels and reinforced concrete beams are no longer regarded as part of the construction industry, but as a branch of manufacturing.

Work on building sites is of necessity less automated than the production of prefabricated components. However, in most industrialized countries construction is fairly capital-intensive, relying heavily on the use of machinery. It also includes a large element of assembly work, putting together prefabricated components delivered to the site.

Some idea of the occupational breakdown of construction workers in the U.K. is given by the returns of the 1971 census, based on a 10% sample.[1] This is shown in *Table 1.2.1*. At that time the total employed population in the U.K. was 23,733 thousand, of whom 8,701 thousand were women. The construction industry employed 1,669 thousand, of whom 97 thousand were women, so that roughly 1 employed man in 10 was engaged in construction. Of the whole employed population 7.7% were registered as 'self employed', compared with 19.3% in the case of the construction industry. The total numbers employed in the U.K. today (Summer, 1984) are, of course, down on the 1971 figures due to the recession, whilst the construction industry has been worse hit than most others.

The most striking thing about *Table 1.2.1* is the large number of occupations which make up the construction industry. Of the 62 occupations listed, only 7 (in italic) individually contribute more than 5% to the total workforce. In addition to those listed the census return shows a further 108 occupations in construction which each contribute less than 0.06% to the total workforce. These are included in the 1.91% shown as not elsewhere described. The largest groups in *Table 1.2.1* are:

Labourers and other unskilled workers (13.24%)
Carpenters and joiners (9.57%)
Construction workers not elsewhere classified (9.45%)
Painters, decorators (9.14%)
Bricklayers and tile setters (6.87%)
Electricians (5.43%)
Plumbers, gas fitters and lead burners (5.2%)

The percentages employed in occupations which are critical to the safety of others or which are inherently hazardous are surprisingly small. The total employed, for example, in steel erection, a very hazardous occupation, is only 0.94%. Similarly crane and hoist operators only account for 0.52%, whilst for earth moving plant operatives the figure is still only 2.64%.

The small percentages of some occupations such as welders and braziers, etc. (9.42%) and draughtsmen (9.41%) in the total workforce appears to be due to the fact that most of these are classified under engineering rather than construction in the census return.

There are a number of important omissions from the table, some of which are included in the more recent data of *Table 1.2.5*, but here appear under 'construction workers n.e.c.'. For example, roof workers, concrete workers, divers, tunnel workers, blasters and others working with explosives are omitted. Safety specialists, often referred to as 'officers' or 'engineers', are also not included as a separate heading.

TABLE 1.2.2. Construction craft operatives employed in the U.K. 1978 (thousands)

Craft	Employed by: Contractors	Employed by: Local authorities & New Towns	Total	%
Carpenters & joiners	107.8	22.2	130.0	15.1
Bricklayers	54.6	10.6	68.0	7.9
Masons	2.8			
Roof tilers & slaters	7.4	1.3	14.5	1.7
Floor, wall & ceiling tilers	5.8			
Plasterers	13.5	3.1	16.6	1.9
Painters (and decorators)	52.9	23.8	76.7	8.9
Plumbers & gas fitters	32.7	11.8	44.5	5.2
Heating & ventilating workers	19.5	0.9	20.4	2.4
Glaziers	3.2	0.7	3.9	0.5
Paviours	2.4	4.3	6.7	0.8
Scaffolders	not shown	0.7	0.7	0.1
other steel erection & sheeting activities	3.9	0.1	4.0	0.5
Electricians	55.5	6.5	62.0	7.2
Mechanical plant & equipment operators	53.6	6.0	59.6	6.9
Other B & C E* crafts and occupations	79.7	17.6	97.3	11.3
All other occupations	200.4	16.1	254.4	29.6
Unskilled labourers		37.9		
Total operatives	695.7	163.6	859.3	100.0

*Building and Civil Engineering

As the figures given in *Table 1.2.1* are compiled from a census return, the occupations given are based on the statements of the employees themselves.

Other information on employment by trades in construction can be found in 'Construction and Housing Statistics' published by the Department of the Environment.[2] *Table 1.2.2* gives the numbers of the various construction craft operatives employed in the U.K. in 1978 by (1) Contractors and (2) Local Authorities and New Towns. The numbers of operatives in most classes is lower than those shown in *Table 1.2.1*, mainly because trainees, apprentices and workers in many small firms were not included. The distribution of workers in various crafts, however, follows a similar trend in both cases. The main trades in *Table 1.2.2* show the following percentages, with the percentages in *Table 1.2.1* given in brackets:

Carpenters and joiners	15.1	(9.57)
Painters and decorators	8.9	(9.14)
Bricklayers and masons	7.9	(7.43)
Electricians	7.2	(5.93)
Mechanical plant & equipment operators	6.9	(3.16)
Plumbers and gas fitters	5.2	(5.2)

With such a vast number of occupations it is interesting to see their distribution in relation to the business or trade of employers. This is shown in *Table 1.2.3* which gives the number of businesses in each speciality, their total numbers of employees and the average number of employees per form. Since it includes 'general builders' with a fair cross-section of

24 Trades and activities in construction

TABLE 1.2.3. Employment in construction firms in the U.K. in 1978

Type of firm	Total employees (thousands)	Number of firms	Average number of employees per firm
General builders	357.0	40081	8.9
Building & civil engineering contractors	221.9	2661	83.4
Civil engineers	80.1	2011	39.8
Plumbers	31.6	8310	3.8
Carpenters & joiners	23.1	6323	3.7
Painters	61.4	12832	4.8
Roofers	23.6	2905	8.1
Plasterers	18.8	2864	6.6
Glaziers	9.3	1374	6.8
Demolition contractors	4.3	475	9.1
Scaffolding specialists	16.0	387	41.3
Reinforced concrete specialists	7.9	353	22.4
Heating & ventilating engineers	55.2	4640	11.9
Electrical contractors	68.4	7757	8.8
Asphalt & tar sprayers	13.8	584	23.6
Plant hirers	37.3	3059	12.2
Flooring contractors	5.6	838	6.7
Constructional engineers	13.9	910	15.3
Insulating specialists	10.8	570	18.9
Suspended ceiling specialists	3.4	367	9.3
Floor & wall tiling specialists	4.1	598	6.9
Miscellaneous	16.3	1181	13.8
Total — average	1083.8	101080	10.7

different tradesmen, the figures are scarcely comparable with those of *Tables 1.2.1* and *1.2.2*. The most striking feature of *Table 1.2.3* is the small average size of the firm. Throughout the whole industry the average number of employees is only 10.7.

This overall average includes the following types of business with higher averages:

Building & civil engineering contractors	83.4
Scaffolding specialists	41.3
Civil engineers	39.8
Asphalt and tar sprayers	23.6
Reinforced concrete specialists	22.4

All the rest by trade or business classification have on average fewer than 20 employees per firm. Noticeably firms specialising in carpentry and joinery, plumbing and painting have less than five workers.

These figures are referred to again in Chapter 4.1, 'Special problems of the industry'. The small average size of firms in construction is considered to be a major reason for its high accident rate.

1.2.2 Comprehensive list of occupations

Table 1.2.4 gives a more complete list of construction occupations as currently found in England and Wales, based on various sources[3,4,5,6]. The

TABLE 1.2.4. Construction occupations, professional, trade, etc. as found in the U.K.

1. *MANAGERIAL, TECHNICAL AND SUPERVISORY STAFF*
   (excluding clerical & secretarial)
   - ARCHITECTS
   - CIVIL AND STRUCTURAL ENGINEERS
   - SPECIALIST ENGINEERS, ELECTRICAL, HEATING, ETC.
     - Specialist draughtsmen, interior designers
   - CONTRACT MANAGERS
     - Quantity surveyors
     - Estimators
     - Buyers
   - CLERKS OF WORKS
   - SITE AGENTS
     - Construction (or building contracts) surveyors
     - Production controllers (or bonus surveyors)
     - Planning (or progress) engineers
     - Site engineers
     - Laboratory technicians
     - General foremen
     - Section foremen
     - Gangers
     - Chainmen (assistants to site engineers)
2. *SAFETY HEALTH, WELFARE AND INSPECTION STAFF*
   - BUILDING CONTROL INSPECTORS
   - ENVIRONMENTAL HEALTH OFFICERS
   - HEALTH AND SAFETY INSPECTORS
   - SITE MEDICAL OFFICERS
   - SAFETY SPECIALISTS (MANAGERS, ENGINEERS OR OFFICERS)
     - Nurses
     - Welfare workers
     - Ambulence men
     - First aiders
     - Canteen workers
     - Guards, watchmen
3. *BUILDING CRAFTS AND OCCUPATIONS*
   - BRICKLAYERS AND STONEMASONS
     - Bricklayers, building construction
     - Bricklayers, chimney construction
     - Bricklayers, furnace construction
     - Banker masons
     - Fixermasons
     - Hod carriers and special labourers
   - CARPENTERS AND JOINERS
     - Carcass work carpenters
     - First fixers
     - Second fixers
     - Shopfitters
     - Bench joiners
     - Woodworking machinists
     - Formwork carpenters
   - ROOFERS
     - Felter roofers (including fixers and potmen)
     - Roof tilers
     - Roof slaters
     - Roof sheeters & wall cladders
     - Mastic asphalters (including potmen and spreaders)
   - CLAY AND CONCRETE WORKERS
     - Clay puddlers
     - Concrete pourers, levellers and vibrator operators
     - Concrete screeders and surface finishers

TABLE 1.2.4. Continued

    PLASTERERS
      Solid plasterers
      Fibrous plasterers (makers and fixers)
    SCAFFOLDERS (including falsework scaffolding)
    STEEL FIXERS
    BARBENDERS
    STEEPLE JACKS
4. *ENGINEERING AND SERVICE WORKERS*
    STEEL RIGGER-ERECTORS
    INSTALLATION ELECTRICIANS
    PLUMBERS
    HEATING AND VENTILATING FITTER-WELDERS
    DUCTWORK FITTER-ERECTORS
    GAS SERVICE FITTERS
    WELDERS, GAS AND ELECTRIC ARC
    PLATERS
    PIPE FITTERS
    ELECTRICAL LINESMEN-ERECTORS
    LIFT AND CRANE ERECTORS AND MECHANICS
    FIRE PROTECTION ENGINEERS (FITTERS)
    THERMAL INSULATION ENGINEERS (LAGGERS)
    TELECOMMUNICATIONS ENGINEERS (FITTERS)
    INSTRUMENT ENGINEERS (FITTERS)
    POWER AND PROCESS PLANT INSTALLATION ENGINEERS (FITTERS)
    CONSTRUCTION PLANT MAINTENANCE MECHANICS
5. *TRANSPORT WORKERS AND PLANT OPERATORS*
   (other than road surfacing)
    DRIVERS AND MOBILE PLANT OPERATORS
      Lorries, tractors and other road vehicles
      Excavators and earth moving equipment
      Site trucks and dumpers
      Mobile cranes, various
    STATIONERY CRANES, HOISTS, GIN POLES, ETC.
      Operators
      Banksmen – slingers
    CONCRETE AND MORTAR MIXER AND PLACER OPERATORS
    AIR COMPRESSOR AND POWER DRIVEN TOOL OPERATORS
    PUMP ATTENDANTS
    TUNNELLING MACHINE OPERATORS
    POST TENSIONING AND PRESTRESSING CONCRETE MACHINE OPERATORS
    PILE DRIVING AND DRILLING PLANT OPERATORS
      Percussive (sheet, etc.)
      Rotary and specialist
6. *ROAD AND GROUND WORKERS NOT ELSEWHERE SPECIFIED*
    LAYERS OF FOUNDATIONS, UNDERGROUND SERVICES
      Navvies, foundation layers
      Timbermen
      Pipe layers (drains, gas, water)
      Cable layers (electricity, telephone, T.V.)
    BLASTING, SHOTFIRERS
    ASPHALT ROAD SURFACING
      Rakers
      Tampermen
      Chippers
      Spreader operators
    PAVING, DRY-WALLING
      Pavoirs rammermen
      Kerb and paving jointers
      Dry wallers

TABLE 1.2.4. Continued

```
      DIVERS AND WORKERS IN COMPRESSED AIR ATMOSPHERES
         Divers
         Tunnel workers
         Caisson and coffer dam workers
         Life support system operatives
   7. FINISHING CRAFTSMEN
      PAINTERS AND DECORATORS
         Industrial painters
         Decorative hangers
         Paper hangers
      GLAZIERS
      CEILING FIXERS
      WALL TILERS
      FLOORING LAYERS
         Floor tilers (including terrazo and mosaic)
         Carpet layers
         Lino layers
         Wood block and timber floor layers
      FENCE AND RAILING ERECTORS
      T.V. AERIAL ERECTORS
   8. ALLIED OCCUPATIONS
      DEMOLITION WORKERS
         Mattock men
         Topmen
      FAIRGROUND WORKERS
      GRAVE DIGGERS
      WINDOW CLEANERS
      BUILDING MAINTENANCE WORKERS (including brick and stone cleaners)
      GENERAL WORKERS
      LABOURERS AND GENERAL CLEANERS
      STOREMEN
```

names of the occupations and their demarcation lines vary even within the U.K., and many differences are found between Scotland and the south of England. Non-British readers should, however, have little difficulty in recognising the different jobs done from the descriptions in *Table 1.2.4*.

The proliferation of occupations in a reducing workforce can be accounted for to some extent by workers carrying out two or more jobs. This has far-reaching effects on the safety performance of the construction industry. The official aim is that all workers, especially those in jobs which are potentially hazardous to themselves or to others, should be properly trained, tested and certified.

This intention tends to be negated, not only because of dual job employees, but also because of the high proportion of workers classified as 'self employed' or 'labour only sub contractors', known as 'The Lump'. Whilst some may be highly skilled and reputable workers, a large number are unregistered casual workers whose main concern is to avoid paying income tax.

1.2.3 More recent data

More recent occupational employment data[7] from the Construction Industry Training Board covers the period October 1974 to April 1982.

TABLE 1.2.5. Construction industry training board. Employment estimates by occupations, thousands, October 1974 and April 1982[7]

Occupation	1974 October Thousands	% of total	1982 April Thousands	% of total	% change from 1974 to 1982
Managerial staff	56.4	5.7	60.5	7.9	+7
Architects, surveyors, engineers	24	2.4	19.5	2.5	+19
Technical staff	29.5	3.0	30.7	4.0	−4
Draughtsmen and tracers	5.3	0.5	3.7	0.5	−30
General foremen & foremen	39.3	4.0	35.3	4.6	−10
Clerical and sales staff	94.5	9.6	86.5	11.2	−8
Bricklayers	58.1	5.9	39	5.1	−33
Masons	3	0.3	2.7	0.4	−10
Carpenters/joiners	121.9	12.4	87	11.3	−27
Painters	54	5.5	38.9	5.1	−28
Plasterers	13.5	1.4	9.4	1.2	−30
Roof slaters and tilers	3.2	0.3	4.2	0.5	+31
Pavoirs	2.7	0.3	1.4	0.2	−48
Miscellaneous craftsmen, excluding mechanical engineering services	10.6	10.8	10.1	1.3	−5
Scaffolders	9.5	1.0	10.2	1.3	+7
Roof sheeters	2.4	0.2	1.5	0.2	−37
Roofing felt fixers	2.7	0.3	2.8	0.4	+4
Floor and wall tilers	1.3	0.1	1.4	0.2	+8
Ceiling fixers	0.8	0.1	1.1	0.1	+37
Mastic asphalters	0.9	0.1	1.7	0.2	+89
Floor coverers	1.5	0.2	1.2	0.2	−20
Floorers	0.2	−	0.4	0.1	+100
Glaziers	2.8	0.3	2.7	0.4	−4
Fencers	0.5	0.1	0.7	0.1	+40
Demolishers	2.2	0.2	1.1	0.1	−50
Steeplejacks	0.7	0.1	0.4	0.1	−43
Cavity wall insulation operatives	0.2	−	0.8	0.1	+300
Demountable partition erectors	0.1	−	0.2	−	+100
Terrazzo workers	−	−	0.1	−	−
Plumbers and gas fitters	35.5	3.6	27.7	3.6	−22
Heat and ventilating engineering workers	19.7	2.0	16	2.1	−19
Other mechanical engineering service workers	7.3	0.7	5.8	0.8	−21
Electricians	55.6	5.7	47.3	6.1	−15
Crane drivers	6.9	0.7	3.8	0.5	−45
Earth moving plant operators	18	1.8	14.3	1.9	−21
Other mechanical plant operators	12.8	1.3	8.9	1.2	−20.5
bar benders and steel fixers	4.1	0.4	2.5	0.3	−39
Steel erectors	0.7	0.1	1	0.1	+43
Concretors	4.6	0.5	2.9	0.4	−37
Gas distribution mains layers	4.9	0.5	2.1	0.3	−57
Plant mechanics	15.1	1.5	14.3	1.9	−5
Other building and civil engineering workers	38.2	3.9	25.7	3.3	−33
Unskilled workers	183.2	18.6	114	14.8	−38
Other occupations	35.9	3.7	30	3.9	−16
Total – all occupations	982.5		769.4		−22

Since the occupational groupings differ from those in *Tables 1.2.1* to *1.2.3*, relevant parts of the recent data are given in *Table 1.2.5*. This shows the numbers of employees (including trainees but excluding self employed persons) in all trades and occupations in UK construction in October 1974 and in April 1982. There has been an overall decline of 22% in numbers employed. The decline is most marked in traditional skilled manual trades. On the other hand, the number of managerial staff, architects, surveyors and engineers, has increased. So too have the numbers employed in certain occupations, some of which (ceiling fixers, mastic asphalters and cavity wall insulation operatives) reflect changing trends in construction techniques.

References

1. BAKER, R., *Manpower Estimates of the Construction Industry*, Construction Industry Training Board, London (1982).
2. DEPARTMENT OF THE ENVIRONMENT, *Construction and Housing Statistics, 1979–1980*, HMSO, London.
3. MANPOWER SERVICES COMMISSION, CAREERS AND OCCUPATIONAL INFORMATION CENTRE, *Leaflets CV 62 to 73 inc.* HMSO, London.
4. BUILDING INDUSTRY CAREERS SERVICE, Various Leaflets, National Federation of Building Trade Employers, London.
5. *National Agreement for Operatives in Heating, Ventilating, Air Conditioning, Piping and Domestic Engineering Industry*, Heating and Ventilating Contractors' Association, London (1979).
6. CIVIL ENGINEERING CONSTRUCTION CONCILIATION BOARD FOR GREAT BRITAIN, *Working Rule Agreement*, Federation of Civil Engineering Contractors, London (1982).
7. BAKER, R., *Internal Information Paper on Employment, etc.* Construction Industry Training Board, London, 10 January 1984.

1.3
Safety and health statistics

Even in the most industrially developed countries it is only fairly recently that meaningful statistics have been kept and published on accidental injuries and occupational diseases among construction workers. This coincides with an increase in the general awareness of the issues of occupational health and safety and improvements in legislation and enforcing inspectorates. Examples include the Health and Safety at Work, etc. Act in the U.K. (1974), the Law on the Improvement of Industrial Accident Prevention in France (1975) and others passed at about the same time in Belgium, Denmark, the Netherlands and the German Federal Republic. These improvements, however, still leave the construction industry as a whole with a far worse accident record than that of manufacturing industry in all the countries concerned, and within the construction industry there are several occupations—steel erection in particular—whose records are notorious and much higher than the average for construction. Thus, despite the improvements that have taken place in developed countries during the 1970s, the record leaves no room for complacency. With greater attention to hazards and safety in construction there seems no reason why the present Swedish trend (see Section 1.3.1) should not set the pattern for all.

Before considering the implications of statistics it is necessary to understand what is meant by the term 'accident'. Even among English speaking safety specialists the word has no universally accepted meaning. The definition preferred by the writer and used here is 'An accident is an unplanned event which has a probability of causing personal injury or property damage'. It follows that accidents can be classified in 4 types depending on their effects:

1. Accidents which cause neither property loss nor personal injury
2. Accidents which cause property loss without personal injury
3. Accidents which cause personal injury without property loss, and
4. Accidents which cause both personal injury and property damage.

Most so-called accident statistics deal only with type 3 and properly speaking are statistics of accidental injuries. Since this itself is a very broad heading, these are generally broken up into those causing fatal injuries, severe injuries and minor injuries, with suitable definitions of each category.

It is an axiom of accident prevention that all accidents have causes and a further act of faith that the great majority are preventable. The cause of an accident may remain latent for a long time before an accident occurs, and to describe a latent accident cause (whether it has caused an accident or not) we use a different word, 'hazard'. A hazard is defined as 'a condition with the potential of causing injury or damage'.

Probably the major part of modern scientific accident prevention is concerned with hazards—specifically discovering them and publicising how and where they arise, how to recognise them and their possible consequences, and making recommendations for eliminating or reducing them. The keeping of reliable statistics of accidental injuries in any occupation is an essential step in this process.

Mere numbers, however, tend to be rather meaningless unless they can be related to everyday experience. Whilst absolute figures for fatal injuries and 'reportable' or 'compensatable' injuries in construction are published annually by a number of countries, their significance is only apparent when they are compared with numbers employed, hours of work, work output, etc. Hence relative figures such as frequency and incidence rates are generally more meaningful than absolute ones. The main source of international statistics in this area is the International Labour Office.

The ILO classifies 'industrial accidents' (accidental injuries) in five different ways[1] following a Resolution adopted by the Tenth International Conference of Labour Statisticians (October, 1962):

1. *According to the degree of injury:*
   Fatal—permanent disablement—temporary disablement and other cases (minor injury)

2. *According to the type of event causing the injury.*
   There are 9 main categories listed below, each with several sub-categories:

   (a) Falls of persons
   (b) Struck by falling objects
   (c) Stepping on, striking against or struck by objects, excluding falling objects
   (d) Caught in or between objects
   (e) Over-exertion or strenuous movements
   (f) Exposure to or contact with extreme temperature
   (g) Exposure to or contact with electric current
   (h) Exposure to or contact with harmful substances or radiations
   (i) Other types of accident, including those not classified for lack of sufficient data

   Two omissions from this list may be noted:

   Being rubbed or abraded by an object, and
   Exposure to blast from an explosion.

   A few countries publish statistics of construction injuries under these headings. From these, the dominant cause of fatal accidents would

seem to be 'falls of persons', followed by being 'struck by falling objects'.

3. *According to the agency*
   Here there are 7 main agencies, such as machines, means of transport, flying fragments, each split up into a large number of categories and sub-categories. It is seldom that this classification is found in published injury statistics.

4. *According to the nature of the injury*
   There are 16 headings in this list, such as fractures, dislocations, etc., which coincide more or less with those used in official U.K. statistics. Such statistics are mainly of interest to first aiders and medical staff, but they should also be studied by safety specialists, particularly when planning protection for personnel.
   The vast majority of injuries suffered in construction are of three types:
   fracture, contusions, and sprains and strains.

5. *According to the bodily location of the injury*
   Here there are 7 main headings, head, neck, trunk, etc., and a number of sub-headings. It is not common to find these shown in official statistics, but the same comments made under (4) also apply.

For the purpose of classifying the risk of different injuries which may be expected in different construction occupations the first and second of these methods of classification are the most useful. A slight modification of (2) as used by the U.K. Health and Safety Executive[2] is used in Section 1.3.2. As already stated, frequency and incidence rates are more meaningful than absolute figures. Here again it is necessary to understand what is meant by these terms.

The ILO has made the following recommendations on frequency rates and incidence rates:

$$\text{FR (Frequency rate)} = \frac{\text{Total number of industrial accidents (of stated severity)} \times 1\,000\,000}{\text{Total number of man hours worked}}$$

Whilst the ILO quote all accidental injury frequency rates (including fatal injuries) on this basis, reportable accidental injury frequency rates in the U.K. are quoted per 100 000 hours worked, and fatal accident frequency rates per 100 000 000 man hours.

$$\text{IR (Incidence rate)} = \frac{\text{Number of accidents per year} \times 1000}{\text{Average number of workers at risk during the period}}$$

In this book the ILO recommendations are followed except where otherwise stated.

In its annually-published statistics for a number of countries, a bare majority quote only frequency rates as defined above, whilst others quote incidence rates only, but unfortunately on one of three different bases:

(a) rates per 1000 man years of 300 days each
(b) rates per 1000 wage earners
(c) rates per 1000 persons employed

An approximate relationship exists between the frequency rate as defined by the ILO and the incidence rate on basis (a) as used by the ILO.

Assuming a 7½ hour working day:

1000 man years of 300 days each = 2 250 000 hours

Hence a frequency rate of 1 would lead to an incidence rate of 2.25, so to convert frequency rates to incidence rates on basis (a) one should multiply by 2.25.

Whilst significant differences are likely between incidence rates quoted on bases (a), (b) and (c), it is difficult to generalise since working hours and reporting practices vary markedly from country to country.

To translate a fatal accident frequency rate (as defined by the ILO and quoted in its published statistics) into more familiar terms, we might consider the chances of a fatal accident for a worker permanently employed in the industry. Such a worker might be employed for 40 years, 250 days per year and 8 hours per day, at the end of which he would have completed 80 000 working hours. A fatal accident frequency rate of 1 (per million hours) then corresponds to a probability of 1 in 12.5 of meeting a fatal accident during a working life in the industry. The chances of a serious injury are a great deal higher, so that an industry with a fatal accident frequency rate of 1 is a very dangerous one to consider for full-time employment. Even a figure of 0.1 is considered by most to be unacceptably high. Much, of course, depends on what other hazards the worker is exposed to during his life and how these affect his life expectancy.

For countries which report injury statistics in construction, fatal injury statistics are generally considered the most reliable. Permanent disablement statistics are seldom reported and the definition in any case may be interpreted in different ways. Many countries, however, report temporary disablement, but on bases other than those recommended by the ILO. In the U.K. it has, in the past, been accidents reported to a doctor and necessitating three or more days absence from work while in Japan the qualifying period is four days off work. France, Italy, West Germany, Switzerland, Sweden and others use compensated accidents rather than reported accidents as the base for their statistics. Since the question of compensation may take years to decide, statistics for temporary disablement are often very delayed. In many countries it has to be recognised that a high proportion of temporary disablements in construction go unreported.

Thus, even within the E.E.C., it is difficult to make comparisons of accidental injury rates in construction (other than fatal injuries) between different countries. The following comment appears in a recent official study on working conditions in construction within countries of the E.E.C. 'For a study of the situations in different countries there is a need for a standard data and information base on general conditions in the European construction industry and a standard system of definitions'.[3]

1.3.1 Comparative international fatal injury statistics in construction

With the points just discussed in mind, the statistics presented here, and taken from section 27 of the 1981 ILO Yearbook of Labour Statistics[4], have been selected on the following bases:

- Only fatal accident frequency rates have been included.
- Only data from countries which have published these figures annually over a complete consecutive eight year period between 1971 and 1980 have been used.

The data given in *Table 1.3.1* is in a condensed form. Instead of quoting figures for each year, the mean and standard deviation for the whole eight year period are given first, followed by the mean for the first and second four years of the whole period. The last column shows whether there has been a rise (+) or fall (−) in the mean fatal accident frequency rate in the second four year period compared with the first. When the difference between these means exceeds the standard deviation for the entire eight year period, this is shown either as 'MF' for 'marked fall', or 'MR' for 'marked rise'. The relatively small variation in FAFR's from one year to the next in most developed countries, as shown by the smallness of the standard deviations, is remarkable even to those used to dealing with the statistics of large populations.

TABLE 1.3.1. Fatal accident frequency rates in construction for selected countries[2] over consecutive 8 year period. Number of fatal accidents per 10^6 man hours worked

Continent	Country	Period	Mean	Standard deviation	Mean Years 1–4	Mean Years 5–8	Marked fall or rise
A. Industrialised countries	**Mean = 0.2560**						
America	Canada	73–80	0.3600	0.0532	0.3550	0.3650	+
	U.S.A.	72–79	0.1585	0.0343	0.1700	0.1470	−
Asia	Japan	73–80	0.1900	0.1000	0.2750	0.1050	MF
Europe	Austria	72–79	0.6140	0.1355	0.7083	0.5198	MF
	France	72–79	0.4004	0.0664	0.4540	0.3468	MF
	German Dem. Rep.	73–80	0.1100	0.0227	0.1225	0.0975	MF
	Sweden	71–78	0.0640	0.0139	0.0700	0.0580	−
	United Kingdom	73–80	0.1509	0.0333	0.1765	0.1253	MF
B. Developing Countries	**Mean = 0.8028**						
Africa	Burundi	73–80	1.2138	1.1148	1.6576	0.7700	−
	Cameroon	71–78	0.9275	0.2849	0.7950	1.0600	+
	Egypt	73–80	0.4363	0.0583	0.4150	0.4575	+
	Tunisia	73–80	0.1853	0.0807	0.2275	0.1430	MF
America	El Salvador	73–80	0.5976	0.6371	0.2325	0.9628	MR
	Panama	72–79	0.4725	0.4062	0.6175	0.3275	−
Asia	Jordan	72–79	0.3270	0.2841	0.1908	0.4633	+
	Korea, Republic of	72–79	1.0050	1.1070	1.4875	0.5225	−
	Malaysia, Sarawak	73–80	1.0600	1.2801	2.700	1.4200	−
C. Transitional countries	**Mean = 0.2026**						
Asia	Singapore	73–80	0.2540	0.1660	0.3650	0.1430	MF
Europe	Malta	72–79	0.1550	0.0989	0.0900	0.2200	MR
	Yugoslavia	73–80	0.1988	0.0344	0.2175	0.1800	MF

The countries are also divided into three groups: (A) Industrialised countries; (B) Developing countries and (C) Transitional countries, whose economies lie somewhere between A and B.

Of the 20 countries which alone qualified for inclusion on this selection basis, the mean fatal accident frequency rates (FAFRs) show a wide variation, from Sweden with 0.064 to Burundi with 1.2138, a nineteenfold range. Even within Group A there is a ninefold range between Sweden and Austria. The figure for Sweden places it almost in a class apart compared with most industrialised countries, and this seems to be due to its high level of safety awareness. An example of Sweden's pioneering work to remove hazards from the workplace is the development by Atlas Copco of pneumatic tools with low vibration and noise levels as experienced by the operator (see Chapter 2.5). The FAFR for mining in Sweden is, however, considerably worse than in construction.

The high FAFRs of Canada, Austria and France in construction are rather surprising; there may, of course, be differences between the bases of their statistics and those of other industrialised countries, such as the possible exclusion of office workers and other low risk occupations from the statistics, but the differences in FAFRs seem too high to be accounted for entirely in this way. Most industrialised countries have shown a marked fall in FAFRs for the second four year period as compared with the first.

Whether this was due to real improvements in safety standards is difficult to establish. It should be considered in the light of declining construction activity during this period in industrialised countries. As a result, fewer people were exposed to hazards. The changing face of construction may also have contributed to this fall in FAFRs. Large companies were growing larger but fewer, whereas the number of small firms, whose standards of accident reporting are poor, was growing.

Some fatal accident incidence rates for industrialised countries, as reported by the U.K. Health and Safety Executive[2] (based mainly on ILO data) are given in *Table 1.3.2*. Whilst the footnotes to this table illustrate the difficulties in making valid comparisons between the rates for different countries, these differences nevertheless appear sufficiently large for the relatively poorer performances of France, the Federal German Republic, Italy and perhaps Canada to be significant.

TABLE 1.3.2. Fatal accident incidence rates in construction in selected countries. Deaths per 1000 employees

Country	Accidents: reported (R) or compensated (C)	1974	1975	1976	1977	1978	1979
Great Britain	R	0.16	0.18	0.15	0.13	0.12	0.12
France	C	0.46	0.43	0.42	0.34	0.31	—
Federal German Republic[a]	C	0.33	0.35	0.39	0.38	0.33	0.35
Irish Republic[b]	R	0.15	0.08	0.09	—	—	—
Italy[a]	C	0.62	0.56	0.54	—	—	—
Netherlands[a]	R	0.08	0.10	0.10	0.10	0.08	—
Canada[c]	R	0.52	0.48	0.41	0.37	0.39	0.39

(a) Based on standard man years of 300 working days. With a 5 day week accidents per actual man year are likely to be some 20% fewer.
(b) Rate per 1000 wage earners.
(c) Including accidents on the way to and from work and deaths arising from occupational illness.

Turning to the developing countries, Group B, the record is far worse and shows little sign of improving. The average FAFR for the nine countries in this group (0.8028) is over three times higher than the average for the eight countries in group A. The record of some other developing countries for which no figures are available may well be worse than those of Group B. The rises in FAFRs shown in several developing countries over the eight year periods may in some cases be more apparent than real and be the result of more complete reporting. Thus apparent rises which were, in fact, due to better reporting were found in the U.K. in the 1950s and earlier.

Personal impressions from visits to construction sites in many developing countries give a more realistic view than any statistics can provide. Vivid descriptions of some of these scenes and experiences are given elsewhere in this book. The reported FAFRs, however, bad though they are, do not give an adequate picture of the total fatalities on any project, since the work is far more labour intensive than in most industrialised countries, and far more people are employed on a given project than would be the case in an industrialised country when much more plant and machinery would be used.

FAFRs for the three countries, Singapore, Malta and Yugoslavia, which, until a few years ago would have been classed as 'developing', provide hope for the future records of those in Group B, since their average FAFR in construction (0.2026) is lower than the average for the industrialised countries in Group A. So far as Singapore and Malta are concerned, this may well be due to the preponderance of medium and large construction firms, and system building in these countries, and the fact that they are predominantly urban with labour forces experienced in modern construction.

The high casualty rates in construction in many developing countries are but a reflection of the general poverty, low living standards and short life expectancies. Even a badly paid and very hazardous job in construction is often preferable to no job at all, and such a job may actually raise the worker's life expectancy by several years. For construction workers in industrialised countries, it is generally in their interests to see improved safety records for their fellow workers in developing countries, if merely to prevent erosion of the relatively better safety standards which they enjoy. Special factors which apply in most developing countries as discussed in Chapter 4.1 add greatly to the difficulties in improving their safety records in construction.

1.3.2 Accidental injury statistics in construction in the U.K.

The U.K. Health and Safety Executive publish biennial reports[5,6,7] on the construction industry, with a number of statistical tables. These give a useful picture for the industry as a whole, but give rather limited data on accidental injury frequency rates for particular occupations in the industry. Special reports which contain further information are, however, available on certain high risk occupations such as steel erecting[9], scaffolding[10] and demolition[11]. The data presented here is for the industry as a whole. *Table*

TABLE 1.3.3. Incidence rates of total and fatal accidental injuries in construction and manufacturing industries and total fatalities in the U.K.[2,6,8] (1973–1979).

Year	Incidence rates per 1000 at risk[(a)]					Fatal accidents in construction
	Construction			Manufacturing industries		
	Total accidental[(b)] injuries	Deaths		Total accidental[(a)] injuries	Deaths	
		All construction	Steel erectors			
1973	35.4	0.216	1.57	37.1	0.042	230
1974	34.6	0.160	1.43	35.2	0.045	161
1975	34.6	0.177	1.14	34.9	0.037	181
1976	35.3	0.153		34.8	0.034	154
1977	33.0	0.131		36.0	0.034	131
1978	34.0	0.122		36.0	0.031	120
1979	31.0	0.117		33.0	0.029	119

(a) Incidence rates quoted by HSE are per 100 000 at risk. Figures here follow ILO recommendations.
(b) Reportable 'accidents' which involve three or more days of lost time.

1.3.3 gives incidence rates for total and fatal accidental injuries in the U.K. in construction and in all manufacturing industries from 1973 to 1979 as well as the total number of fatalities in construction. It also gives fatal accident incidence rates for steel erectors only for the first three years.

Whilst the incidence rates for all reportable accidents in construction and in manufacturing industries are roughly equal, the incidence rate for fatal accidents is still about four times as high in construction as in manufacturing industries despite a marked fall during the period. The fatal accident incidence rate for steel erectors was about eight times the figure for construction as a whole for the three years for which figures are available.

Table 1.3.4 gives the distribution of fatal accidents in construction by causation over two three-year periods, 1974–1976 and 1977–1979, and

TABLE 1.3.4. Distribution of fatal accidents in construction by causation[6,7,8] (1974–76 and 1977–79), and total for 1980–82.

	1974–1976		1977–1979		1980–1982*
	Number	%	Number	%	Number
Falls of persons	217	43.8	200	54.0	
Lifting equipment	59	11.9	35	9.5	
Non-rail transport	65	13.1	34	9.2	
Falls of materials	54	10.9	32	8.6	
Machinery	9	1.8	10	2.7	
Excavations	27	5.5	21	5.7	
Electrical	30	6.0	16	4.3	
Tunnelling	2	0.4	7	1.9	
Fires and explosions	9	1.8	3	0.8	
Striking an object	1	0.2	1	0.3	
Poisoning and gassing	14	2.8	3	0.8	
Handling	1	0.2	–	–	
Not otherwise specified	8	1.6	8	2.2	
Total	496	100.0	370	100.0	443

*Changes in the classification of causation which occurred in this period make comaprison with previous periods difficult.

total figures for 1980–1982. After a welcome fall during the second three-year period, the rise over the last period is particularly disturbing since the number of workers employed in the industry had fallen. The poor economic conditions in the industry over this period appear to have led to more chances being taken.

'Falls of persons' represents by far the main cause and accounted for over half the total number over the second three-year period. Lifting equipment, non-rail transport and falls of materials each account for nearly 10% of the total, whilst excavations and electrical each account for approximately 5%.

Table 1.3.5 gives the occupations of fatally injured workers over the three-year period 1977 to 1979.

TABLE 1.3.5. Occupations of fatally injured workers (1977 to 1979)[7]

Trade/occupation	Number of fatalities
Labourer	76
Roofing worker	52
Painter	37
Managerial/supervisory/technical	30
Driver	28
Demolition worker	24
Steel erector	21
Scaffolder	21
Joiner	12
Steeplejack	9
Bricklayer	8
Tunnel worker	6
Welder/plater	6
Cladder	5
Electrician	4
Lift engineer	4
Rigger	4
Others	23
Total	370

It is unfortunately very difficult to give realistic fatal accident frequency rates for the occupational groups listed in *Table 1.3.5*. This is partly because of the difficulty in finding realistic figures for full-time employment in these occupations, and partly because the casualties reported were not all working in the trade or occupation shown at the time of the fatal accident. Thus, some of the 'roofing workers' shown in *Table 1.3.5* were carpenters and some were plumbers working on a roof. However, some general idea of the more hazardous occupations can be formed by comparing *Table 1.3.5* with the occupational census returns shown in *Table 1.2.1* and with the Department of the Environment's figures shown in *Table 1.2.2*.

By making such a comparison we can make the following assessment of fatal accident frequency rates for some of the trades, i.e. roofers about 2.0; painters 0.16; labourers and others 0.13; bricklayers 0.10; carpenters and joiners 0.03; and electricians 0.02. This highlights the wide range of the risk factor in the different construction occupations.

TABLE 1.3.6. Distribution of fatal accidents by type of work[7], 1977 to 1979

Building operations		Works of engineering construction	
Construction		Tunnelling	14
Industrial buildings	53	Dams and reservoirs	1
Commercial & public	37	Bridges and viaducts	2
Flats & dwellings	35	Pipelines and sewers	20
Maintenance	110	Docks, harbours & inland navigation	8
Demolition	26	Waterworks and sewage works	2
		Steel and reinforced concrete	25
		Sea defence and riverworks	5
		Roads and airfields	26
		Others	6
Totals:	261		109

Although no employment figures are available, the numbers engaged in the following occupations (which appear in *Table 1.3.5*) appear to have been relatively small, so that their fatal accident frequency rates were correspondingly high—steel erectors[9], demolition workers[10], scaffolders[11], steeplejacks and tunnel workers.

Finally, *Table 1.3.6* gives the distribution of fatal accidents by the type of work during the same three-year period. The most striking feature here is the large number of men killed in building maintenance—110 out of a total of 261 for building operations and 370 for all construction workers. Most of these were falls of painters and roofing workers. Improper use of ladders and inadequate protection against falls for roof workers were dominant causes.

1.3.3 Statistics of occupational diseases in construction in the U.K.

The figures given here are extracted mainly from section 10 of the H & SE publication 'Health and Safety Statistics 1979–1980'[2] which is based on data supplied by the Department of Health and Social Security (DHSS).

The only figures available which apply specifically to construction workers are the numbers of those suffering from prescribed industrial diseases who qualify for injury benefit. Whilst there were at the time fifty such occupational diseases which so qualify, plus pneumoconiosis and byssinosis which qualify for disablement and death benefit but not injury benefit, the qualification only applies when the worker suffering from the disease is, or has recently been, engaged in an occupation or conditions of work in which there is a recognisable risk of contracting the disease. Occupational diseases and the health of construction workers are discussed in more detail in Chapters 2.1 and 2.2.

These limitations eliminate most of the prescribed diseases for most occupations in construction, and leave only three main groups of occupational diseases for which figures are available: (1) beat hand, beat knee and beat elbow; (2) inflammation of tendons of hand, and (3) non-infectious dermatitis. Published figures cover the years from June,

TABLE 1.3.7. Construction workers qualifying for injury benefit from prescribed industrial diseases[8]

Prescribed disease	Year				
	74/5	75/6	76/7	77/8	78/9
Beat hand, beat knee & beat elbow	130	133	127	127	116
Inflammation of tendons of hand, etc.	292	280	295	303	246
Non-infectious dermatitis	611	553	521	451	424
Others	12	9	18	17	15
All prescribed diseases	1045	975	961	898	801

1974 to June, 1979, since the DHSS statistical year starts on the first Monday in June. The figures are given in *Table 1.3.7.*

Beat hand, etc. is far more common among construction workers than in manufacturing industries, but the incidence is higher still in mining. The incidence of the other prescribed diseases among construction workers is roughly the same as the average for all manufacturing industry.

Whilst these are the only figures which apply specifically to construction, the numbers of death certificates throughout the country as a whole which mention specified asbestos-related diseases are worth quoting since many are bound to have been construction workers[2]. The figures are given in *Table 1.3.8.*

TABLE 1.3.8. Death certificates mentioning asbestos-related diseases 1968 to 1979[8]

Year	Asbestosis	Mesothelioma
1968	80	154
1969	78	159
1970	87	194
1971	94	179
1972	108	210
1973	107	222
1974	138	233
1975	147	263
1976	191	309
1977	184	330
1978	195	391
1979	167	428

Both totals include death certificates mentioning both asbestosis and mesothelioma, so that some deaths have been counted twice. The increasing trend in both cases is probably due to better recognition of the diseases in the deceased persons rather than to real increases in incidence of these diseases.

In prevention terms occupational health statistics have to be treated with caution and interpreted with care as their relevance is not as immediate as those for accidental injuries. This is due to the intervening time lapse between contracting a disease and the resulting disability being sufficiently serious to be statistically relevant (see Section 2.2). For example, the figures in *Table 1.3.8* can easily give the impression that there is a continuing growing problem of exposure of individuals to asbestos. The number of deaths in 1979 in which asbestos related diseases were mentioned now exceeds the number of reported deaths from accidental injuries at work in all manufacturing, construction and other industries together. Such circumstances need to be carefully investigated before conclusions are drawn. In this case as in many others the onset of the particular disease could predate the statistics by anything up to thirty years. In the meantime more stringent controls have been implemented. However, for a number of years to come statistically significant figures may continue to appear because of this time lapse. The misleading impression this creates can lead to time and effort being wasted on achieving higher standards of control for virtually no additional benefit. More important may be the smaller figures for newly recognised occupational diseases which may be the tip of a future iceberg.

Taking the health of construction workers as a whole, there seems little doubt that the figures given in the official statistics quoted above do not reflect the widespread illhealth, e.g. loss of hearing, back disorders and hernias, among construction workers which are referred to in Part 2. Only the worst cases qualify for injury benefit, still leaving an army of 'walking wounded'.

Health statistics among construction workers in other countries are not quoted here since little reliable statistical information is readily available.

1.3.4 Risk comparison with other activities

In considering hazards of any kind it is necessary to maintain some perspective so that we do not find ourselves concentrating too much effort on minor hazards whilst ignoring more serious ones. Kletz[12] has attempted to present a comparison of the risks of various activities by considering how long a person might be expected to live if this was the only risk to which he was exposed. The figures given by Kletz, plus others estimated from statistics presented previously, are given in *Table 1.3.9*. They only take into account the risks of fatal injury or fatal illness and the risks of any lesser injuries and illnesses are ignored (but which are, of course, much higher).

TABLE 1.3.9. Life expectancy for a U.K. citizen if exposed only to risk specified

Risk	Life expectancy years
Smoking 40 cigarettes a day	100
Riding motor cycle for 10 hours per week	300
Steel erecting for 2000 hours a year	750
Railway shunting for 2000 hours a year	1100
Drinking one bottle of wine a day	1300
Driving a car for 10 hours per week	3500
Working in construction industry (average risk) for 2000 hours a year	5000
Working in the steel industry for 2000 hours a year	6000
Working in manufacturing industry (average) for 2000 hours a year	20000
Travelling in a train for 500 hours a year	40000
Staying at home for 2000 hours a year (man aged 16–65)	50000
Being struck by lightning	10000000
Being hit by falling aircraft	50000000

Some of the risks shown here, such as smoking 40 cigarettes or drinking a bottle of wine per day, are commonly regarded as voluntary risks, although both are usually cases of drug addiction which the victim can only break at the expense of some (temporary) physical or mental disturbance.

Table 1.3.9 should leave us in no doubt that the hazards of the construction industry, particularly high risk occupations such as steel erection, roofing, painting and scaffolding, should not be ignored by any socially-conscious society.

References

1. *International Recommendations on Labour Statistics*, International Labour Office, Geneva (1976).
2. HEALTH AND SAFETY EXECUTIVE, *Health and Safety Statistics, 1979–1980*, HMSO, London.
3. EISENBACH, B., 'Bewertung und Vergleich Vorhandener Daten und Bezugsgrossen zur Bestimmung der Arbeitsbedingungen in der Bauindustrie in der Landern der Europaischen Gemeinschaft, Eine Studie im Augtrag der Europaischen Gemeinschaft', Bundesanstalt fur Arbeitsschutz und Unfallforschung, Dortmund, FDR (1981).
4. *1981 Yearbook of Labour Statistics*, International Labour Office, Geneva (1982).
5. HEALTH AND SAFETY EXECUTIVE, *Construction, Health and Safety, 1977–78*, HMSO, London.
6. *ibid* 1979–80.
7. *ibid* 1981–82.
8. HEALTH AND SAFETY EXECUTIVE, *Fatal accidents in construction*, 1977, HMSO, London.
9. HEALTH AND SAFETY EXECUTIVE, *Safety in Steel Erection*, Report on a sub-committee of the Joint Advisory Committee on Safety and Health in the Construction Industries, HMSO, London (1969).
10. *ibis Safety in Scaffolding*.
11. *ibid Safety in Demolition*.
12. KLETZ, T.A., 'The Mathematics of Risk', Paper given at the 1981 annual meeting of the British Association for the Advancement of Science.

1.4

General trades and their hazards

This chapter describes briefly the work and hazards of the following trades:
1. Bricklayers and stone masons
2. Carpenters, joiners, formworkers, shop fitters and wood-working machinists
3. Painters and decorators
4. Plumbers, heating and ventilating engineers, gas services engineers, welder and gas cutters
5. Electricians
6. Plant operators, including pile driving operators, slingers and riggers
7. Concreters, plasterers
8. Glaziers, cladders
9. Wall and floor tilers
10. Lift engineers
11. Others: storekeepers, labourers and general workers.

Certain occupations, such as steel erection, scaffolding, roof work, demolition, tunnel work, groundwork and diving have been excluded from the above list. These have high fatal accident frequency rates in relation to the trades already listed[1]. For this reason these 'high risk' occupations are considered separately in the next chapter.

Health hazards, which are touched on in this chapter, are discussed in more depth in Part 2.

1.4.1 Bricklayers and allied crafts

1.4.1.1 Bricklayers

Bricklayers work mainly out of doors. Much of their work is done on or around scaffold platforms. Thus they are exposed to the hazards of falling and being hit by falling objects (*Figure 1.4.1*). Bad weather, especially

Figure 1.4.1 Bricklayers are exposed to the hazards of falling, to the weather and other dangers.

winds, the strength of which increases with height, adds to these hazards. Bricklayers are also exposed to:

   bruises, contusions and cuts from the materials they handle;
   eye injuries from dust and chippings when grinding, cutting and chipping bricks;
   dermatitis and skin ailments from contact with some of the mortars they handle;
   'slipped disc' and other back and muscle injuries from handling heavy and awkward loads, often in cramped or unnatural postures.

In spite of these hazards, bricklaying appears to be one of the safer building trades.

Furnace construction bricklayers who work with high silica-content bricks in rather confined spaces are particularly prone to silicosis[2].

1.4.1.2 Stonemasons

The craft of bricklaying evolved from that of the stonemason. Today in most countries there are fewer stonemasons than bricklayers. Much of their work is now concerned with the maintenance of old buildings.

There are two main sub-categories of this craft:

   banker masons who shape the stone, both with traditional hand tools and modern power tools, usually at ground level and under cover; and
   fixer masons who lay the stones on site with mortar.

Both banker masons and fixer masons face the hazards of bruises, contusions and back and muscle injury through handling heavy objects. Banker masons are more exposed to the hazards of dust (including silicosis) and flying chips. Fixer masons are more exposed to the hazards of weather, insecure scaffolding and contact with mortar and injurious constituents in it (lime, chromium compounds).

1.4.2 Carpenters, joiners and allied crafts

1.4.2.1 Carpenters and joiners

Joinery is mostly carried out on a bench in a covered building. It consists of making cupboards, drawers and fixtures of wood or other materials on which wood-working tools can be used (e.g. some plastics, chipboard and asbestos cement board). Most joinery used for building today is factory made and delivered to site.

Carpentry is carried out on site. It consists of cutting, shaping and fixing the wooden structural parts of a house, e.g. floor beams and boards, staircases, banisters, sills, lintels and mullions of windows and door openings, hanging doors and windows, etc. In some countries, such as Norway and Canada, where timber is cheap, the whole carcass of the house may be built out of wood by the carpenter.

Whilst some craftsmen are exclusively carpenters and others are exclusively joiners, many craftsmen combine both trades and are known as carpenter-joiners. They are constantly exposed to the hazards, including noise and vibration, of using hand and powered woodworking tools, as well as the hazards of wooden splinters, nails and other sharp objects. Both are also exposed to respiratory hazards from the dust of some of the materials cut or otherwise shaped, including asbestos, polyurethane plastics and many woods. Their skins may be exposed to dermatological hazards from solvents and chemicals used in adhesives and treating woods. Wood treated against rot, mildew and insect and rodent infestation often contains highly toxic compounds of copper, chromium, zinc, arsenic and tin, as well as phenols. Woods such as pitch pine are poisonous. Thus splinters of wood which enter the skin may cause septic wounds and wood dust may cause pneumoconioses. These toxic hazards are discussed in more detail in Chapter 2.4. The carpenter working outside on scaffolding is exposed to similar hazards as bricklayers and masons.

1.4.2.2 Formworkers (shuttering carpenters)

The formworker is a joiner-carpenter who specializes in the construction of timber moulds for pouring concrete on site (*Figure 1.4.2*). He may also

Figure 1.4.2 The formworker is exposed to most of the hazards of carpentry and joinery as well as contact with cement and release agents.

have to erect wooden supports or falsework to support the formwork and the concrete in it until the structure is self-supporting. The formworker is exposed to most of the hazards common to carpentry and joinery. His skin may also suffer from contact with timber impregnated with cement and release agents.

Designers of formwork must ensure that their designs comply with recognized standards, such as the recommendations of the Bragg report on falsework[2] and BS 5975 Code of Practice for Falsework. The formworker must work in accordance with instructions and design requirements. Many serious accidents involving death and injury to other building workers arise from the collapse of formwork and falsework under load (see Section 1.5.3).

1.4.2.3 Shop fitters

The shop fitter is a carpenter-joiner who specializes in the fitting out of display counters and windows in shops. He may also combine the crafts of glazier, lighting engineer, electrician and decorator. Thus he is exposed to the many hazards of all these trades.

1.4.2.4 Woodworking machinists[3]

The woodworking machinist cuts, shapes and prepares wood in a machine shop or shed on various machines, such as saws, planes and morticers, for assembly by a joiner or semi-skilled worker. Most of this work is now done in factories rather than on site, but some of the work which would be done by a joiner on a small construction site is done on a larger site by the woodworking machinist in a special shed. Woodworking machinery includes many different types, all of which have hazards which require special training, guards and precautions. These hazards are recognized by the Woodworking Machines Regulations 1974 (SI 1974, No. 903 as amended by SI 1978, No. 1126) which specifies both general and special safety and health requirements, including protection of workers against excessive noise. This is discussed in Chapter 2.5. Because of the small numbers of woodworking machinists employed on construction sites the special hazards of their machinery are not discussed here.

Some woodworking machines, particularly circular saws, are, however, widely used by other woodworking trades on construction sites. These have many hazards, especially those of 'kick-backs' which cause many injuries.

Woodworkers may also be exposed to:

- cuts and lacerations from wood splinters and when sharpening and setting-up their tools;
- dust and flying particles from grinders and sanders with risk of injury to eyes and lungs.

All woodworkers employing machinery require special instruction in their safe setting-up and operation.

1.4.3 Painters and decorators

Painters often combine their profession with decorating, although they are treated separately here. Of the two occupations, the available U.K. statistics indicate painting to be the more hazardous. Some would include it among the specially hazardous occupations discussed in Chapter 1.5. The reasons for this are discussed below.

1.4.3.1 *Painters*

Painters, particularly on outside work, are exposed to the hazards of falling from heights, often from ladders or inadequate scaffolds or improvised temporary structures. A high proportion of their outside work consists of maintenance.

Most buildings have to be repainted many times during their lives. The short time that the ladder or scaffold is needed in any one place results in many risks being taken. Painters are also exposed to toxic and dermatological hazards both through the solvents, pigments and other constituents of their paints and from the fumes of the chemicals used in removing old paintwork.

Table 1.3.5 showed a total of 37 fatalities among painters in the U.K. over the years 1977–1979. This may be an underestimate, since some painters appear to have been classified under other occupations, such as roofing workers.

In 1979 there were a total of 1530 notifiable accidents involving painters and decorators, of which 578 were due to falls from heights (*Figure 1.4.3*).

Figure 1.4.3 In 1979 there were 1530 notifiable accidents involving painters and decorators in the U.K., of which 578 were due to falls from heights.

Table 1.2.2 shows a total of 76,600 painters (probably including decorators) employed in the U.K. This suggests a fatal accident incidence rate (FAIR) of about 0.2.

Of 18 reported fatalities among painters in 1977, at least 18 were caused by falls:

4 from internal roof structures
3 from defective tower scaffolds
2 through roof glazing or skylights
2 from a bosuns chair or cradle
2 from single planks or inadequate working platforms
2 from ladders
1 through an unguarded opening.

At least half of these fatalities occurred on industrial premises. The main problem therefore is one of safe access. To quote the 1979–80 HSE report[1]:

> "In this highly competitive field the conscientious firms who do plan for proper access frequently find their tender undercut by less scrupulous firms. All too often, clients accept the lowest quotation without considering the safety of the painters or of others who may be involved. What is needed is for the client to consider the safety precautions proposed in the tenders before awarding the contract."

It is clear from this that the safety problem is, to a large extent, economic. This is discussed further in Part 4. The problem is far worse in developing countries as the following report in the *Times* (15 November, 1982, p. 7) by Andrew Thompson from Buenos Aires shows:

> "Senor Juan Castro, a grocer in the neighbourhood of Forencio Varela, has had to start doing odd painting jobs to survive. 'When I get one of these jobs, I take on some kids to help me', he said. 'Last week I took on three. The three of them were barefoot, I had to go out and buy them shoes'."

Painters are more exposed than most other construction workers to health hazards from the materials they use. These are discussed in more detail in Chapter 2.4. They also face fire and explosion hazards from the use of LPG and other fuels for blow lamps and from flammable solvents and thinners (see Chapters 3.1 and 3.2).

A former occupational disease of painters associated mainly with the use of lead-based paints was known as 'painters cholic'[4]. Typical symptoms were stomach cramps and sickness. Though the use of lead in paints is now rare, most painters today still suffer from irritation of the eyes, nose and throat, dizziness and stupor caused by inhaling solvent vapours.

Other special health problems may arise from:

burning or dry rubbing surfaces treated in the past with paints containing lead, arsenic, chromium and other toxic materials;
handling two-pack paint systems for epoxy, polyurethane and other synthetic resin paints (skin and respiratory allergies);
handling rust-inhibiting chemicals (possible eye damage and skin burns);

stone, brick and cement cleaners (possible skin burns and ulcers, especially if hydrofluoric acid is used);
paint strippers

These problems are accentuated by the use of mechanical equipment—rotary wire brushes and sanders, descaling pistol guns, burning torches and mechanical spraying equipment.

1.4.3.2 Decorators

The work of the decorator is mainly indoors. He applies wall and ceiling papers and other coverings. He is less exposed to falling risks than the exterior painter. Nevertheless, he is at risk from falling when working on high walls and ceilings in entrance halls, auditoria, lobbies and stairwells. Like the painter, much of his work is done from ladders and temporary scaffolds, trestles or improvised supports. He is often exposed to dermatological hazards from the materials he works with, for example, stripping and cleaning fluids and adhesive formulations.

1.4.4 Plumbers and allied crafts

1.4.4.1 Plumbers

Traditionally the plumber was involved with lead piping. Now he works with a variety of materials—copper, steel and several plastics. He is generally responsible for the installation and maintenance of all systems handling water, inside buildings. Sometimes he is required to weld steel. Plumbers in former times were exposed to the toxic hazards of lead fumes. Today, when little lead is employed in plumbing, the plumber faces respiratory hazards of other toxic materials, e.g. PVC, polyurethanes and the vapours formed from them when they are heated. He may also be exposed to dermatological hazards from the solvents and jointing compounds he has to use. These toxic hazards are discussed in Part 2, particularly Chapter 2.4.

Much of the plumber's work has to be carried out in confined spaces, in cupboards or beneath floorboards where it may be difficult to provide adequate ventilation to remove toxic fumes and vapours resulting from the work.

Plumbers are required to use a wide variety of hand tools and portable powered tools, including blow lamps and soldering irons. These have their own hazards (see Part 3).

Plumbers in construction are generally less exposed to the risks of falling than, say, roofing workers, although the installation and maintenance of roof guttering and occasionally lead flashing on roof-wall intersections carries these risks.

1.4.4.2 Heating, ventilating and air conditioning engineers

Much of the work of these craftsmen consists of fitting air ducting in new buildings to provide clean air at a controlled temperature to the different rooms. The work involves the use of hand, and sometimes portable, power tools and spot welding equipment. Today it is probably no more hazardous than the work of the carpenter, although severe respiratory risks may arise where an old heating system with asbestos insulation is being removed. The hazard is specially serious in countries which still use asbestos insulation.

1.4.4.3 Gas services engineers

The gas services engineer is usually a trained specialist employed by the Gas Board or supply company to install piped fuel gas and gas burning appliances in buildings. In many cases, the work is done by plumbers. The work itself is not particularly hazardous, but unless it is done properly and in accordance with relevant codes, there will be serious risks to the users of the building. As well as the danger of explosions from leaks, there is also the danger of carbon monoxide poisoning through incomplete combustion, inadequate flues and poor ventilation. Similar hazards exist where liquefied petroleum gases supplied in transportable containers are used. The point to be emphasized here is the need to ensure that only trained men should be allowed to do this work.

1.4.4.4 Welders and gas cutters [5]

Site welding is used for structural steelwork, balustrading and pipework such as gas transmission and distribution mains. A great deal is done on offshore platforms, process and power plant. Some welding of steel reinforcement may be found on site, but this is generally done in factory conditions off-site.

Both electric and oxyacetylene gas welding are used for steel. Oxyacetylene and oxy-arc gas cutting of steel is also widely used, particularly in demolition work.

Welding and gas cutting are also carried out under water by divers, but acetylene cannot be used in deep water because it explodes spontaneously under pressure. Hydrogen is used in these situations.

Sitewelders are subject to all the usual hazards of welding for which proper eye and skin protection is essential, as well as those of working at heights and in bad weather. Accidental damage to gas cylinders and fittings, gas hoses and electric leads and cables is far more difficult to control on a building site than in a permanent welding bay. Unless their work is carefully controlled they can be a source of fire hazards and personal injuries (e.g. sparks, globules of molten metal, explosions and ultra-violet radiation) to other workers.

Fire and explosion hazards are discussed in Part 3.

Gas cutters and welders employed in demolition and repair work may be exposed to the toxic fumes of lead (from old lead paint) and other metals (such as cadmium, chromium, cobalt, nickel, vanadium and zinc). The respiratory protection needed for such work needs careful assessment; regular medical examination of these workers is also needed. The physical and chemical health hazards of welding are discussed in Chapters 2.4 and 2.5.

When gas cutting is used in demolition, the completion of a cut through a reinforcement rod or structural steel member may release heavy objects which fall to the ground from considerable heights. It may also lead to the collapse of the whole structure or part of it. Such work, therefore, has to be carefully planned and controlled to prevent accidents involving the gas cutter, other workers and members of the public and adjacent property.

1.4.5 Electricians

The main hazards of electricians are electric shock and flash burns through contact with, or short-circuiting of, electric cables. They have a heavy responsibility to other construction workers and to the ultimate users of their appliances in protecting them from similar risks. Virtually all electrical hazards can be eliminated or reduced to negligible proportions by thorough training of electricians, by compliance with the appropriate codes and regulations and by the use of materials and appliances which conform to these codes[6].

The present position in many developing countries faced with the need to expand electrical supply with inadequate resources is often very hazardous. Sometimes the standards of the buildings themselves make it difficult or impossible to electrify them safely, even assuming that safe wiring and fittings could be afforded and provided (see Sections 1.1.4 and 1.1.5).

Installation electricians have to be jacks of all trades, cutting holes in walls, floors, etc. and making good afterwards. They are also exposed to the risk of falling when working on ladders, temporary platforms and

Figure 1.4.4 The electrician must always be prepared for unexpected hazards

supports. A sudden electric shock (though slight and otherwise harmless) may cause a man to lose his hold on the ladder, etc. so that he falls and injures himself.

Temporary electric wiring on building sites for the portable power tools used by other craftsmen has been the cause of many electrical accidents. It is now a general requirement in the U.K. that a reduced supply voltage of 110 volts with the midpoint earthed be used for all such temporary wiring. This greatly reduces the risk of fatal electrocution injuries. Electrical hazards are discussed in more detail in Chapter 2.5.

1.4.6 Plant operators and related trades

The term 'plant operator' covers the operation of cranes, trucks, excavators, bulldozers, dumpers, road rollers, fork lift trucks, and other mobile machines used in construction. Most of these jobs have considerable hazard potential to other workers and property, as well as to the operators themselves. Because of this, operators of most of these machines require special training with the emphasis on safety. In most cases the operator should be above a minimum age, and be tested and certified as competent to use the particular machine before he is allowed to start work on it. For vehicles which have to use roads, care is needed to ensure that all Road Traffic Regulations are complied with.

The main hazards are those of:

collision, tipping and overturning;
over-stressing, with damage to the machine;
dropping loads being carried;
contact of cranes with overhead live electrical cables;
contact of excavators with underground services.

Some machines are excessively noisy and/or vibrate severely, causing health hazards to their operators from which they must be protected.

Some concreting machines lift, transport, spread and apply concrete pneumatically under compressed air pressure. They can cause serious injury to anyone in the path of the emerging stream of wet concrete. Since concrete sets fairly quickly, these machines are subject to blockages. Serious accidents have occurred by trying to clear these while the machine was still under pressure.

1.4.6.1 Pile driving operators

Operators of percussive types of pile driving machines deserve special mention if only because of the intolerable noise they often have to endure. This is not only a hazard to the hearing of the operator, but also to others who are obliged to be in the proximity. The operator should be protected by ear muffs. Noise as a health hazard and quieter methods of piling are discussed in Chapter 2.5.

1.4.6.2 *Slingers*

These men are exposed to the hazards of falls and falling objects, and other injuries arising from the fastening and release of loads from slings, hooks and ropes. Mistakes may cause injury to others, and property damage. Thus it is essential that they be well trained and experienced. They usually work as members of a team with crane drivers, banksmen and other workers. Their own safety is dependent on the skill of others in the team, especially the crane driver.

Riggers involved in steel erection are subject to the special hazards discussed in Section 1.5.1.

1.4.7 Concreters and plasterers

1.4.7.1 *Concreters*

The main responsibility of the concreter is to mix and place the concrete, though in industrialised countries today most mixing is done by machine, often some distance from the site. Sometimes he has also to make and assemble the formwork and reinforcement. He uses special tools such as vibrating pokers and shuttering vibrators which act on the wet concrete. This allows trapped air bubbles to escape and ensures that the concrete fills the mould completely and leaves no voids round the reinforcement. He also has to level and tamp the concrete. The work involves many critical factors which affect the setting and subsequent strength and integrity of the concrete. These are essential to the subsequent load-bearing capacity of the building or works. The concreter needs to be thoroughly trained in all these aspects or must work under close, trained supervision. The work has many typical construction hazards—falls, material handling, vibration exposure. It also has distinct dermatological hazards since cement is strongly alkaline and usually contains small quantities of chromium compounds which are harmful to the skin (see Chapter 2.4).

1.4.7.2 *Plasterers*

The plasterer works mainly indoors on walls and ceilings and is less exposed to the weather and heights than the bricklayer or roofing worker. Sometimes, however, he may do stucco work or apply rendering on outside walls. This can involve the risk of falling from a temporary work place. The same risk arises when working indoors on high walls and ceilings in halls, lobbies, etc. The plasterer's work is arduous and requires continuous muscular exertion. This is often in very cramped and unnatural bodily positions. Slipped discs, hernias and strained muscles and ligaments are hence common.

The plasterer carries out partition work as well as plastering, using various types of artificial boarding. He or a labourer or demolition worker has to rip out old and defective plastering which may contain asbestos and other injurious materials. The plasterer is exposed to respiratory, dermatological and eye hazards from the materials he works with. These include a variety of proprietary plasters, whose composition is often unknown. Eye protection is frequently necessary for plasterers.

1.4.8 Glaziers and cladders

1.4.8.1 Glaziers

The glazier handles, cuts and fits glass panes and panels in windows, doors, partitions, showcases, screens, etc. He is exposed to the risk of injury from the material he handles which has sharp edges and is easily broken. Cuts may be deep and sever veins and arteries. Fatal wounds from glass are, however, fairly rare.

The glazier is also exposed to falling hazards, especially when replacing broken window panes at first storey level and higher. One reason for this is that windows are normally made so that the panes can only be fitted from the outside. The carrying and holding of large window panes on a ladder on a windy day is also very risky.

The glazier is sometimes exposed to toxic hazards through splinters of glass containing lead, arsenic and other toxic elements which may penetrate and lodge in his skin. He may run dermatological risks from the putties and other materials which he handles. Finally, he may also injure himself with the tools he uses.

1.4.8.2 Cladders

Cladders are mainly engaged in fixing panels of concrete, aluminium, asbestos cement, glass, glass reinforced plastic, stainless steel and other materials to the outside walls of buildings.

The work is potentially quite hazardous and requires careful planning, proper equipment, and well-trained and experienced workers. Often essential parts of the scaffolding used for constructing the carcass of the building interfere with the cladding operation. They thus have to be removed or modified before the cladding can be applied.

The panels are often quite large and until secured in position they must be carefully braced, propped or temporarily supported to counteract wind pressures. Sometimes cartridge-operated fixing tools are used to secure panels to walls or frameworks by sharp hardened-steel pins. They are discussed as an explosion hazard in Part 3. Their use must be carefully controlled.

1.4.9 Wall and floor tilers

Wall and floor tilers work mainly indoors with ceramic or plastic tiles and tile adhesives. The work is fairly arduous and sometimes has to be done in cramped or unnatural body positions, with risk of back or muscular disorders. Most of the work is done at fairly low levels, where the falling risk is low. There may sometimes be toxic or dermatological hazards from the materials handled, which include dirt and fragments of old surfaces and tile adhesives.

Some tile adhesives contain flammable solvents[7] whose vapours are heavier than air and accumulate at low levels in poorly ventilated rooms and spaces. These have caused a number of flash-over fires in which tilers and others have been badly injured. Such fire hazards are discussed further in Part 3.

1.4.10 Lift engineers

The engineer who installs and maintains lifts in building is usually highly-trained and employed as a sub-contractor. Thus he should be thoroughly familiar with the hazards of his work. These include the risks of falls, being trapped between the lift and counterweights or crushed under them, as well as those of lifting heavy objects. The safety of the ultimate users of the lift depends critically on the soundness of his work.

1.4.11 Other categories

1.4.11.1 Storekeepers

Most construction sites other than the very smallest have a store for construction materials and tools used, with a storekeeper in charge. His duties involve a great deal of manual handling of different materials, some of which may be toxic or inflammable. He is principally exposed to the hazards of manhandling loads. He is also at risk of being struck from falling objects which are being unloaded from trucks and lorries, or placed on or removed from high shelves or racks in the store.

Storemen are responsible for checking and inspecting building materials delivered to site, as well as tools and equipment returned to the store and re-issued later. They sometimes have to repair them before re-issue. The storeman has a heavy responsibility for the safety of workers who have to use the tools, equipment and materials issued.

1.4.11.2 Labourers

Much of the manual handling and unskilled manual work (e.g. digging) involved in the forementioned activities is carried out by labourers. They are exposed to all the hazards of the activities concerned, often without any proper appreciation of them. In developing countries which use labour-intensive construction methods, the majority of the construction workers are often labourers. Rarely are they equipped with protective hats, footwear and other safety equipment. The scaffolds they work on are generally substandard.

The accident rates among unskilled labourers on construction sites in developing countries are usually high, although it is seldom possible to find reliable statistics.

1.4.11.3 General workers

General workers and odd-job men may be involved from time to time in any of the forementioned activities and exposed to the same hazards. They are less familiar with the tasks, the risks involved and the safety procedures to be followed than a trained tradesman. Their chances of an accidental injury are thus higher.

References

1. HEALTH AND SAFETY EXECUTIVE, *Construction, Health and Safety, 1979–1980*, HMSO, London.
2. EISENBACH, B., *Evaluation and Comparison of Existing Data and Criteria for the Definition of Working Conditions in the Construction Industry in the European Community* (original in German), Bundesanstalt fur Arbeitsschutz und Unfallforschung, Dortmund, West Germany (1981).
3. HEALTH AND SAFETY EXECUTIVE, *Guidance Note PM 21*, *Safety in the use of wood working machines*, HMSO, London.
4. HUNTER, D., *The Diseases of Occupations 2nd Ed.*, English Universities Press, London, (1957).
5. SANDERSON, P. G., 'Welding Operations', Chapter 13 in *Industrial Safety Handbook*, editor Handley, W., 2nd ed., McGraw Hill, Maidenhead (1977).
6. FORDHAM-COOPER, W., *Electrical Safety Engineering*, Butterworths, London (1979).
7. HEALTH AND SAFETY EXECUTIVE, *Guidance Note EH 7, Petroleum based adhesives in building operations*, HMSO, London.

1.5
High risk occupations

Construction includes several occupations which repeatedly show higher fatal accident frequency rates than those of the industry as a whole. Many of these have been the subject of special investigations in the U.K. by joint working parties in which the Health and Safety Executive has played a guiding role. The principal ones are steel erection, scaffold erection and dismantling, roof work, demolition, excavation and tunnel work. Divers and steeplejacks form smaller groups in this category.

1.5.1 Steel erection

Steel erection (*Figure 1.5.1*) has one of the highest incidence rates of fatal and serious injuries of any in construction. In the U.K. there were about

Figure 1.5.1 Most fatal accidents in steel erection result from falling and too few workers have as much thought for their safety as these men 200 metres above Berlin (Popperfoto).

58 High risk occupations

TABLE 1.5.1. Fatal accident incidence rates in steel erection compared with all construction and manufacturing in the U.K.[1,2,3] **(fatalities per 1000)**

	Year									
	1970	1971	1972	1973	1974	1975	1976	1977	1978	1979
Manufacturing industry (factory processes)	0.048	0.046	0.041	0.045	0.045	0.039	0.034	0.034	0.031	0.029
Construction	0.188	0.206	0.201	0.231	0.167	0.193	0.153	0.131	0.122	0.117
Steel erectors only	2.00	0.71	1.57	1.57	1.43	1.14	unknown	1.29	about 1.0	

Note: Data from 1970 to 1975 from Ref. 1. Data from 1976 to 1979 from Refs. 2 and 3.

7000 full-time steel erection workers in the early 1970s and an annual total of about 1,200 reportable accidental injuries. Of these, over 20 were fatal and about 400 were serious (Group 1) injuries. *Table 1.5.1* compares accidental injury incidence rates for steel erectors with those for manufacturing and construction as a whole in the U.K.

It is surprising that no specific regulations for controlling the safety of steel erectors exist in the U.K., although they are subject to all the requirements of the Construction (Working Places) Regulations, 1966. The attitude has been that it is accepted as a dangerous manly occupation by a fairly small group of workers who 'know how to look after themselves'. They may even enjoy the hazards of their work, like Surtees' description of fox hunting as having 'all the excitement of war and only half its dangers'.

Steel framework sometimes forms part of the integral structure of walls and buildings. Sometimes it forms parts of open structures for the support of roads, railways, roofs, plant, equipment and pipework. It is used thus in bridges, steel mills, oil refineries, chemical works, power stations, railway stations, aeroplane hangars, etc. Similarly it may form part of a single-storey building or shed, or of a multi-storey structure.

The popularity of structural steel (as compared to reinforced concrete) is due mainly to the speed with which it can be assembled. This speed of erection is one of the main underlying reasons for its hazards. The steel framework can usually be erected much faster than the scaffolding and safe working platforms needed for the steel erectors to work from.

The special U.K. committee which studied the cause of accidents in steel erection reported the causes of all accidents and fatal accidents over a period of four years[1]. These are given in *Table 1.5.2*.

As expected, most fatal accidents were caused by falls from over 2 metres (27 out of 30 deaths). These, however, only caused 21% of the total reported accidental injuries. A variety of other causes were responsible for the remaining 79% of accidental injuries—struck by materials *or* steelwork, 17%; falls from below 2 metres or on the flat, 14%; strains from handling steelwork, 13%; trapping, 11%.

The authors of the report consider the attitude to hazards that has grown up within the industry to be the principal cause of its high fatality rate. "The concept that an erector should be able to look after himself in dangerous situations has so pervaded the attitude of man and management to steel erection work that it has radically affected the way trade practice

TABLE 1.5.2. Accidents as percentage of study group, and fatalities ( ) (actual numbers in study group) to steel erectors.

Item	Causation	1970	1971	1972	1973	Average (total)
1	Falls over 2 metres					
	(a) from steelwork	11 (5)	12 (4)	15 (5)	11 (3)	12 (17)
	(b) from ladders	5 (1)	7 —	7 (2)	6 —	6 (3)
	(c) from platforms, etc.	3 (3)	2 —	4 (1)	3 (3)	3 (7)
2	Falls below 2 metres and on flat	15 —	16 —	17 —	9 —	14 —
3	Struck by materials	8 —	7 (1)	7 (1)	8 —	8 (2)
4	Struck by steelwork during erection	7 —	9 —	9 —	11 —	9 —
5	Strains from handling steelwork	14 —	13 —	12 —	13 —	13 —
6	Use of tools	4 —	5 —	4 —	6 —	5 —
7	Trapping	11 —	11 —	10 —	11 (1)	11 (1)
8	Crane/plant failure	1 —	1 —	2 —	4 —	2 —
9	Eye injuries	3 —	2 —	2 —	2 —	2 —
10	Burns	2 —	2 —	1 —	1 —	2 —
11	Nails, etc. in feet	2 —	1 —	1 —	1 —	1 —
12	Miscellaneous	14 —	12 —	9 —	14 —	12 —
Total fatalities in study group		(9)	(5)	(9)	(7)	(30)
Total fatalities in year		14	5	11	11	41

has developed, to the detriment of safety. Reliance should not be placed on the skill of the worker to protect himself from injury, but all practical steps should be taken to remove or prevent the risk".

The HSE's 79–80 Construction Report[2] classes the basic hazards in steel erection in three broad categories:

1. Falling from the structure under erection, or from access ladders;
2. Instability of partially-erected structures; and
3. Materials being dropped while working at heights.

The HSE's main comments and precautions recommended against these hazards are summarized below.

1.5.1.1 Falling

Reliance on the provision of safety belts for steel erectors has proved ineffective since they are rarely worn and used. This is partly due to:

the lack of suitable anchorage points,
the fact that safety belts are not designed to carry tools and bolts and have to be worn over the normal steel erector's belt,
the restriction of safety belts on the mobility of the erector.

Hence safety belts (or harnesses) are now considered the last means of fall protection. More emphasis is placed on the use of lightweight mobile tower scaffolds and hydraulic platforms. Where ladders have to be used for access, greater control is needed to ensure that the ladders used are long enough and properly tied or footed.

Further remedies against falling are given subsequently under 'General Precautions'.

1.5.1.2 Instability of partially-erected structures

Erectors frequently fail to appreciate the instability of structures under erection and fail to provide guy ropes or props for columns, etc. Hence they collapse as a result of:

high winds,
being struck by loads handled by cranes,
through men working on ladders leaning against them.

Even when a column has been guyed, some of the guys are often slackened or removed to facilitate erection of other parts of the steelwork, thereby reintroducing instability.

Two Indian experts[4] recommend that the first four adjacent columns at the corners of a grid should be erected and guyed using open pockets for the holding down bolts. With the guys in position the columns should be aligned and plumbed whilst the horizontal and diagonal interconnecting members are erected and fixed in position. The holding down bolts are then grouted in when this part of the structure is rigid, level and plumb. The erection of the steelwork may then proceed row by row. In the U.K. there is a preference for casting the holding down bolts into the foundations before erecting the columns, then using shims for levelling and plumbing.

Whichever method is adopted, the instability of the partially-erected structure must be appreciated. The structure must be adequately guyed or propped until it is plumb, level, has sufficient cross members in position and joints bolted up so as to be mechanically stable.

1.5.1.3 Falling material

When men are working at considerable height with bolts, nuts, spanners, podgers and other tools, the risk of dropping metal objects, sometimes sharp and pointed, onto someone standing below is considerable. Even safety helmets have been known to fall and injure someone below.

Wherever possible the work should be planned so that no one has to work beneath steel erectors. Barriers and notices should be used at lower levels to keep people away. Special measures, such as the use of nets, hard hats and safety shoes, or the erection of protective scaffolds and fans, are required when it proves impossible to exclude people from these areas.

1.5.1.4 General precautions and recommendations

Besides the above precautions against the three main hazards of steel erection, there are several other general precautions which should be taken to improve the safety of steel erection.

Architects, structural engineers and engineering draughtsmen, besides progress and planning engineers on site, can do a great deal to design hazards out of steel erection. By carefully 'going through the motions' to be performed by steel erectors and other workers when the work is being designed and planned, risks can be identified and appropriate precautions introduced. This involves close collaboration at the planning stage between the designers, the main contractor and the steel erection contractor.

A few examples follow:

- Sections of steel frames for shed-type buildings can often be pre-assembled more efficiently and safely on the ground and then lifted by crane into position rather than being assembled in mid-air.
- The ground should be well consolidated and the sub base prepared to ensure the safety of mobile cranes and enable mobile scaffold towers or other mobile access equipment to be used by erectors.
- Columns should be designed with temporary supports (welded bearing blocks, cleats, etc.) for beams which have to be erected and attached to them.
- The placing of ladders against steelwork during erection should be anticipated. Lugs should be incorporated in the design and fabrication of the steelwork to facilitate safe and easy attachment and removal of erection ladders—by crane if possible.
- Horizontal access between parts of structural frames, including permanent walkways with handrails, should be provided at as early a stage in erection as possible. Temporary gangways with handrails and toeboards should, where needed, be provided for access to steel structures.

Other general points are:

- Steel erectors, riggers, crane operators and banksmen should undergo thorough pre-selection (including medical examination), training and certification.
- The importance of teamwork and of effective communication between members of teams is stressed.
- Safety helmets and appropriate safety boots should be worn by everyone on a site where steel erection is in progress.
- Steel erection should not be carried out in unsuitable weather or bad light, or when the structure is covered with ice or snow or slippery from other causes, such as wet paint.
- Contracts and payment for steel erection should be free of clauses or penalties which increase the risk of the work being undertaken under unsafe conditions.
- Erectors should be provided with proper drawings of the structures being erected, together with detailed instructions which cover the order and method of erection.
- Erectors should be included in discussions relating to the order and method of erection. This ensures that the system of work is fully understood and possible problems eliminated.

The principal recommendations from the (U.K.) Joint Advisory Council's report 'Safety in Steel Erection' are given in Appendix B to this book.

1.5.2 Scaffolding (including falsework)

Sound scaffolding is essential to safety in construction. The words of a British government report[5] published in 1907 still apply: 'We are convinced that serious dangers are due to insufficiency of plant and the use of defective materials for scaffolding, coupled with unskilled and careless workmanship in erection of scaffolds'.

The term 'scaffold' is a very old one, and its meaning has altered. Chaucer spoke of 'skaffauts', meaning mobile towers equipped with battering rams used for assaulting castles. Shakespeare referred to the gallery at the Globe Theatre as 'scaffoldage'. Today, in the U.K. the term 'scaffold' in its legal sense[6] means a temporary structure and working platform for construction workers and their materials (e.g. bricks), and includes stairs, ladders, gangways, guardrails and toeboards which form part of the structure, but not lifting appliances or structures used to support lifting appliances or other plant or equipment.

With this narrow meaning of the term 'scaffold' another term 'falsework' has been coined to describe temporary structures used to support parts of buildings under construction. Examples are the supports used for constructing brick and stone arches and the formwork for reinforced concrete floor slabs, lintels, etc. Falsework is defined as 'any temporary structure used to support a permanent structure during its erection and until it becomes self-supporting'[7]. In some countries the term 'access scaffold' is used for our narrower definition of 'scaffold' and 'support scaffold' is used for falsework.

1.5.2.1 Wood and bamboo scaffolds

From BC 2000 to about 1920 AD wood and bamboo, generally in the form of poles, were used almost exclusively for the construction of both access and support scaffolds. Sawn timber also came to be used after the industrial revolution. Since 1920 there has been almost a total switch in the U.K. and many industrialised countries to the use of metal tubes (mostly 2 inch steel as well as aluminium) for the structural members of scaffolds. These are used in conjunction with metal couplers and other fittings. Timber is now only used in the U.K. for working platforms, toeboards, trestles, gantries, soleplates and ladders.

This change has come partly from economic and partly for safety reasons. There is in the U.K. today no official code of practice specifically for timber scaffolds. But both German and American standards for scaffolding, DIN 4220 and A.10.8 (American National Standard, 'Safety Requirements for Scaffolding') give details on the design and construction of timber as well as metal scaffolding.

One exception in the U.K. is the use of trestle scaffolds (mainly by decorators, indoors). There is an anomaly in the present law which allows these to be used as working platforms without guardrails and toeboards for heights up to 15 feet (4.57 metres). All other working platforms more than 6.5 feet (2 metres) from the floor or ground must have guardrails and toeboards.

The bamboo scaffolding used by early Chinese bricklayers, as shown in contemporary illustrations, had neither ladders, guardrails nor toeboards. The wooden working platforms were narrow and spanned considerable distances between transoms. These scaffolds appear to have been deficient in bracing, with poor stability, and only suited to low heights and light loads. Construction methods throughout much of Asia remained relatively static until the last century. Even today one sees many scaffolds very similar to the early Chinese ones. Bamboo scaffolding has, however, been developed for use today on modern high rise buildings of 30 storeys and more. A single square lattice of 3 inch bamboo poles lashed together by 'rattan' is used with larger poles for tower lifts. A speaker at an Indian seminar on construction safety described the situation thus[8]:

> "It is very common in our country to see traditional bamboo or bally scaffolds erected even for the construction of tall buildings and structures. They present a dreadful sight and the veritable forest of bamboo is hardly safe for such tall structures, and workmen on them are in constant fear for safety. Long exposure to weather produces distortions in the bamboo and the coir rope and the scaffolding goes out of alignment and plumb and the resulting crooked structure can hardly be deemed safe to work on. These scaffoldings are built with almost scant regard to safety requirements with regard to working platforms, guardrails, toeboards and accesses, etc.".

Other speakers proposed that wooden and bamboo scaffolds should be restricted to, say, 15 metres in height. Western engineers who have seen these scaffolds will agree.

It is only recently in the industrialized countries that the public has become concerned about deaths and injuries in construction, and the importance of safe scaffolds. The International Labour Office recommends that scaffolding made from bamboo and wooden poles and sawn timber should be designed to similar or even more stringent standards than those in force for tubular metal scaffolds[9].

Thus the ILO recommends two guardrails (against our one) as well as toeboards for working platforms.

ILO publications[9] reveal many differences in practice when using sawn timber, wooden or bamboo poles for scaffold structures. The simplest type of timber scaffold is the single ladder scaffold (*Figure 1.5.2*). The deck supported by the rungs of a single row of ladders is, however, very narrow and restrictive. An alternative form using pairs of ladders (double ladder scaffold) to support transoms which carry a wider deck is obviously safer (*Figure 1.5.3*). Ladder scaffolds need to be particularly well braced as well as anchored to the building at every storey.

The next stage of complexity in scaffolding involves the use of timber poles to construct a framework on which a deck can be placed. Details of ledgers, splices and bases for pole scaffolds are shown in *Figure 1.5.4*. The ILO strongly recommend[9] that pole uprights be let into the ground from 0.6 m to 1 m and secured against further sinking by resting them on sole plates. When this is done the poles will generally need special treatment to protect them against rot and termites, etc. Squared timber scaffolds should only be made by trained carpenters to drawings prepared by qualified

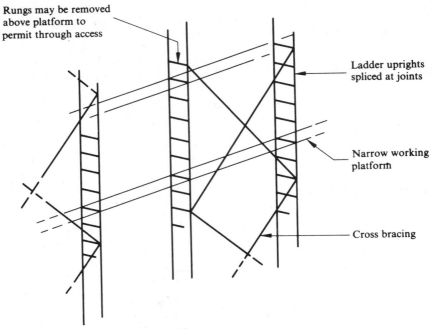

Figure 1.5.2 The single ladder scaffold.

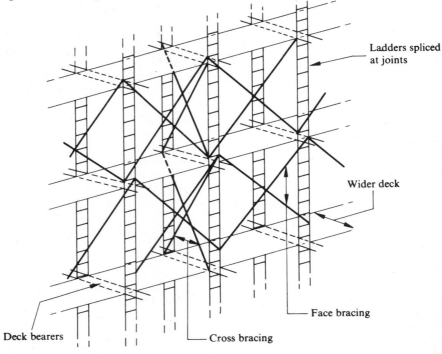

Figure 1.5.3 Double ladder scaffold providing wider deck.

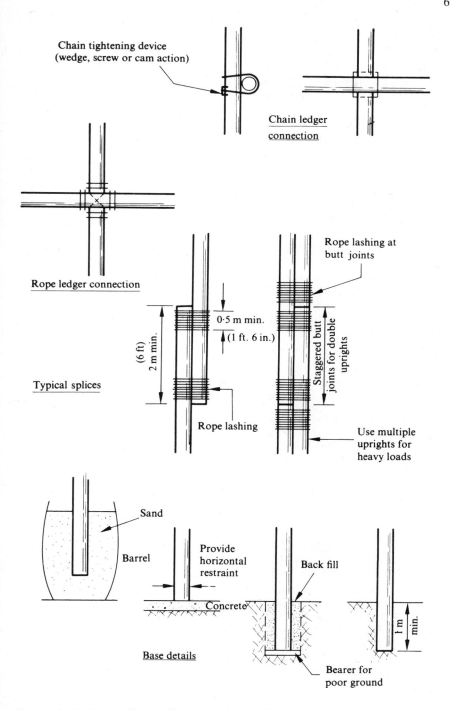

Figure 1.5.4 Ledgers, splices and bases for pole scaffolds.

persons who can do the necessary stress calculations. Such structures may be used as work scaffolds for loads up to 400 kg and should be assembled by bolting.

Among serious disasters involving the collapse of timber scaffolding was the Nagarjunsagar Dam disaster in India (1969) when the entire scaffolding, more than 60 metres high, collapsed, killing a number of workers.

1.5.2.2 Tubular Metal Scaffolding and Falsework[10] (See BS 1139, 5973, 5974, 5975 and GS 15)

Tubular metal scaffolding and falsework are generally made from the same or similar components. Both should only be erected by trained scaffolders.

In the U.K. a scheme known as the Construction Industry Scaffolders' Record Scheme is organised by the Construction Industry Training Board. The CITB run courses in basic and advanced scaffolding and issue and control individual training record cards. Scaffolders are grouped into three categories, trainee, basic scaffolder and advanced scaffolder. To qualify for the two latter groups they must have completed the appropriate course and period of practical experience.

In considering the hazards of scaffolding, one has to distinguish between those to:

- the scaffolder or erector of the scaffold;
- the people who later use it; and
- people, including the general public, who may be exposed to dangers during its transport, erection and dismantling or accidental collapse.

A steel scaffold coupler is a hard and heavy object. When thrown off the back of a lorry it can cause serious injury to the head of a passing pedestrian. (The writer had this painful experience in a London mews several years ago.)

Tubular metal scaffolding is nearly all erected manually. Here the scaffolder often still works from the partially-erected scaffold, before ladders, hand rails, platforms and boards have been fixed in position. In more progressive firms he is expected to work from a minimum of two board wide platforms. These will not have guardrails or toeboards unless required in these positions as part of the complete scaffold. In any case he is more exposed to the risks of falls and being hit by falling objects than the person who subsequently uses the scaffold. These risks are almost in the same category as those for the steel erector.

Safety helmets with chin straps should be worn when erecting tubular metal scaffolding, particularly when others are working at higher levels. There are many instances where safety harnesses on static or roving lines should also be used.

According to the HSE's 79–80 annual report[3], 'scaffolders disregard their own safety, and even such relatively simple requirements as the provision of properly secured ladders are ignored. Scaffolders still take chances by clambering up the outside of the scaffold or sliding down standards, and single-board working platforms are often encountered.

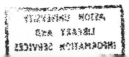

Scaffolding (including falsewo[rk]

Figure 1.5.5 A young roofing worker, who was not wearing a safety helmet, died as a result of injuries when scaffolding collapsed into a shopping street in Faversham, UK.

league'.

Scaffolders are also exposed to the risk of collapse of the scaffold they are erecting. Over recent years there have been an average of over 3000 such accidents every year in the U.K. (*Figure 1.5.5*). Independent and putlog scaffolds generally collapse fairly slowly. Collapses generally result from a combination of faults. Insufficient ties to the building is a common cause. Heavy support or falsework scaffolds usually collapse under load due to insufficient bracing, often the result of inadequate design. Such collapses are sudden and cause several fatalities.

Several scaffold collapses have also occurred during demolition work, due to premature or unexpected collapse of the building of structure to which the scaffold was tied. These are referred to again under 'Demolition' (Section 1.5.4).

Table 1.5.3 gives names, descriptions and uses of several different types of tubular metal scaffolds made of straight tubes joined by couplers. The tubes and fittings have various names, given in *Table 1.5.4*, depending on their purpose and position in the scaffold. Platforms are built of varying widths and load-bearing capacities, depending on the purpose they will be put to (light duty for painting, extra heavy duty for building masonry walls). Lifts between ledgers will also vary according to the purpose.

Scaffolding should be designed by experienced scaffold designers who know the use to which the scaffold will be put and the loads involved. They should be put up by scaffolders who have been through a proper scaffolders' training course such as those run by the CITB. Sufficient and correct scaffolding materials of the types needed must be made available

TABLE 1.5.3. Some common types of tubular metal scaffolds

Type	Description of main features	Uses and notes
Independent tied scaffold	Two rows of standards and ledgers carry transoms over the width of the scaffold. The transoms support the platform boards. Movement of the scaffold is prevented by ties to the building.	For heavy duties (masons scaffold) and cases where wall or building is unsuitable for load-bearing or does not allow insertion of putlogs (e.g. glass walled buildings). All independent tied scaffolds are classified by type of use, standard spacing, lift height and other features.
Putlog scaffold	Platforms are supported between a single row of standards with ledgers and the building itself by horizontal putlogs inserted into the building fabric. Platforms allowed at each level of ledgers subject to load restrictions, but usually only one platform boarded out.	Restricted to use with load-bearing walls which allow the insertion of putlogs, e.g. brick and some masonry walls. Principally a bricklayer's scaffold.
Birdcage scaffold	Wide independent scaffold, usually rectangular, built up from several rows of standards on base plates and usually sole plates on floor of building. Often tied or braced to walls of building. Single working platform.	To provide access to ceilings, walls, soffits in factories, public halls, cinemas and other large buildings.
Truss out scaffold	Independent tied scaffold projecting from building or structure and entirely dependent on building for support.	For light work on upper storeys of tall buildings in busy street and other cases where it is impracticable or inadvisable to erect a scaffold from ground level.
Cantilever scaffold	Similar to truss out scaffold except that scaffold is erected on beams cantilevered out from building.	Similar to those of truss out scaffold but generally allows greater loads as result of beam supports.
Slung scaffold	Working platform suspended at fixed height below load-bearing brackets or bearers or from structural members of roof or over-head structure. Suspended by wire ropes or tubes without means of raising or lowering.	To provide access to any type of ceiling or internal roof where it is impracticable to build up from the ground or clear access below is required.
Suspended access scaffold	There are three types, all suspended from outriggers or tracks, and all having manual or power operated mechanisms for raising and lowering.	For painting, glazing, scaling, cleaning and light repair work on tall buildings above busy street or above intervening intrusions which make it impracticable to build up from ground.
a) Hinged	Allows two horizontal platforms at different levels with sloping gangway.	
b) Independent c) Cradles	Single platform—rigid structure. May be fixed and able to move up and down only, or travelling if suspended on tracks and able to move both horizontally and vertically.	

TABLE 1.5.3. Continued

Type	Description of main features	Uses and notes
Tower scaffolds	Small independent scaffolds with single working platform, not tied for heights up to 9.6 m, but tied to structure from height of 9.6 m up to maximum of 12 m. There are two types:	For painters and others doing light work on the sides of structures for heights up to about 13 m from the ground. Height limited to 3 × minimum base measurement for external use and to 3.5 × minimum base measurement for internal use.
a) Fixed	on four base plates	
b) Mobile	on four locking castors	
Steeplejacks scaffold	See Section 2.3.7.	

TABLE 1.5.4. Some terms used for parts of tubular mtal scaffolds

1. Tubular members

Brace	Diagonal member providing stability.
Bridle	Tube slung between putlogs adjacent to the wall across a window, door or other opening to provide support for intermediate putlogs.
Butt	Very short tube.
Buttress	Inclined tube assemblies fixed between scaffold and ground to increase stability.
Guardrail	Horizontal tube fixed to standards at side of platform to prevent persons falling.
Ledger	Horizontal tube parallel to face of structure joining standards and used to support transoms or putlogs.
Physical tie	Compression and/or tension member fixed from scaffold to structure.
Puncheon	Vertical compression member bearing neither on ground nor on structure.
Putlog	Tube with one end flattened for insertion into horizontal joints between bricks or masonry.
Raker	Inclined thrust member (similar to buttress).
Reveal tube	Tube wedged between window reveals to provide reaction for a tie.
Standard	Vertical compression member.
Transom	Horizontal tube spanning onto ledgers and supporting boards of platform where present.

2. Fittings and timber

Baseplate	Metal plate incorporating loading pin for distributing load from standard or other compression tube.
Baseplate-adjustable	As above, but embodying some form of length adjustment.
Castor	Swivelling wheel attached to bottom of standard.
Coupler	Metal connecting fitting for joining two tubes.
Frame	Support for platform or beams on which platform may be laid, usually hanging from wire rope.
Joint pin	Short rod used to attach two tubes end to end, fitted internally.
Platform	Comprised of wooden boards or steel plate supported by transoms or putlogs.
Reveal pin	Threaded pin used to tighten reveal tube.
Sleeve coupler	Coupler for connecting two tubes end to end externally.
Sole plate	Thick wooden section resting on soil or fragile floor to support baseplates and spread load.
Toeboard	Board fixed on edge at side or end of platform to prevent falls of materials, tools and persons.

on site. A percentage of spares is needed to cater for contingencies, and parts which fail to pass inspection. The scaffolder must inspect all parts carefully before starting to erect. He must reject damaged or sub-standard items, or allocate them to some less critical duty, e.g. sub-standard boards used as duckboards.

Standard designs are available for a number of purposes. Quite often, however, a special design has to be made. It is essential to establish this point at an early stage when the scaffolding is planned. The technical aspects of scaffold design and construction are all essential to safety. Since these are well documented elsewhere and form an essential part of the scaffold designer's and scaffolder's training, they are not discussed in detail here.

Tubular metal scaffolding is designed with a high safety factor, generally 4 or 5. Thus there are usually a number of faults or deficiencies before a failure occurs. Unfortunately far too much scaffolding is sub-standard and fails to comply with the regulations. Whilst the regulations stipulate that all scaffolding should be inspected weekly by a competent person such inspections only start when the scaffold is complete (faults and all). By this time it often becomes very difficult and expensive to rectify the faults. Anyone with only a modest knowledge of the regulations[6] and British Standards (see list at end of chapter) can find faults in many of the scaffolds he sees erected. The most serious common faults and the ones most likely to cause accidents and collapse of tubular metal access scaffolds are the following:

- Failure of foundations. These are often due to soil washing away after the scaffold has been erected and in use for some time. Sometimes there is a deeper cause such as uncompacted soil or a cavity (cellar, etc.) below the surface.

    This is largely a matter of training and experience. Inadequate foundations and bases for scaffolds are one of the commonest and most obvious faults to be seen.
- Failure to tie in the scaffolding adequately to the building.
- Removal of ties by unauthorised persons to carry out particular jobs.

    Ties in or through window openings have an unfortunate history. Through ties (*Figure 1.5.6* and *1.5.7*) are often unacceptable when the building is under construction as they interfere with plastering and decorating. When the building is occupied the occupants are exposed to wind and rain. Reveal ties, on the other hand, are seldom strong enough to be relied on exclusively. They depend very much on the general state of the building. They also require regular checks and adjustment to counter swelling or contraction of timber packing in wet or dry weather. If tightened too far they may cause cracks in the building. It is recommended (BS 5973:1981) that they should not exceed 50% of the total number of ties in a scaffold. Properly drilled-in anchor fittings are preferred. It is, however, difficult to fit these securely if the bricks are unsuitable or if the wall is covered by cladding, pebble dash or some facade finish.

    The best solution is to design and build anchor points for scaffolding into the wall at the beginning. These should be protected by removable plastic plugs.

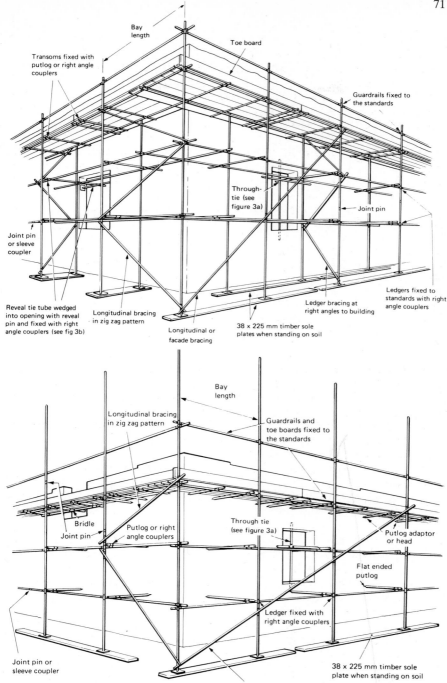

Figure 1.5.6 Typical independent tied (top) and putlog (bottom) scaffolds erected on substantial timber soles and with secure guardrails and boards.

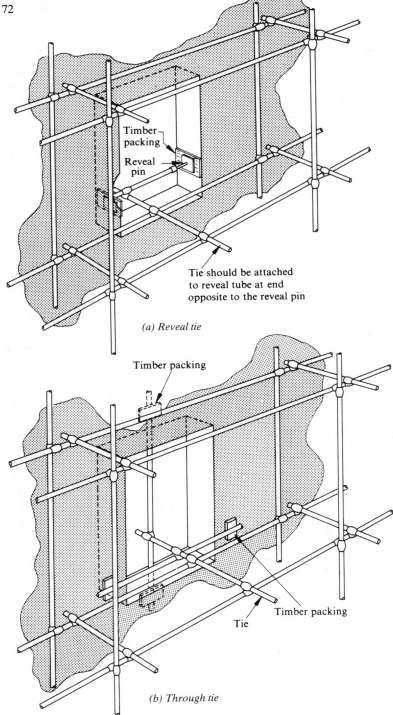

Figure 1.5.7 (a) Reveal ties and (b) through ties are technically acceptable but a nuisance to occupants or other tradesmen (British Standards Institution).

The main contractor on a site where several sub-contractors need to use the same scaffold for different purposes must ensure that any necessary modifications to the scaffold, such as changing the positions of ties and working platforms, are carried out and checked by scaffolders.
- Failure properly to install or counterweight the outriggers of suspended or cradle scaffolds has led to a number of falls, usually fatal, of painters, window cleaners and others.

  This is a technical fault caused by lack of proper training and understanding by the scaffolders. To prevent counterweights or outriggers being moved when the scaffold is not in use, these should be formally inspected each time before the scaffold is used. If movement is detected no one should be allowed on the scaffold until it has been rectified. To ensure safety, a permit to work system should be used. The person inspecting should sign the 'permit to work' and work should not be allowed without such a signed permit.
- Collapse or overturning of lightweight scaffold towers, particularly mobile ones, when being climbed or moved.

Many publications dealing with scaffolds show pictures of a mobile tower scaffold with a vertical ladder attached to one end which has to be climbed externally. These can be overturned by a heavy man climbing the ladder when the scaffold is out of plumb. This type is now being superseded by mobile tower scaffolds that have steps which can be climbed internally. Some have trap doors on platforms through which the worker climbs.

1.5.2.3 Lightweight and modular metal scaffolding (proprietary systems)

Lightweight aluminium tube and frames, of various shapes—'H', rectangular and triangular, as well as other modular systems which can be simply fitted together, without loose couplers, are now increasingly being used. Their main danger is that they are too easy to erect by untrained people, who often have no knowledge of the basic principles of scaffolding and fail to inspect the parts before using them.

There are further dangers in mixing the use of steel and weaker aluminium tube. Aluminium tube is easily damaged by tightening a coupler designed for steel tube onto it and accelerated corrosion occurs at a junction of a steel and aluminium tube.

Modular elements often have welded joints which can crack. Such cracks may be hidden by a coat of paint. As there can be many such joints in a single element, the task of inspection is correspondingly onerous.

1.5.2.4 Mobile access platforms

Several types are now being used in place of conventional scaffolds:

- hydraulic-arm platforms mounted on lorries or as towable trailer units;
- continuous mast scaffolds;
- telescopic masts;
- scissors lift 'flying carpet' platforms.

Provided they are well designed, constructed and maintained, used by properly-trained people and their bases properly supported, they are generally safer than conventional scaffolds. They contain a number of built-in features such as 'fail safe' hydraulic systems which safeguard against overloading.

1.5.2.5 The joint advisory council report on scaffolding

Further information is given in a report by a sub-committee of the Joint Advisory Council on Safety and Health in the Construction Industries[11]. Its principal recommendations are given in Appendix B.

1.5.2.6 Support scaffolds. Falsework

Falsework may be constructed from scaffold tube and fittings, adjustable metal props and timber supports, or from proprietary systems.

There have been several spectacular failures of falsework in the U.K. and elsewhere in the early 1970s on bridges and viaducts under construction or repair. These led to the appointment of a special U.K. committee to study the problem and to the issue by the HSE of the Bragg report on Falsework[7].

The main conclusions of this report were that falsework is all too seldom properly designed and controlled. Rule of thumb methods are too often applied. As a result the structure is inadequate for the loading to be applied and lacks lateral and longitudinal stability.

To ensure better standards of falsework design, special training courses on falsework were set up.

Judged by a recent survey by the HSE[2], the work of the Bragg Committee is slow in bringing improvements.

These problems are world wide. We quote the same Indian authority[8],

> "Although no single predominent cause of failure has been established, there is sufficient evidence regarding a potential cause of failure—namely lack of lateral stability. Inadequate or misplaced bracings and improper tying back of the structure make it unstable".

And again:

> "The quality of timber generally used for formwork and scaffolding is very poor and it twists, warps badly and quickly and deteriorates under the stress of the heat and contact with wet concrete. It is seldom possible to evaluate its strength and therefore rule of thumb methods are adopted in constructing the timber centering".

Of the remedies proposed, one of the most important is that 'the designer should ensure that the drawings and instructions are sufficient and easily understood by the erector. Particularly for large scale centering jobs like for bridges, etc. it should be made compulsory for the contractor to submit proper calculations with detailed working drawings and sequences of construction for the centerings for approval of a competent professional engineer before erecting'[8].

1.5.2.7 *Steeplejacks' scaffolds*

Steeplejacks' scaffolds, which are supported by and built up from the spire, chimney, monument, etc. under repair, are exempt from certain aspects of the U.K. Construction (Working Places) Regulations, 1966. They are discussed briefly under 'Steeplejacks' in Section 1.5.7.

1.5.3 Roof work

Roofs are the most essential part of any building in providing protection against the elements. Before discussing the hazards of roof work, the various types of roof and the reasons for their choice are considered.

Roofs with a pitch of less than 10° are considered to be flat[6];
those with a pitch of 10° to 30° to have a shallow slope[6];
those from 30° to 50° to have a moderate slope; and
those with a pitch greater than 50° to have a steep slope.

Flat roofs are most popular in warm and dry countries, where they can provide additional living space. Shallow sloping roofs are popular in climates with moderate snowfall, where the snow tends to remain on the roofs and provides extra insulation until a strong thaw occurs. Roofs with a steep slope are most popular in very wet and windy climates, and also in climates with heavy snowfall which might overload flat or gently sloping roofs. For the same reason, fragile roofs are less common in cold climates with heavy snowfall where they are more prone to collapse under the weight of the snow.

In the U.K. and other countries with temperate climates there is an increasing tendency to use fragile sheeting on widely spaced rafters or purlins on shallow sloping roofs.

Roofs with moderate and steep slopes are more commonly tiled using fairly closely-spaced battens, which when new are generally strong enough to stand a man's weight. The danger of sliding or falling off a sloping roof obviously increases with the slope of the roof.

Roof work has always been one of construction's most hazardous activities. A quick look at *Tables 1.3.4* and *1.3.5* makes this clear. There were an average of 27 roofing workers killed each year in the U.K. between 1977 and 1979. In 1978 there were approximately 8500 roof tilers and slaters employed. If the 27 casualties were all members of the 8500 roof tilers and slaters the FAIR in this group would have been over three. Since some of the casualties were painters, carpenters, plumbers, labourers and other craftsmen who were working on roofs when they fell, the FAIR probably lies between one and two for full-time roofing craftsmen. Nearly all the serious and fatal injuries resulted from falls. These occur repeatedly in the same old and obvious ways—from edges and eaves and through fragile roof materials.

In view of this predominance of falls, other hazards of roof work, such as dermatitis and respiratory infections through handling birds' droppings, tend to be overlooked.

A high proportion of roofing accidents occur during maintenance, particularly on or within the roofs of industrial premises.

Two special safety problems of roof maintenance are:

short duration of work, with high cost of safety measures compared with the cost of the job,
deterioration or masking of the roofing materials used.

An HSE publication[12] gives details of 26 fatal roof accidents in 1977. These involved 11 roofing craftsmen, five painters, as well as joiners, plumbers, labourers and an architect. It seems that 17 of these fatal accidents occurred during maintenance.

The recorded falls were of three classes:

roof edge falls,
falls through fragile roofing material,
falls from the internal structure of roofs.

1.5.3.1 Roof edge falls

There were six fatal roof edge falls, from heights ranging between 3.65 m and 14.5 m. At least two of these occurred during maintenance. Two victims were plumbers, two roofing workers, one an architect and one of unknown occupation. Three or four of the roofs were known to be flat, one shallow pitched, and one or two sloping at more than 30°. In only one case was the victim working on the slope or ridge of a roof. This man fell whilst sitting on the ridge and holding a rope to protect another man who was sitting on the eaves to examine a gutter. The main lesson is the need for barriers at roof edges, both on flat roofs and sloping roofs. These are needed at the edges of unfinished roofs under construction, as well as on completed roofs, particularly on flat roofs and roofs of shallow slope. The dangers of falling off the edge of a flat roof seem every bit as great as from a sloping roof. This may be because of the greater ease and confidence with which people move about on flat roofs. Men handling large roofing sheets or skylights are specially vulnerable to wind gusts and eddies which may literally blow them off a roof which has no guard rail at the edges.

1.5.3.2 Falls through fragile roofing material

"Fifteen men were killed as a result of falling through fragile roof materials, ten during maintenance. Of these, nine were roofing workers, two joiners, one a painter, two handymen and one a labourer. Nine fell through asbestos cement sheets, five through plastic roof lights and one through glazing."[12]

Practically all the accidents were caused either through the total lack, or inadequacy, of crawling boards, roof ladders and staging. The number of professional roof workers who succumb to the temptation to risk their weights on asbestos cement and PVC sheets without roof ladders or crawling boards is surprising. In one or two cases, however, the roof lights in stronger corrugated steel roofs had been painted and were disguised.

Roof work 77

Figure 1.5.8 A high proportion of roofing accidents occur during maintenance, particularly on roofs of industrial premises (HMSO).

1.5.3.3 *Falls from internal roof structures*

"Five men were killed through falls from internal roof structures, four of them painters and one a scaffolder. It seems that all five accidents occurred during maintenance. Nearly all were due to inadequate scaffolding or misuse of scaffolding. One, a painter, was actually wearing a safety belt but had not secured it."[12]

1.5.3.4 *The special problems of roof maintenance*

The two special problems mentioned earlier are discussed below.

Short duration of job
The problem is perhaps best appreciated by taking an example familiar to most of us, that of replacing a few broken tiles on a moderately sloping roof after a severe gale. The client, a working family man with a large mortgage is mainly interested in finding a local tradesman with a reputation for sound work who will do the job at an 'economic' price. Needless to say the 'economic' price covers only the use of a single ladder without roof ladders or crawling boards. The fact that these precautions are in the U.K. legal requirements of the Construction (Working Places) Regulations, 1966, is quietly ignored. Expediency, therefore, wins the day. Before any roof work is started one or two people bring ladders, climb on the roof, inspect the job and give estimates. Having survived this ordeal a bargain is struck between the owner-occupier and the lowest bidder. Would we expect that the man who actually does the job will have any more protection than his boss who risked his neck to get it in the first place?

It is surprising that more people are not killed through repairing roofs of domestic dwellings. Statistics suggest that industrial buildings present a greater menace.

Fragile roofing materials
Much of the lightweight sheeting used for modern industrial buildings, asbestos cement, PVC and other plastics, woodwool and wired glass spans considerable distances between rafters or purlins. In such circumstances the materials are not strong enough to carry a man's weight concentrated on one heel, particularly if he stumbles.

Asbestos cement sheeting is specially brittle. PVC and other plastic roofing sheeting becomes embrittled on prolonged exposure to sun and rain. Even metal sheeting, such as corrugated galvanized iron, can corrode as a result of condensation underneath, as well as the weather outside, and collapse under a man's weight. Hence, unless a roof is known to have been designed to carry people's weight and is maintained to that standard, it should not be trusted to do so. Men's weights should otherwise always be distributed by roof ladders or crawling boards or on adequate staging with guard rails and toeboards at the sides of the roof.

The masking of weak areas of roofs is another cause of fatal falls. Corrugated asbestos sheets have sometimes been used to replace galvanized iron, and then painted over so that the two are indistinguishable. Glass and plastic skylights have also been painted over, covered with bitumen or become discoloured.

1.5.3.5 Precautions

HSE Guidance Note GS 10[13] details precautions which should be taken in roofing work to minimize falls.

Edge protection
There are many different cases to be considered, depending on the slope of the roof and on the type of building and roof construction. No single type of edge protection can be applied in all cases. Under U.K. regulations it

should incorporate a guardrail at a height of 0.913 to 1.15 m (3 ft to 3 ft 9 ins) from the edge of the roof and a toeboard or equivalent protection, such as the upstand of an existing roof edge. The toeboard should stand at least 152.5 mm (6 ins) above the level of the roof, or more if necessary so as to leave a gap not wider than 762 mm (30 ins) between the top of the toeboard or upstand and the guardrail. The guardrail and toeboard must be strong and secure enough to resist any likely impact from a man running on, or sliding off the roof.

For strong flat roofs intended to be walked on, such as those made of reinforced concrete beams, a type of portable edge protection dependent for its security and location on the use of precast concrete counterweights may be used (*Figure 1.5.9*).

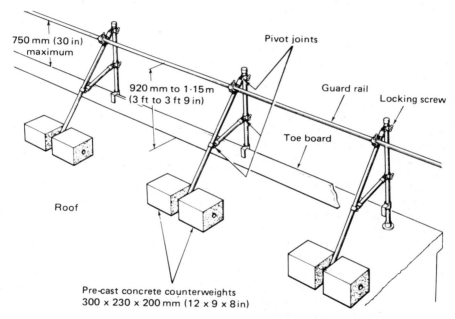

Figure 1.5.9 Free-standing counterbalanced frames may be used on strong flat roofs.

For roofs with a sufficiently strong upstand at the edge, the guardrail may be attached by short vertical scaffold poles to the upstand itself. (Care is needed here as many apparently solid parapets have little resistance to a horizontal blow.)

For single storey buildings with roof edges not more than 16.5 feet (5 m) above ground level, the guardrail and toeboard may be supported by scaffolding erected from the ground, with proper base and sole plates and bracing (*Figure 1.5.10*).

For multi-storey buildings with sufficient window openings on the top storey, it is sometimes possible to erect a guard rail and toeboard by cantilevering or trussing out scaffold tubes from the top storey.

80   High risk occupations

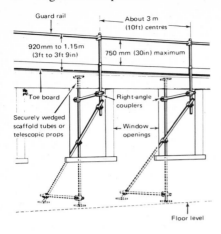

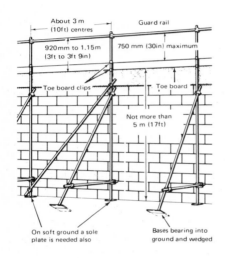

Figure 1.5.10 Scaffolding erected from the ground can be used on buildings where the roof is not more than five metres above ground level.

Brackets are also available for supporting roof edge protection guardrails and toeboards from the columns or edge beams of framed buildings, or from substantial floor slabs (*Figure 1.5.11*).

It is fairly obvious that edge protection should not be attached to gutters, fascia or soffit boards, which would not be strong enough to resist the overturning moment of a man's body striking the guard rail.

Proper edge protection for roof workers can generally be provided at a reasonable cost during construction of a building as an integral part of the scaffolding used for construction. The position of the top scaffold platform in relation to the roof battens, roof ladder or other form of roof structure onto which a man has to step, should be between 9.25 inches and 11.8 inches (235 and 300 mm) below it. This provides a reasonable step-up distance. If a tower scaffold is used for access to the roof it must be securely tied to the building and arranged so that there is no danger of a man falling between the tower scaffold and the building.

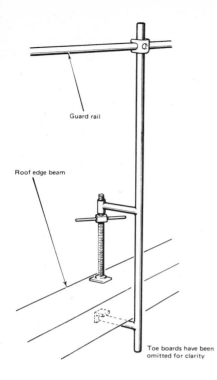

Figure 1.5.11 Edge protection clamped to edge beam of framed building.

The problem is more difficult during maintenance. Most roofs will require maintenance work several times during the life of a building. The problem should, therefore, be anticipated when the building is designed and built. This may take the form of:

erecting permanent walkways with guardrails at roof edges,
erecting permanent walkways along roof valleys and over ridges,
providing means of securing standard guardrails and toeboards, and
providing permanent anchorages for lines used with safety harnesses or belts.

Protection around fragile roof areas
Protection is not only required against falling from roof edges; it is also necessary around fragile roof surfaces (including roof lights). These sometimes extend to the entire roof (with the exception of permanent walkways with guardrails which should have been provided in the design of the building). The law in the U.K. requires notices to be prominently and permanently displayed at the approaches to fragile roofs. (A roof which becomes fragile through deterioration will present a problem.)

For large flat roofs where work has to be done in limited areas, it may be more appropriate to provide adequate fencing round the areas where work has to be done, rather than fit edge protection round the whole roof area. Fragile roof areas and roof openings must in any case be separately protected by barriers or strong covers. These must either be securely fixed or have a notice attached warning of the hazard.

Before any roof is used as a means of access or as a place of work during any operation, whether it be construction, repair, maintenance or demolition, it is essential to identify parts covered by fragile material and decide on the precautions to be taken.

Sloping roofs
On moderately sloping roofs whose covering is fragile or of uncertain strength, at least two roof ladders or crawling boards should be used. A man can then support himself on one ladder while moving the other ladder to a new position. Roof ladders and crawling boards must be properly made and suitable for the job in hand, be strong enough to support men between the supports for the roof covering, and be secured against movement. Safe anchorage of roof ladders and crawling boards is particularly important. Top anchorage should not depend on the ridge capping, but bear on the opposite slope by a specially made ridge iron. Roof ladders are often lashed to the tops of access ladders, with a man standing on the access ladder near the junction. This is better than leaving the roof ladder quite unsecured, but it creates hazards, especially when the roof ladder has to be moved. Even the lifting, erection and securing of a roof ladder when the only access to the roof is a portable extension ladder presents serious safety problems (see Chapter 2.9).

For work on steeply sloping roofs, special working platforms or 'cripples' should be used. Work on steeples is discussed under the hazards of steeplejacks in Section 1.5.7.

Hipped roofs present difficult problems in securing adequate anchorage for crawling boards and roof ladders. Gable roofs without a parapet at one or both ends, where the roof is carried over the wall to form a verge, may present a present a problem in providing edge protection.

Special shapes
Similar problems are encountered on curved roofs or roofs of special shape, which may require the construction of special platforms for construction and maintenance workers. These problems can sometimes be solved by the use of an aerial platform, e.g. one mounted on a vehicle with a hydraulically-operated boom. Before deciding on the use of this method, the supervisor must:

    check that there is adequate unimpeded access,
    ensure a solid base for the platform vehicle,
    check that all parts of the roof to be worked on are within the safe range of the boom.

Lifting appliances (see HS PM 27)
Another hazard sometimes encountered in roof work arises from the use of small lifting appliances—gin wheels and scaffold cranes—near the edges of roofs. One problem is that of providing adequate and secure counterweights for the appliance. A second is that of providing adequate edge protection for workers at all times when material is being raised or lowered. Here the HSE recommend that if a guardrail and toeboard cannot always be maintained in position, any worker who has to be near the edge to help move the load or to signal should be protected by a suitably anchored safety harness or belt.

1.5.3.6 Work in adverse weather

Other hazards of roof work arise from adverse weather conditions—ice, snow, rain and wind. Sudden changes in the surface may also occur in moving from shaded areas to areas exposed to the sun. Contracts for roof work must not penalise the contractor who suspends work when the weather renders it specially dangerous. The handling of large sheets can be dangerous even in quite moderate winds. Care is needed to ensure that work materials brought to roof level and not immediately required impose no excessive local loads on the structure, and that they are also adequately lashed down or otherwise secured so that they cannot be blown away.

1.5.3.7 Electric cables (see HSE GS 6)

Further hazards arise in roof work from electric cables which enter the building at roof level, particularly those bringing in electrical supplies at supply voltages. Details of any such cables should be obtained by the contractor before starting work. If the work cannot be done without risk of contact with them, arrangements should be made to have the electricity supply disconnected for the duration of the work. The greatest hazard is when erecting or removing metal ladders and scaffolding. Electrical hazards are discussed in more detail in Chapter 2.5.

1.5.3.8 Access ladders

Ladders used for access to roofs should wherever possible be secured at the top. Frequently they are lashed to eye bolts screwed into the fascia board which is often none too secure. If possible, the ladder should be secured to a brace inside a window, or to one of the rafters. Ladders are discussed in more detail in Chapter 2.9.

1.5.4 Demolition[14,15,16,17,18] (see also BS 5975)

The hazards of demolition, the types of business (general, dismantling, explosives and civil engineering), and the main categories of workers (top men, mattock men and general labourers) were discussed briefly in Chapter 1.1. Demolition is generally regarded as a dirty, dangerous and unhealthy occupation. The demolition worker faces a variety of hazards:

Falling. He sometimes works near the edges of tall buildings with no working platform and no anchorage for a safety harness.
Being hit or trapped by falling objects.
Excessive noise, from demolition balls, pneumatic drills, explosives and falling masonry.
Vibration, especially from hand-held pneumatic tools.
Respiratory hazards, from dust which may contain toxic constituents, such as asbestos and silica.
Flying particles, causing eye and skin injury.
Fires and explosions, especially when demolishing tanks which contained oils or chemicals.
Weather. Demolition proceeds in all weathers.

Demolition work is difficult and skilled. It requires knowledge of the methods and materials of construction of the structure being demolished.

Demolition proceeds at a rapid pace. Workers must constantly appraise the strength, stability and weak points of partly-demolished structures. They must know:

- how and in which direction things will fall; and
- whether the collapse of one wall or arch will cause adjacent walls to collapse.

1.5.4.1 Statistics and the Joint Advisory Council Review

The hazards of demolition in the U.K. were reviewed by the Joint Advisory Committee on Safety and Health in the Construction Industries in the early 1970s. Their report[14] was published by the Health and Safety Executive. Its main recommendations are given in Appendix B. The injury statistics given in *Tables 1.5.5, 1.5.6* and *1.5.7* are taken from it.

TABLE 1.5.5. Reported accidents in demolition and all building operations

Year	Demolition of buildings			All building operations			Demolition accidents % of all building	
	Fatal	Total reported†	Fatal % of total	Fatal	Total reported†	Fatal % of total	Fatal	Total
1969	21	809	2.6	174	34982	0.50	12.1	2.3
1970	19	687	2.8	138	30938	0.45	13.8	2.2
1971	15	580	2.6	144	26683	0.54	10.4	2.2
1972	14	718	1.9	132	27524	0.48	10.6	2.6
1973	15	663	2.3	148	27991	0.53	10.1	2.4
1974	16	614	2.6	113	25831	0.44	14.2	2.4
1975	16	559	2.9	121	26555	0.46	13.2	2.1
Total	116	4630	2.5	970	200504	0.48	12.0	2.3

†The total reported accidents shown here are thought to be considerably lower than those which disable the worker for more than three working days, perhaps by as much as 50%. In any event, they do not include accidents to self employed men who constitute about 20% of the construction labour force.

Table 1.5.5 gives the number of reported accidental injuries in demolition, and all reported accidents in building operations from 1969 to 1975. It clearly shows that the chances of an accidental injury proving fatal are far higher for demolition than for building as a whole.

The number of workers employed in demolition at any one time is somewhat uncertain. Taking the Department of Employment's figure of about 6000[15] in the mid 1970s, the FAIR for demolition workers would have been about 2.75 between 1969 and 1975. The accidental injury incidence rate would have been over 100. Thus the average full-time demolition worker had a 1 in 7 chance of an accidental injury every year.

Since 1975 there has been a drop in the annual toll of demolition workers, with an average of eight fatalities per year for the three years 1977 to 1979[2]. This still corresponds to a fatal accident incidence rate greater than one.

TABLE 1.5.6. Reported demolition injuries related to type of building involved (fatalities in brackets)

Type of building	1970	1971	1972	Year 1973	1974	1975
Industrial	278 (8)	220 (7)	284 (6)	234 (8)	265 (8)	204 (7)
Commercial and public	164 (9)	153 (4)	177 (3)	206 (5)	144 (4)	146 (6)
Blocks of flats	7 (–)	13 (1)	14 (1)	14 (1)	8 (–)	14 (–)
Dwelling houses	133 (–)	98 (2)	136 (1)	129 (1)	131 (1)	103 (2)
Other building operations	105 (2)	96 (1)	107 (5)	80 (–)	69 (3)	92 (–)
Total reported accidental injuries	687 (19)	580 (15)	718 (14)	663 (15)	614 (16)	559 (16)

TABLE 1.5.7. Analysis of reports of fifty fatal accidents to men employed in demolition

Type of accident	Number of cases
Collapse of part of building meaning a collapse not expected by workmen involved, or a deliberate collapse before every person in a safe position	18
Material falling	4
meaning (a) part of a building, after being prised off, falls on to a person, or	2
(b) part of a building being thrown down catches on clothing pulling man down	2
Men falling from a height	17
meaning (a) from a working position or means of access, or	9
(b) through fragile roof covering or through roofs when stripping sheets, or	4
(c) through openings in floors	4
Plant accidents meaning negligence in use of or alteration to plant such as vehicles or cranes	7
Miscellaneous or unknown cause	4
Total	50

Table 1.5.6 gives the distribution of demolition injuries between the types of building involved between 1970 and 1975. Most occurred during the demolition of industrial, commercial and public buildings. The FAIRs among workers so engaged must have been exceptionally high—perhaps 4 or 5—which corresponds to a chance of about one in six of a fatal accident during a working lifetime. The hazards appear to increase with greater wall heights, floor loads and floor spans, and greater use of steel and reinforced concrete in construction.

Table 1.5.7 gives an analysis of reports on 50 fatal accidents to demolition workers.

1.5.4.2 Need and features of safe systems of work[16,18,19,20]

The high injury rates among demolition workers point to the need to consider the hazards before work starts and to plan safe systems of work. Often, however, the time gap between the release of the building, etc. for demolition and the start of new work on the site is very short. Many important points need to be considered in planning for safe demolition and serious hazards can easily be missed due to time constraints, resulting in planning degenerating into improvisation with all its dangers.

86   High risk occupations

In planned demolition the work should fall into the following phases:

Precontract survey
The Contract with its various requirements
The demolition itself

These different aspects of planned demolition are considered next.

1.5.4.3   Precontract survey

A survey of the premises and surrounding site should be made before a contract is placed for the demolition. This is generally done by an architect on behalf of the site owner(s). The survey has two main purposes:

- To ensure that property on surrounding sites is protected during demolition;
- To ensure that the demolition can proceed safely.

The contract should be based on the recommendations of the survey. The survey normally covers a wide range of items. Only those which particularly relate to the safety of the workers are listed below.

- Details and location of all public services on site must be known and shown on a large scale plan. They are mostly buried and include drainage; electric cables; gas pipes; water pipes; telephone, radio and TV lines and others such as hydraulic pressure and district heating. The survey should state which of these have to be cut off, plugged or diverted, and which need protection during demolition. The survey should make clear who is responsible for doing each of these things.
- In the case of factories, power stations, oil refineries, etc. details of any special services, raw materials and product lines should be stated and treated similarly.
- The age and condition of the building or structure to be demolished. The structural design should be examined to determine what parts depend on each other for structural stability.
- Identification of whether the structure is dangerous and thus likely to pose special problems in demolition.
- The presence of asbestos (e.g. in thermal insulation) should be checked. If present, its type should be determined, by analysis if necessary. Demolition work involving the handling of asbestos should be treated as a specialist operation. Those doing it must be properly instructed, protected (by appropriate clothing and respirators) and supervised.
- Items which present special hazards during demolition must be indicated. These include reinforced concrete, pre-stressed members, structural steelwork, cantilevered balconies, canopies and arches and external staircases and ladders.
- Basements, cellars, wells and storage tanks, especially those which have contained toxic or flammable materials.
- Previous use of buildings. They may still contain toxic, flammable or radio-active materials. For example, one of the authors found glass

carboys of acid and cyanide solution on premises just about to be demolished. If the carboys had broken and the contents had mixed, a lethal gas cloud capable of killing many people would have been released.
- Any radio-active sources, such as old lightening conductors and special level gauges.
- Local road access for cranes, demolition vehicles and plant, and details of roads and footways which may have to be closed (with appropriate permission) during demolition.
- Suitability of ground around the building or structure for cranes, mobile platforms, scaffolds, etc.
- Details of fans, hoardings and protective scaffolding needed during demolition.
- Details of temporary support for the structure needed to enable demolition to proceed.
- Recommended means of disposal of rubbish (subject to Control of Pollution Act, 1974 and Control of Pollution (Special Wastes) Regulations, 1980).

1.5.4.4 *The Contract and its requirements*

Standard forms for tender and contract are available in the U.K. from the Royal Institute of British Architects, the Institute of Civil Engineers and the National Federation of Demolition Contractors. These may be useful to help draw up contracts in countries where no such forms are available. Special contract requirements will undoubtedly arise as a result of the pre-contract survey. These may include provision for sealing branch drains with cement and slurry, the use or avoidance of explosives in demolition, the use of cranes to lower heavy steel or reinforced concrete beams and columns and the preliminary removal of glass from all windows. In circumstances where special precautions need to be used the methods of demolition to be employed may also have to be stated in the contract.

Sealing of drains is very important to avoid dangers of toxic and sometimes flammable sewer gas escaping to the site, as well as rat infestation. Respiratory protection may be needed when sealing drains. Special precautions are needed if drains or other confined spaces have to be entered. These are described in Chapter 2.7.

The contract should also require the contractor to insure against injury to his workmen and third parties.

Demolition contracts generally preclude sub-letting the contract without the written agreement of the client. There is a danger that a demolition contract or parts of it may be sub-let to a roving band of self-employed demolition sub-contractor with very low safety standards.

The contractor must comply with all statutory requirements, obtain all necessary permissions and comply with various conditions, such as the shoring up of adjacent buildings.

1.5.4.5 *Safety requirements during demolition*

Whatever methods are used, a number of safety requirements must be met. Some general ones are listed below.

1. *Competence of Supervisor*
Demolition needs close supervision. The supervisor should be experienced in the methods to be used and their problems. He should be familiar with the methods of erecting different types of building—reinforced concrete, steel frame, wooden and pre-fabricated structures. Thus he will be able to recognise the structural details of buildings to be demolished and take or initiate remedial action in the event of problems.

2. *Demolition Workers*
Demolition workers should be trained and experienced in their work. Young persons should be restricted to low levels or to clearing-up. All workers should wear at least hard hats, shoes with protective toecaps, gloves and goggles, and any other special protection prescribed for the job.

3. *Supervisor's Inspection*
Notwithstanding the pre-contract survey, the supervisor should carefully inspect the building to be demolished and adjacent structures and check that the methods proposed by his management are suitable and safe. He should then prescribe safe methods of working in writing in full detail.

4. *Essentials of Working Method*
The method must ensure not only the stability of the building or structure under demolition so long as they remain standing, but also that of adjacent properties. This may often require the erection of bracing shores or other support to achieve this object. Alcoves, balconies, cornices, arches and roof trusses must always be secured against inadvertent or premature collapse.

5. *Passageways, Floors and Stairs*
Passageways should always be kept clear. Floors and stairs should not be overloaded with rubble. So long as stairs are used, the bannisters should not be removed. When floorboards are removed from supporting joists or girders, these should be covered by strong and secured planks if anyone still has to work or move on them.

6. *Danger zones*
Where danger zones are created, they should be barricaded and indicated by warning notices or permanently guarded.

7. *Throwing Down Material*
So far as possible, material should not be thrown down, but enclosed chutes should be used (*Figure 1.5.12*). When there is no practical alternative to throwing, a loud warning should be given and precautions taken to ensure that no one is likely to be struck. This should include erecting barriers or posting look-outs. Demolition workers should wear close fitting clothes with no loose belts, ties, etc. which could catch in material which they throw down.

8. *Stagewise Demolition*
It is generally preferable to bring a building down in stages starting at the roof and leaving the removal of floor supports at any stage until the last. No material should be left supported or hanging dangerously.

9. *Demolition of Reinforced Concrete*
This should only be undertaken by persons with expert knowledge. Some of the special risks are discussed in Section 2.3.4.10.

10. *Use of Demolition Balls*
When using a demolition ball, the effects of the blows on the rest of the

Figure 1.5.12 Precautions should be taken to protect the general public and enclosed chutes should be used to bring down loose material (HMSO).

structure and the foundations must be anticipated. It is generally recommended that a demolition ball should only be dropped on a target and not swung against its side. Some exceptions may be made if the crane is sufficiently strong and its driver specially trained and appreciates the limitations of the crane-ball combination.

1.5.4.6 Demolition by knocking down and taking to pieces

The following general precautions apply:
1. *Safe Working Place*
A safe working place is always needed. Where necessary, scaffolds should be erected. Walls which are being knocked down should not be stepped on, nor when beams or girders are being taken down or tipped over.
2. *Work from Tops of Walls*
The ILO recommend that work should only be done from the tops of walls if they are more than 50 cm wide and less than 3 m high[19]. Workers knocking down walls must be protected against falls by suitable means, such as roping. Stricter rules apply in the U.K.[6] These require that there

should be a safe place of work which is guarded on all open sides where a person could fall more than 1.98 m (6 ft 6 ins).

3. *Demolition of Panelled Walls*
Masonry should be taken down before a panelled wall is demolished, taking care that the panelled wall is still stable.

4. *Demolition of Foundations and Underground Structures*
Before underground structures are knocked down, the adjacent ground must be secured against falling-in.

5. *Undermining of Structures*
Structures should not be demolished by undermining or cutting through them because of the danger of premature collapse.

6. *Use of Bars and Racks*
Bars and racks should not be used where there is any danger that the wall, etc. may fall on the wrong side and cause danger.

1.5.4.7 *Demolition by pulling and pushing down*

The following general precautions apply:

1. *When to Pull Down*
Buildings should only be pulled down when they are in an unsafe condition and cannot be knocked down.

2. *Rope for Pulling Down*
Only strong wire ropes should be used for pulling down. These must be long enough to enable them to be pulled from a safe place.

3. *Fastening of Ropes*
Ropes are best fastened to the building to be pulled down by independent mobile ladders to reduce risks to the workers.

4. *Danger Zone Evacuation*
Pulling down should not begin until everyone is clear of the danger zone. This zone includes the area where a rope might whip round if it broke.

5. *Plant for Pushing Down and Operator Training*
Only power plant specially designed for pushing down buildings, with strong all-round protection for the operator (especially windscreens) should be used. Such plant may incorporate a pusher arm or a mechanical rake. Operators should be specially trained before being employed on demolition work.

1.5.4.8 *Demolition by explosives*

The following general precautions apply:

1. *Experience and Training of Explosives Engineer*
The explosives engineer should have sufficient experience in demolition by explosives and have served as an assistant for a considerable period in this task. Experience of shot firing in quarrying or underground construction is not enough.

2. *Normal Explosives Precautions*
All customary precautions in the use of explosives should be taken. Shots should be covered so as not to endanger the neighbourhood. Local and structural conditions must be considered when fixing the sizes of charges. It

is often advisable to divide the structure to be blasted into a number of sections, applying suitable charges to each section.

3. *Drilling Shot Holes*

Where the parts to be demolished are unstable, it is safer to drill the shot holes electrically than pneumatically because of the danger of vibration, including a premature collapse.

4. *Dangers of Cast Iron*

Great care is needed to check the presence of cast iron and to avoid placing charges near it, since it easily shatters into shrapnel.

5. *Barricading*

Special care is needed to barricade the area being demolished and round the firing point, and to prevent the entry of unauthorised persons.

6. *Electrical Firing*

Shots should, as a rule, only be fired electrically.

7. *Freeing Firing Point and Removing Barricades*

The engineer must personally satisfy himself that no dangerous conditions have been left or created after the charges have been detonated before barricades are removed. Any danger zone found must be barricaded until a new shot has been placed and fired or the situation rendered safe by some other means.

1.5.4.9 Demolition of structures with special risks

Several common problems are discussed below. It is, however, impossible to cover every case. Demolition supervisors and workers need always to be on the alert to special risks which they have not encountered before.

1. *Chimneys*

Tall chimneys should only be demolished by those experienced in their construction. Chimneys should not be demolished from the top of the wall. Inside scaffolds or outside bracket scaffolds should be used for knocking down chimneys. As demolition proceeds they should be progressively lowered. The platforms should be at least 10 inches (25 cm) but not more than 4 ft 11 ins (1.5 m) below the top of the wall. Before material is thrown down inside the chimney, an opening should be made in the base of the wall for removing it. No work should be allowed above this opening while such material is being removed. Where conditions permit, chimneys may be demolished by deliberate collapse or explosives. The direction of fall must then be fixed in advance.

2. *Steel Structures*

Those responsible for demolition should know how the stresses are distributed and the order in which members should be removed to prevent collapse. Supports and guys should be used to prevent temporary instability. A look-out should be kept for damaged, insecure or partly cut steelwork. This should be removed or made safe. Heavy steel and reinforced concrete beams and columns should be supported by a crane and lifted clear during demolition and not dropped. Before cutting or releasing any steelwork precautions must be taken to safeguard against a sudden twist, spring or collapse. Where gas cutting or welding is done on structures coated with lead paint, the welders should wear respirators with suitable filters to protect against lead fumes.

92 High risk occupations

3. *Pre-stressed and Reinforced Concrete*

Pre-tensioned and post-tensioned concrete creates special problems due to the forces which may be released when it is demolished. For this reason demolition should only be done in accordance with an engineer's instructions. As far as possible pre-tensioned beams should be removed intact from the structure and broken-up on the ground. Post-tensioned beams present more difficult problems which require special knowledge. Reinforced concrete slabs should be cut parallel to the main reinforcement. Where necessary, inspection holes should be drilled to determine the position and direction of the reinforcement.

4. *Oil, Chemical and Gas Works and Storage Tanks*

A special survey should be made of any oils, chemicals and explosives which may still be present, and a signed statement by the owner or occupier prior to demolition should be made. Specialist advice is needed to determine the right precautions before demolition begins. Tanks which contained toxic, flammable or gum-forming materials need special care. The tanks should first be freed of all toxic and combustible material by displacement with water, steaming out, and/or passing air or inert gas through them. The atmosphere inside the tank should be tested for flammable vapours and oxygen before any hot work is done. These precautions are, however, insufficient if the inside of the tank has an adherent deposit of gum, tar, wax or heavy oil. It is then best to fill the tank to the roof with water. The roof is then removed by cold cutting. The water level is next lowered by about £ ft 8 ins (0.5 m) at a time, and the sides are cut down to the water level under close supervision. If hot cutting is used, the cutters must be protected against fires and toxic gases released during cutting.

Tanks which have contained leaded petrol may contain highly toxic residues. Expert advice and suitable protective clothing and respirators are needed.

5. *Entry into Pits, Basements, Cellars, Old Wells, Tanks and Other Enclosed Spaces*

The atmosphere in these may be toxic, flammable or deficient in oxygen. Entry should not be allowed without a written permit signed by the supervisor stipulating that certain precautions have been taken. These include the isolation of the enclosed space from all sources of toxic and flammable gases, the ventilation of the space and the testing of the atmosphere by a competent person using proper testing equipment. The hazards are discussed in more detail in Chapter 2.4.

1.5.4.10 Trends in safer demolition

There are several current trends in industrialized countries which should make for safer and more economic demolition.

1. *Use of Mobile Work Platforms and Man-riding Skips*

These reduce the need for men to work on or suspended from structures under demolition.

2. *Greater Use of Explosives*

This has followed from improved techniques in the use of explosives. The use of explosives generally make it necessary for neighbouring premises to be evacuated while they are used. The process, however, may be so fast that they prefer this inconvenience for a short time to the prolonged exposure to noise, dirt, etc. over a longer period which is characteristic of other methods of demolition.

3. *Greater Use of Specialized Demolition Techniques*

These include the use of hydraulic and chemical bursters and thermic reactions using oxygen lances or chemicals.

In the U.K. moves are afoot to improve the legal regulations to ensure that:

(a) the laws relating to demolition can be presented in a single publication;
(b) a precontract survey of the property to be demolished is mandatory and a written record of the survey is available for inspection;
(c) the present duty to notify the HSE of the commencement of demolition should apply to work lasting 3 weeks and longer, rather than the present 6 weeks or more; and
(d) local authorities should have powers to extend the areas around demolition projects from which the public are temporarily excluded in order that quicker and safer demolition methods (e.g. explosives) may be used more extensively.

1.5.5 Excavation[2,17,18,19,20,21,22,23] *(see also BS 6031)*

Excavation in building and civil engineering is a vast subject ranging from the digging of a small hole to plant a tree or for a soak-away drain to a deep canal for ocean-going tankers. In building work the majority of excavations are for foundations or cable ducts and pipe runs. Civil engineering, on the other hand, is associated with deeper excavations, including tunnels and caissons which are discussed in Section 1.5.6. Apart from these exceptions, we try to cover the hazards of the whole field of excavation in this chapter.

The physical work of excavation is done mainly by ground workers who carry out all work below ground level. This can include hand digging the excavation as well as pouring foundations and laying drains and services. Mechanical aids, such as excavators, have replaced hand digging in many cases, leaving only side trimming and bottoming out to manual methods. The hazards considered in this section apply mainly to the ground workers and to a lesser extent to the plant operators outside the excavation.

Ground work was treated as a high risk construction activity in HSE's report 'Construction 1979–80'[2]. This was the result of the high proportion of accidents to ground workers which proved fatal.

In digging excavations to provide for the foundations and services of the future, we should never forget that the land area of this planet is limited and its history of human occupancy a long one. All excavations are to some extent a journey into the past and sometimes turn up the unexpected such as unexploded bombs, long forgotten mine workings and even buried treasure.

Not all the hazards of a new excavation can be anticipated, but judging by the record, the vast majority of them can be avoided by thorough

TABLE 1.5.8. The main hazards of excavation

No.	Description
1.	Falls of persons into excavation (including children)
2.	Collapse of sides of excavations and falls of material excavated onto persons in excavation.
3.	Collapse of adjacent building or structure into excavation.
4.	Collapse of temporary structure made to support sides of excavation.
5.	Persons in excavation struck by parts of machine, falls of spoil from excavator buckets and other objects dropped on them.
6.	Striking underground electrical cables with resulting flash burns and electric shock.
7.	Striking and breaking other underground service lines—gas (fire and explosion hazard), water (flooding), and sewage (toxic gases).
8.	Fire and explosion from heavier than air flammable gases and vapours (especially LPG) entering excavation.
9.	Poisoning or asphyxiation from heavier than air gases such as hydrogen sulphide or carbon dioxide present in the ground itself, or entering the excavation from outside.
10.	Poisoning from carbon monoxide produced from torches, burners, etc. used in excavation with insufficient ventilation, or from exhaust gases produced by plant and machinery used in connection with the excavation, including pumps for dewatering.
11.	Toxic and radioactive hazards from the ground itself, usually resulting from its previous occupancy.
12.	Flooding with risk of drowning.
13.	Accidental explosions through the use of explosives in excavation.
14.	Unexploded bombs.
15.	Fall through bottom of excavation into disused mine shaft or other cavities in the ground.
16.	Slipped disc, bursitis, muscular disorders resulting from work in cramped positions.
17.	Weil's disease, transmitted by rats; tetanus.

preliminary surveys and careful planning both of the technical and safety aspects.

Different aspects of excavation are covered by several U.K. regulations, particularly the Construction (General Provisions) Regulations, 1961, Regulations 8–14. They are also included in the Code of Practice BS 6031:1981 which has a special section on safety procedures.

A brief listing of the main hazards of excavation work is given in *Table 1.5.8.*

On medium and large scale excavations, a pre-contract survey is necessary before work starts on the drawing board. In civil engineering, consulting engineers are often involved in this process. The survey will involve testing and examination of the ground sometimes to depths of 30 metres or more to determine its load bearing capabilities and the type of foundation, etc. required. This information will also be needed in defining the scope of excavation contracts, preparing tender documents and in estimating the costs of excavation when tendering.

The same information which provides the basic data for designing the permanent works can be of inestimable value in identifying safe methods of excavation. Whilst on larger jobs pre-contract surveys are the rule this is not necessarily the case on small building works. As a result excavation often starts without any proper survey. This leads to last minute improvisation resulting all too often in an accident which should and could have been avoided.

The importance of a pre-contract survey has been stressed in the case of demolition (Section 1.5.4) and equally in the case of excavations such surveys are an essential prerequisite.

A number of points involved in the surveying, planning and execution of excavations are considered next.

1.5.5.1 Site investigation

The nature of the ground should be investigated before excavation to determine its type and natural angle of repose, wet and dry. The ILO in a guide to safety in building work[19] divides ground into three broad classes:

- non-cohesive ground or light soil, such as sand or gravel, whose natural angle of repose when dry is usually 45° or less;
- cohesive ground or heavy soil, such as stiff clay, whose natural angle of repose is about 60°;
- rock, whose natural angle of repose varies from about 80° for loosely bonded or light rock, to 90° for tightly bonded heavy rock. Rock may, however, have steeply sloping clay planes which may fail with resultant collapse.

1.5.5.2 Site records, including underground hazards

A search must be made for all relevant site records before planning an excavation. They must be checked for type of soil and any changes made to it, old mine shafts, wells, springs, buried tanks and underground culverts, etc.

1.5.5.3 Underground cables and services

A clear legal obligation to protect workers from buried electric cables is contained in Regulation 44(1) of the Construction (General Provisions) Regulations, 1961. This is a real and serious risk. According to the HSE[2] an average of 600 people are injured, some fatally, due to this cause, and the staggering figure of 20 000 or more buried cables are also damaged every year.

The injuries received are mostly flash burns to the hands, face and body from the explosive arcing, rather than electrocution.

Public utility companies and other owners of underground cables and services should be informed of the proposed excavation and its exact location at an early stage and should be requested to provide locations and details of their cables, pipes, etc. The Electricity Board will, where possible, make available a copy of a dimensioned plan showing the locations and depths of any of their buried cables in the proposed work area. They may also assist the contractor to plan a safe working procedure.

The line of any cable found to be within or adjacent to the area to be excavated should be marked with chalk or paint on paved surfaces and with wooden pegs in green field locations. Details of any likely to be exposed during the excavation should be specially noted. Decisions will have to be taken as to whether any need to be turned off or disconnected during the excavation, and whether any will need to be rerouted because of the excavation.

One U.K. publication[21] by the National Joint Utilities Group gives a number of recommendations on the avoidance of danger from underground electric cables. Another[22] describes cable locating devices available and advises on their selection and use.

Most, but not all, buried cables in the U.K. belong to one of the Electricity Boards and there is seldom any visible indication of the presence of a cable below ground. Whilst some cables are covered in earthenware pipes or by tiles many are laid directly in the ground, generally at depths of 0.4 to 1 metre below the surface.

Cables may have a protective layer of steel wire, tape, hessian or PVC overlaid on aluminium or lead sheaths. Buried cables are easily damaged by hand held tools such as pneumatic drills, picks and forks, as well as by excavating machines.

Trial holes should be excavated to locate any electric cable believed to lie in the area of excavation. This should be by hand digging methods (spade or shovel) and special care must be taken when close to the presumed line of the cable. Mechanical excavators or power tools other than for breaking paved surfaces should not be used within 0.5 m of the indicated cable line. This applies to the excavation proper as well as to the trial holes.

The position and depths of cables shown on drawings are not always reliable since reference points (such as kerb lines) may have altered. In addition, cables may have been reinstated during a previous excavation or the ground level may have changed.

Cable locating devices are very useful in confirming and determining the actual positions of cables even when drawings exist. They are essential if records have been lost and for emergency work which has to be done before records can be obtained. In residential and urban areas one should always assume that there may be a cable in the area to be excavated.

There are three main types of cable locating devices:

- live cable or 'hum' detectors;
- very low frequency radio detectors, which detect dead, as well as live, cables, including long metallic pipes;
- transmitter-receiver intruments.

Some instruments combine more than one method.

The first type is the simplest to use, although it will not detect unloaded cables and may fail to detect lightly loaded low voltage cables. Cable locating devices are not infallible, but they will remove much of the risk and uncertainty when excavating near buried cables. They should also be used as work proceeds and when a recorded cable is being approached.

Great care is needed when approaching and exposing a cable, and only spades and shovels should be used. Exposed cables should be protected from damage. If the exposed length exceeds 1 metre the cable should be supported with slings or props, and planks should be placed over cables crossing a trench. Any cables lying on the bottom of an excavation should be protected by wooden planks or other material which will not damage the protective sheath of the cable.

If a cable is damaged, only slightly, the Electricity Board (or other owner) should be informed at once, and people should be kept well clear until it has been made safe.

1.5.5.4 Effect of excavation on adjacent buildings

The effects of the excavation and the operations used in making it on the stability of adjacent property, especially the foundations, have to be assessed and appropriate precautions taken.

1.5.5.5 Flooding

The risks of flooding from heavy rain, rivers, tidal waters, ground water, surface and underground streams and other water sources must be carefully considered. The floors of excavations should slope down to sumps at low points where drainage pumps can be used.

1.5.5.6 Side protection

The dangers arising from the sudden collapse of unsupported or inadequately supported sides of excavations can scarcely be overemphasized. Few people recognise the danger of being buried alive in a trench two metres or less deep. The fact that a man is standing in an excavation with his head at or above ground level does not mean he will remain in this position if a side collapses. It is more likely that he will be knocked over by the falling soil. Once buried and unable to breathe, his chances of survival are small.

The need to support or batter the sides of excavations in all but hard rock, or to protect workers in excavations in some other way, e.g. inside mobile shields, is thus imperative. Following the ground investigation, decisions must therefore be taken as to how the sides of the excavation are to be protected. All necessary materials must be obtained and delivered to site. All foundation excavations and service trenches deeper than 1.2 m (4 ft) should have their sides supported or protected by sloping or battering. Exceptions can sometimes be made in the case of hard rock, when it becomes clear to experienced engineers on site (as the excavation proceeds) that the rock is solid and has no dangerous cleavage planes which could cause the side of the excavation to collapse.

If the sides of the excavation are to be battered, the maximum angle in the U.K., where frequent changes in the water content of the soil are common, is usually taken as 45°, less in the case of non-cohesive soil. Higher angles may be justified for very cohesive ground, and more particularly in countries with dry climates.

Apart from the wide variety of soils and moisture contents, other factors which must be taken into account include:

- adjacent ground loads
- vibration
- nearby traffic
- the effect of excavation on ground stability.

Trench supports may be of timber or steel. One of the most reliable (if expensive) methods is to drive interlocking steel piling on either side of the trench prior to excavation. The piles must be driven far enough below the bottom of the trench to ensure that they are firm and secure. A quiet method of piling may be needed (see Section 3.5.1) to protect both workers and local residents.

When a trench is hand dug it is not difficult to either batter the sides or to install supports while excavation progresses. The problem of installing supports in safety is less easy when a mechanical excavator has been used to dig a narrow trench to its full depth in one pass. This may call for the use of a movable metal cage or frame as deep as the trench and somewhat narrower within which the man installing the supports can work.

In long machine-excavated trenches a sliding trench shield which can be moved by the back hoe of the excavator is sometimes used instead of fixed supports when laying pipes and cables. It is then, of course, essential to ensure that nobody is allowed in the trench outside the protection of the shield.

Timber supports are more liable to failure than steel ones, since they suffer from swelling or shrinkage depending on the moisture content, as well as splitting. The quality of timber used for supporting the sides of trenches is also often very poor.

1.5.5.7 Fencing

Adequate barriers must be provided and placed as close as possible to all accessible sides of excavations where other workers could fall more than 2 metres (6 ft 6 ins) (Regulation 13). Notices and warning lights will also be required if the excavations are accessible to the public. Special care is needed where the excavation is not in an enclosed and fenced site area from which the public is excluded.

1.5.5.8 Excavated material

The (often temporary) disposal of excavated material must be considered when planning the work. Before deciding to leave it in spoil heaps near the excavation, those in charge must be sure that there is space to stack it without interfering with other operations, yet leaving sufficient unexcavated ground clear between the edge of the excavation and the edge of the spoil heap. The very minimum distance is 0.6 m. The officially recognized depth of the excavation depends on the distance of the spoil heap from its edge, and for excavations up to 6 m deep, the following rules apply in the U.K.[23]

- if the spoil heap is within 1.5 m of the edge of the excavation, the depth of the excavation should be measured from the top of the spoil heap;
- if the spoil heap is more than 1.5 m horizontally from the edge of the excavation but less than the actual depth of the excavation, the official depth should be taken as the actual depth plus half the height of the spoil heap.

- if the spoil heap is further horizontally from the edge of the excavation than the actual depth of the excavation, it may be assumed to have no effect on the sides of the excavation, and the official depth is taken as the actual depth.

1.5.5.9 Unidentified or suspicious objects

A procedure should be laid down for the reporting of unidentified or suspicious objects found as a result of soil probes, instrument searches and during the excavation itself, and for appropriate action to be taken. Thus, if an unexploded bomb is found, the police should be informed immediately. All work in or near the excavation should be suspended, and people kept well away (by barriers and warning notices) until the object has been investigated by the police and made or pronounced safe.

1.5.5.10 Use of explosives and noisy plant in excavation

Special consideration must be given to their effects on adjacent property and residents as well as site property and workers if explosives, vibratory drills, piling hammers, noisy dumpers, etc. are to be used during excavation. Noise and vibration problems are discussed in Chapter 2.5.

1.5.5.11 Access to, exit and escape from excavations

Excavations should be provided with adequate ladders projecting at least five rungs or 1.07 m (3 ft 6 ins) above the edge and be securely fixed for persons working in the excavation to enter and leave and escape quickly in an emergency, such as flooding, fire or the entry of gases or vapours. Workers should not have to enter or leave an excavation by climbing the supports for the sides.

1.5.5.12 Risks of toxic and flammable gases in excavations

The risks of flammable, toxic and asphyxiating gases from various sources entering and accumulating in excavations need to be assessed. LPG and acetylene could enter from cutting, welding or heating equipment used in or near an excavation. Carbon monoxide could enter from the exhaust of internal combustion engines, including portable pumps in or near an excavation. Carbon dioxide, hydrogen sulphide and occasionally methane are sometimes found in the ground or in underground water and seep into an excavation. Wells and shafts must be treated with special caution.

Procedures should be laid down for testing the atmosphere in excavations for oxygen, flammable gases and hydrogen sulphide, as well as ventilating the excavations. Positive ventilation with air fans drawing clean air from outside the excavation may be required to maintain a safe atmosphere, particularly when burners, welding and cutting torches and internal combustion engines are used in excavations. The hazards associated with toxic, asphyxiating and flammable gases are discussed in more detail in Parts 3 and 4 under Health and Fire Hazards respectively. Special care and ventilation are needed when gas torches or burners or internal comnbustion engines have to be used in an excavation.

1.5.5.13 Toxic and radioactive materials in soil

The possibility of toxic or radioactive materials in the soil (generally from previous site occupancy) needs to be considered and expert advice and assistance (e.g. from the HSE) sought if there is any reason to suspect their presence. Soil contaminated with highly toxic or radioactive materials may have to be removed by a specialist contractor before excavation work can start.

1.5.5.14 Biological hazards in soil

The risks to ground workers of tetanus (from a special bacterium in the soil), Weil's disease (from rats' urine) and other biological hazards should be assessed and, where suspected, monitored by bacteriological examination of soil samples. Where risk of infection exists, consideration should be given to immunization (e.g. by suitable inoculation). Ground workers should be instructed in personal hygiene and proper washing and adequate toilet facilities must be provided for them. First aid boxes with antiseptics and sterile dressings for treating wounds and abrasions should be readily available.

1.5.5.15 Risks to ground workers from cranes, mobile plant and other site operations

These risks need to be carefully assessed and appropriate rules and procedures laid down.

Vehicles should be operated with special care when near the edge of an excavation to avoid disturbing the ground. Precautions, such as special barriers, should be used to prevent dumpers or tipper trucks back filling excavations from falling in themselves.

Sometimes machinery, such as cranes and excavators, has to be used close to and in connection with an excavation while someone is working in it. If the driver is unable to see inside the excavation and anyone in it, an experienced signaller should be used.

Personal protective devices are discussed in Chapter 2.8, but as minimum protection ground workers should wear hard hats and protective footwear.

1.5.5.16 Risks to timbermen engaged in supporting the sides of excavations

A special procedure is needed for the work of timbermen to ensure that they are not exposed to the collapse of sides of excavations whilst fixing timber supports. The details will depend on the type of supports used. Sometimes it is necessary to provide a metal cage or shield inside or behind which the timberman can work safely.

1.5.5.17 Statutory inspections

Regulation 9 of the Construction (General Provisions) Regulations, 1961 makes the following requirements for inspection and examination of every part of any excavation, shaft, earthwork or tunnel where persons are employed:

- they should be inspected by a competent person at least once every working day when people are working in them;
- the face of every tunnel, the working end of every trench more than six feet six inches deep and the base or crown of every shaft shall be inspected by a competent person at the start of every shift;
- no one should be employed in any excavation, shaft, earthwork or tunnel unless a thorough examination has been made by a competent person:
  (a) of parts and particularly timbering and other supports near where explosives have been used in a way which may have affected their strength and stability;
  (b) of parts and timbering and other supports that have been substantially damaged and where there has been an unexpected fall of rock, earth or other material;
  (c) of every part within the last seven days, except for timbering or supports erected within the last seven days;
- the results of every such examination shall be made on Form 91, Part 1, Section B on the day of the examination and signed by the person making it;
- certain exemptions are made, particularly if the sides of an excavation are battered rather than supported, and in the case where the only person in the excavation is the man engaged in supporting the sides.

1.5.6 Tunnelling and caisson work[20,24] *(see also BS 6164)*

Tunnelling forms branches of civil, mining, as well as military engineering and is of great antiquity. Thus the walls of Jericho had almost certainly been undermined. Why else did they collapse at a mere fanfare of trumpets? Only the civil enginering branch and its hazards are discussed here. There has, however, always been some exchange of ideas and experience between these three fields of engineering.

Caisson work is included with tunnelling partly because many techniques of tunnelling, particularly the use of a cylindrical shield were derived from caisson work. Work in compressed air chambers, first used in pneumatic caissons was later used in sub-aqueous tunnelling. Decompression sickness, the hazard of all work in a compressed air atmosphere, is known also as 'caisson disease', as well as 'the bends' and 'divers' palsy'.

A brief acquaintance with the history of tunnelling[24], particularly in the 'heroic age' of the 19th century, reveals a horrific toll of human life comparable only to that of military operations. Thus Edwin Chadwick, writing in 1866 of the construction of the Woodhead tunnel (on the Sheffield–Manchester line) commented:

"Thirty-two killed out of such a body of labourers and one hundred and forty wounded, besides the sick, nearly equal the proportionate casualties of a campaign or a severe battle. The losses in this one work may be stated as three percent of killed and fourteen percent of wounded. The deaths (according to the official returns) in the four battles, Talavera, Salamanca, Vittoria and Waterloo, were only 2.11 percent of privates; and in the last forty one months of the Peninsula war the mortality of privates in the battle was 4.2 percent, of disease 11.9 percent."

The accident record of this tunnel was by no means exceptional, and even well into the 20th century fatality rates of 10 or more per kilometre of tunnel dug were common.

Tunnels other than those used in mining and military operations serve a variety of purposes, which are summarized in *Table 1.5.9*.

The hazards discussed here apply to all types, but particularly to types (1), (2) and (3).

Tunnels often have enormous strategic and economic importance which may indeed be quoted to justify the human risks taken in building them.

TABLE 1.5.9. Types and purposes of tunnels in civil engineering

Type	Purpose
1. Roads Railways Canals Pedestrian passageways	improved communications and transport of people and goods
2. Urban and industrial irrigation Hydro-electric	conveyance of water
3. Sewers	conveyance and disposal of sewage
4. Large sub-terranean chambers for:	
power stations (hydro-electric)	economic
underground railway systems	economic
underground storage of petroleum, commodities, etc.	economic and protection from enemy action
underground factories civilian protection	protection against nuclear attack and extremes of climate

Recent examples are the tunnels under the Suez Canal at Ismalia and Port Suez and the Russian-built tunnel in the Salang pass in Afghanistan. British fears of invasion through a channel tunnel have reinforced the economic and technical reasons against one.

Most large irrigation schemes in arid climates depend on tunnels at some point of the canal systems (e.g. the Snowy river scheme in Australia, qanats in Iran, Orange river scheme in South Africa).

Tunnels have to cut through or cross various types of natural and man-made obstacles. The principal ones are hills and mountains on the one hand, and rivers, canals and limited sea crossings, on the other. They are built in all types of soil, from hard rock to sand and mud.

Relatively soft sedimentary rocks without fissures, such as sandstone, are the easiest to tunnel. River muds present the most serious problems as such tunnels have to be lined immediately behind the cutting face and there is the risk of mud and water bursting into the tunnel. Such tunnels require the use of a shield and compressed air chamber at the cutting face. The object of using compressed air is to increase the pressure in the tunnel to balance the forces created by the weight of mud and water and eliminate the risk of sudden inrushes flooding the tunnel.

Hard rocks require much energy, explosives and make greatest demands on cutting tools.

Rock pressures under mountains can be extremely high. Water contained in fissures encountered when tunnelling through mountains is an ever-present hazard as it is usually at high pressures and enters the tunnel with high force and velocity

Inflammable methane gas, poisonous hydrogen sulphide, as well as other gases, are frequently encountered when tunnelling and may enter the tunnel unexpectedly as the tunnel advances.

Rock temperatures found when tunnelling through mountains can reach 50 °C or higher, which is well above human blood temperature.

Silica dust is produced from most acid rocks and is a constant health hazard to the tunnel worker.

1.5.6.1 *Principal hazards*

The tunnel worker faces a number of lethal hazards, which may overwhelm him suddenly and unexpectedly and from which the possibilities of escape or rescue are restricted. The principal ones are:

1. Drowning
2. Interment
3. Injury from falling rock
4. Asphyxiation (through insufficient oxygen in atmosphere)
5. Fire (from methane and timber in tunnels)
6. Poisoning (from toxic gases, principally hydrogen sulphide, oxides of nitrogen and carbon monoxide from explosive charges)
7. Respiratory disease from dusts, especially silicosis
8. Decompression sickness for workers in compressed air compartments
9. Explosions (from premature firing of explosive charges)

In addition, the tunnel worker is exposed to a number of disabling and debilitating health hazards:

10. Excessive noise, leading to hearing loss
11. Vibration, principally from hand-held pneumatic rock drills
12. Bursitis (principally beat knee) is common among miners
13. Bronchitis and pneumonia, e.g. from frequently transferring from hot tunnel atmosphere to cold mountain air.

This is not to suggest that tunnel workers are exposed to these hazards all the time. Most, however, are liable to be exposed to several of them on any particular job.

The hazards listed above do not, as a rule, just strike a single worker but several members of a working team simultaneously. Neither are they hazards which an individual worker can avoid by taking special care. The contractor in overall charge of the operation can do a great deal to reduce them, by careful preliminary surveying, by choice of method, plant and machinery, by monitoring the tunnel atmosphere, by supervision and training of workers and by providing all necessary personal protection (clothing and devices).

Safety in tunnelling is, in short, largely a technical problem and depends in large measure on the competence of the contractor and his engineers. Safety in tunnelling, however, has its price, which those who commission the work and award the contract are not always willing to pay. The contractor who gets the job on the basis of the lowest tender may then find himself the victim of circumstances. The situation may be aggravated by bonus or penalty clauses for completing the work early or late.

1.5.6.2 Some historical examples[24]

A classic example of the economic disincentives to safety was the building of the St. Gothard tunnel. This was commissioned by the Swiss Central Railway Company to take the Zurich Milan railway. The contract was won by Louis Favre of Genoa with a bid of £1.9 million in the face of competition from six other firms. He agreed to complete the tunnel in eight years and had to put up £320 000 which he would forfeit if the work was not finished within one year of the deadline. There were also other bonuses and penalties for early and late completion. Work was started in 1872.

Favre and his men worked in appalling conditions. Out of a total work force of about 4000 at any time, 310 died and 877 were seriously invalided by accident and disease. Favre himself died in 1879. His firm lost their deposit of £320 000, as well as £590 000 by which their costs exceeded the contract price. His family were ruined financially.

Injuries, many fatal, were caused by accidental explosions, falls of rock, and mishaps with machinery. Many died from disease, silicosis, bronchitis, pneumonia and 'miners' anaemia', and from a parasitic worm. The tunnel ran into high pressure water, high temperatures (over 40 °C) and pockets of foul gas, while ventilation was poor and inadequate even for clearing the toxic gases produced by blasting.

The Great St Bernard tunnel and the Mont Blanc tunnel were constructed some eighty years later. The same hazards as were found in the St Gothard tunnel were met on both and very similar methods used—a cycle of drilling, blasting and mucking out with three shifts working. The techniques however were better and safer. On the Great St Bernard tunnel, 17 men were killed and 800 injured (for a distance of 5.5 kilometres), whilst on the Mont Blanc tunnel 23 men were killed for a distance of about 11 kilometres.

Whilst techniques and machinery used in tunnelling are continuously being improved and reduce its hazards, *ad hoc* solutions to particular problems are constantly having to be found. In the Telecote tunnel under the Santa Ynez mountains in California, started in 1950, unusually high temperatures of about 45 °C were found. These seemingly impossible

Figure 1.5.13 Italian miners in steel helmets work among the wooden roof props of the Great St. Bernard Tunnel (Popperfoto).

working conditions were aggravated by jets of hot water and pockets of hydrogen sulphide gas. To counter the temperature problem the first contractor filled mining skips with cold water and brought the miners to the work face with their bodies immersed. After a short working spell, they returned to their cold baths to recuperate.

1.5.6.3 Development of sub-aqueous tunnelling techniques

The use of sealed tunnel chambers filled with compressed air for driving tunnels through soft soil under rivers developed from the use of compressed air for caisson sinking. This had been used since 1830 after the invention of a suitable air lock by the British engineer Thomas Cochrane.

An early cylindrical tunnelling shield, invented by Peter Barlow also based on the vertical cast iron caissons then in use was driven horizontally to build a foot tunnel under the Thames at Tower Hill. The shield prevented the periphery of the tunnel at the work face from collapsing and allowed the tunnel to be lined with masonry or shaped cast iron segments immediately behind it.

Meanwhile Haskin used compressed air to fill the Hudson river tunnel under construction in 1879 in an attempt to prevent water from seeping into it. An air lock through which men, tools and equipment had to pass to enter the tunnel was fitted at the bottom of the shaft.

Although the use of compressed air can prevent or reduce the seepage of water into the tunnel, it brings the new hazard of air being blown out of the tunnel through the soft soil and the water above it. This can lead to the rapid depressurisation of the tunnel and an inrush of water into the tunnel through the hole made by the escaping air. During completion of the Hudson river tunnel in 1904 a workman attempting to stop a blowhole in the tunnel with a bale of hay was actually blown into and through the hole and shot like a cannonball through the silt of the river bed and fifteen feet of water to the surface of the river. He was rescued unharmed.

J.H. Greathead combined Barlow's use of a cylindrical metal shield with Haskin's use of compressed air at the work face to produce the greatly improved Greathead shield, first used to build the London Underground tunnel between the Monument and Stockwell in the 1880s. This shield had an airlock built into it. Most subsequent engineering developments of sub-aqueous tunnelling are based on the Greathead shield.

Machines are now widely used for excavating the working face and removing the spoil. With the development of silicon carbide tipped drill bits, machines are available which can be used in hard rock as well as soft, and can obviate the use of explosives with all its hazards and uncertainties.

Various other techniques are used to prevent leakage of water and gases into tunnels under construction. These include injecting cement under pressure behind the lining and various methods of freezing the soil (for temporary stoppage).

1.5.6.4 *Accident prevention in tunnelling*

Two general factors which greatly affect the safety of tunnel work are:

- the position of the tunnel and the natural geological features of the terrain; and
- the techniques, machinery and skills employed.

In addition, the attitude of management in the contracting firm, of the local community and its safety authorities and of the client for whom the tunnel is built are critically important to safety.

The most difficult and hazardous tunnels being built in the world today are probably in third world countries, in high mountains such as the Andes, the Himalayas, the Pamirs and the Hindu Kush. Life and labour in these surroundings are generally considered cheap. Shortage of capital may preclude the use of the most suitable modern machinery. Thus the methods used tend to resemble those used in Europe in the first half of this century, with its accompanying toll of life.

The following features seem to be critically important to the safety and health of tunnel workers[20]:

(a) The geology of the ground through which the tunnel is to be built should be very thoroughly and competently surveyed before the precise route is chosen and before actual tunnelling work starts. Such

surveys are expensive, involve the extensive drilling of boreholes, the use of sophisticated instruments to detect stresses, water and gases in the rocks, extensive analyses of rock samples, and skilled evaluation of the results.
(b) For tunnels beneath high mountains, complete preliminary surveying is impossible. Some chances on the geology have to be taken. Further surveys have then to be undertaken whilst the tunnel is under construction by driving pilot headings in advance of the main tunnel. Those driving the pilot headings are, of course, exposed to the greatest risks.
(c) The tunnel and its ventilation system must be carefully designed, taking into account the maximum likely rates of entry or production of the various toxic and flammable gases liable to be present, and the permissible 'threshold limit values' of these gases for human exposure.
(d) Pumps and drainage for removal of water need to be designed on similar principles.
(e) Lighting must be adequate.
(f) All electrical equipment should be 'safe' for the particular types of explosive atmosphere which may be encountered.
(g) Electric lighting and power supplies for pumps, ventilation and critical machinery should be assured at all times. If there is any doubt about the reliability of the main electrical supplies, emergency generators which will 'cut in' immediately in case of power failure should be installed adjacent to the shafts or other entries to the tunnel. Special care in safeguarding these supplies is needed when the tunnel is built in a remote or mountainous area, which may be subject to avalanches, severe storms, snowdrifts, earthquakes, etc.
(h) The use of wood and other combustible materials in a tunnel, particularly one where methane may be present, should be eliminated or kept to the barest minimum in quantity.
(i) The tunnel itself, its lining and all shafts, services and equipment used in its construction, must be inspected at frequent intervals and maintained to a high standard.
(j) The atmosphere in the tunnel to which miners are subjected should be continuously or frequently monitored by sampling and analysis for flammable and toxic substances (e.g. hydrogen sulphide, methane, carbon monoxide, oxides of nitrogen and silica dust). This must be done by a properly trained person with suitable experience in industrial hygiene. Dangerous conditions should be promptly reported, not only to the senior engineer present but also to the director of the contracting firm with special responsibility for health and safety, and also to any official inspectorates involved with these matters.
(k) Appropriate protective clothing and special devices such as ear plugs or muffs, should be provided to workers after careful selection on ergonomic lines. Hard hats, rubber boots with toe-protectors, safety glasses and suitable gloves or gauntlets and tunics are nearly always the minimum essentials. Sufficient self-contained breathing apparatus for escape through toxic or flammable atmospheres should be available for all workers present at any time.
(l) Proper medical and first aid facilities are essential. This involves

pre-medical inspection of all tunnel workers and regular subsequent inspection and prompt investigation of the causes of abnormal incidences of any diseases or accidents among workers. It means the provision of a properly equipped first aid room with trained full-time attendants, together with trained part-time first aid men and ambulance men among the workers.

(m) Properly trained rescue teams, with self-contained breathing apparatus, resuscitation apparatus and shovels and other tools for rescuing workers buried by rock falls are also essential.

1.5.6.5 Work in compressed air atmospheres

This aspect of tunnel and caisson work has probably received most medical and legal attention, partly because it is an entirely artificial situation and partly because of the particular medical hazards associated with it.

In the U.K. work in compressed air atmospheres is covered by 'The Work in Compressed Air Special Regulations, 1958'.

Fit people can work in air compressed to up to 3.5 atmospheres (55 pounds per square inch) without harm provided they are suitably prepared, compressed and decompressed afterwards. The main danger lies in the fact that nitrogen is more soluble in the blood and body tissues at higher pressures than at atmospheric pressure. Thus when a person who has been working in compressed air is suddenly decompressed to atmospheric pressure, the dissolved nitrogen in his blood forms small bubbles. These cause damage to body tissue, such as joint lesions, bone necrosis, interfere with blood circulation, and even cause heart failure. A key to safe working in compressed air is gradual decompression afterwards at such a rate that the dissolved nitrogen in the bloodstream and tissues can be released through the lungs without separating and forming bubbles in the body.

In addition, most people feel discomfort when the air pressure changes due to pressure on the eardrums. Relief may be obtained by chewing, but the rate of pressure change must be carefully controlled to avoid discomfort and injury.

Combustion proceeds more rapidly in compressed air than at atmospheric pressure and smouldering objects readily burst into flames. Hence smoking and the use of flames are prohibited in compressed air chambers. Fire and explosion hazards are considered in Section 3.2.4.3.

A number of other precautionary measures must be taken[20].

1. *Physical Condition of Workers*

A pre-employment medical examination of all compressed air workers is needed, followed by regular re-examination at least once a year. Any worker suffering from a head cold, sore throat, ear ache or other similar ailment should report this to his supervisor, and if otherwise fit for work, be given alternate employment until his ailment is cured.

2. *Provision of Air Locks*

Air locks, which may form part of the compressed air chamber or bulkhead, must be provided through which all workers must pass for compression and decompression. No materials or equipment other than hand tools should be taken through these air locks, known as man locks.

They must be soundly constructed, of adequate dimensions and fitted with suitable entrance and exit doors. They should be kept clean and at an agreeable temperature.

3. *Fittings in Air Locks*

Every air lock must be fitted with dual pressure gauges to show the pressure inside the lock and in the pressurised working chamber, both to those inside the lock and to the attendant outside. They should also be provided with two-way audio and visual communication between the attendant and those inside. Two clocks should be provided, one for those inside and one for the attendant.

4. *Notice*

A notice showing precisely the precautions to be taken during compression and decompression should be fixed to every man lock.

5. *Pressure Control and Regulation*

Effective controls for regulating the air pressure in the lock must be supplied. These are normally operated by the attendant.

6. *Aattendant*

A skilled and trained attendant must be in charge of every man lock.

7. *Conditions in Air Lock*

Maximum air pressure should not exceed 3.5 atmospheres (55 psi) for normal purposes, and 1.2 atmospheres (18 psi) for most routine work. The air temperature and humidity in an air lock must be controlled. The maximum allowable wet bulb temperature in an air lock is 26.5 °C.

8. *Rules for Compression and Decompression*

Precise rules must be laid down for both procedures.

The rules for compression must ensure that compression takes place gradually and in stages to prevent discomfort. If any person feels discomfort during compression he must signal to the attendant who must take remedial action.

Rules for decompression must take the following factors into account (among others):

> The period of exposure to compressed air.
> The pressure to which persons have been subjected.
> Possible differences in both when several persons are being decompressed simultaneously.
> Possible use of 'decanting' procedures, whereby a person is rapidly decompressed in the air lock, followed promptly by rapid compression and correct decompression in another chamber.
> Possible use of 'stagewise decompression' whereby the pressure is reduced stepwise with intervals between each step instead of slowly and smoothly. This method was introduced by Professor Haldane in 1905 as a result of extensive research and has been mainly used for diving. Most modern rules are derived from the rules then laid down and known as 'Haldane's rules'.

9. *Medical Lock*

A separate medical lock should be provided at a suitable place for treatment of any persons working in compressed air found suffering from 'the bends'. This must be properly equipped and in the charge of a person trained in first aid and especially the treatment needed in such cases.

10. *Identification Tags*

All workers in compressed air should wear a tag showing they have been working in compressed air and other necessary information such as the location of the medical lock on site provided for them. This precaution is needed in case the worker becomes ill after work which usually results from too-rapid decompression.

11. *Arrangements with Hospitals*

Appropriate arrangements should be made with local hospitals for the transfer and treatment of injured workers

12. *Rest Room, Washing, Changing and Locker Facilities*

A suitable rest room should be provided where compressed air workers can rest after decompression. They should also be provided with proper washing, changing and locker facilities.

13. *Inexperienced Workers*

Workers inexperienced in working in compressed air chambers should be closely supervised, and accompanied by an experienced person when first passing through an air lock.

14. *Door Opening Between Air Lock and Compressed Air Working*

So far as possible the door between the compressed air working and the air lock should be kept open to facilitate emergency evacuation of persons in the working chamber.

1.5.7 Diving[25,28]

Natural unassisted diving has been practised for thousands of years for sponge in the Mediterranean and pearls in the Indian and Pacific Oceans. Attempts to obviate human limitations by the use of diving bells (especially for salvage) appear to have begun in the sixteenth century. Modern diving dates from the early nineteenth century with the development of the closed diving dress with helmet, air pump at the surface and flexible compressed air tube by Augustus Siebe. This was first used to salvage property from the Royal George in 1837. The first self-contained diving dress with its own supply of compressed air was designed by W.H. James in 1825.

Diving serves a variety of purposes, salvage, military, marine and archaeological research and fishing, as well as construction.

Modern diving is essential for underwater construction, inspection and surveying (*Figure 1.5.14*). In the last two decades it has received particular impetus from the challenge of offshore oil production and the construction of rigs and submarine pipelines.

Divers carry out a number of 'trades' under water[28]. All have their own hazards and require special training, even when carried out on terra firma. These include:

Figure 1.5.14 A diver inspecting one of the legs of Cormorant 'A' production platform in the North Sea. (Courtesy of Shell Photographic Library).

(a) cutting steel and other metals by oxy-hydrogen cutters, the oxy arc system and thermic lances. The last-mentioned method, which involves feeding high pressure oxygen through a tube packed with steel and magnesium rods, can even be used to cut concrete.
(b) welding, generally electric with a coated electrode. To get best results, this is often done inside a dry habitat on the bottom.
(c) use of high pressure water jets, for dispersing mud and sand, cleaning, burying pipe and cables and even cutting timber, stone and concrete.
(d) use of explosives for dispersing rock or wrecks and for cutting channels and excavations.
(e) use of cartridge-fired bolt guns for fixing plates, repairing piles, etc.
(f) use of hydraulic and compressed air tools.
(g) use of underwater film and video cameras.

The diver is exposed to the special hazards of these trades on top of the many hazards of diving. Diving is thus a highly specialized and potentially hazardous activity requiring the ultimate in training techniques, inspection and maintenance.

The high risks of a diving accident proving fatal are illustrated by accident figures published by the Department of Energy for North Sea oil operations. Between 1969 and 1979 the number of fatal accidents to divers (22) exceeded the number of serious accidents (18)[26].

Diving in waters subject to U.K. control comes under the Diving Operations at Work Regulations, 1981. An explanatory guide to them is published by the Health and Safety Executive. The main subjects covered by the Regulations are:

Diving Contractors
The Diving Supervisor
Divers
Diving Teams
Diving Rules
Qualification of Divers
Certificate of Medical Fitness to Dive
Plant and Equipment
Maintenance, examination and testing of plant and equipment.

It should be appreciated that any attempt here to cover in detail the safety aspects of so specialized a subject would only be misleading.

The U.K. Regulations which have to cover the extreme conditions of Britain's territorial waters in the North Sea are more comprehensive than those of most countries. Thus the notes should be of interest to non-British readers who may soon be facing similar problems.

1.5.7.1 Diving contractors

With a few carefully stated exceptions a clearly defined organization referred to as 'the contractor' must be responsible for every commercial diving operation and for ensuring that all requirements of the regulations are complied with. The need to ensure that a particular organization is responsible was only too clear in the trial following the incident described later in this section.

The duties of the contractor for each diving operation include:

- the appointment of one or more qualified diving supervisors with appropriate experience as divers and supervisors to control the operation directly.
- to issue diving rules (described later) to be followed in the operation.
- to provide a diving operations log book in which certain details of each operation must be recorded.
- to ensure that all plant and equipment, including that required by the Regulations, is ready for immediate use.
- to prohibit the use of compressed air as the breathing mixture at depths over 50 metres. (Diving at depths greater than 50 metres must be done by the closed bell method using an oxygen-helium breathing mixture.)
- to ensure that the diving is carried out from a safe and suitable place.
- to ensure that emergency services are available, including facilities for transferring divers safely under pressure to a pressurized treatment chamber when saturation diving is used or the depth exceeds 60 metres. (In saturation diving, the divers live in special chambers on the surface, pressurized with the appropriate breathing mixture close to the pressure on the seabed when not actually diving. They then transfer under pressure to their bell on the surface and descend directly to the seabed to work. This avoids the lengthy

period needed for decompression between dives and the risk of decompression sickness.)
- to ensure that there are effective means of communication between divers, diving supervisors and emergency services.

1.5.7.2 Diving supervisors

The diving supervisor must ensure for each diving operation for which he is responsible that:

- the plant and equipment has been properly maintained and inspected and covered by a valid inspection certificate issued within six months of the date of the operation;
- each member of the diving team has received a copy of the relevant parts of the diving rules;
- he does not dive whilst supervising any diving operation.

1.5.7.3 Divers

A diver may not take part as such in any diving operation unless:

- he has a valid certificate of training issued by the HSE or by a person or body approved by the HSE;
- he has a valid certificate of medical fitness issued by an approved doctor or by the HSE;
- he is competent to carry out the work which he is called upon to perform in that operation.

Further, he must inform the diving supervisor if he is unfit or if there is any other reason preventing him from taking part as a diver;

- he must maintain a personal log book containing his signature and photograph;
- he must record specified details of the operation in his log book which must be countersigned by the diving supervisor;
- he must present his diving log book to the doctor who examines him and certifies his fitness to dive;
- he must keep his log book for at least two years from the date of the last entry.

1.5.7.4 Diving team

- There must be enough divers and other competent persons (the 'diving team') to ensure that the operation can be undertaken safely and all plant, equipment, etc. necessary for safety can be operated;
- in addition to the diver or divers who will actually be diving, there shall be a stand-by diver on all operations where the water is more than 4 ft 11 ins (1.5 m) deep;
- where a diving bell is used, the stand-by diver shall descend in the bell and remain in it to monitor the diver(s) who leave it and be ready to assist in the event of an emergency;
- where no bell is used, the stand-by diver shall be in immediate readiness to dive. However, if two divers are in the water

simultaneously and each near enough to communicate with and assist the other, each may count as the stand-by diver for the other;
- an extra diver on the surface, as well as the stand-by diver, is needed in these three cases:

  where diving stops are required for routine decompression;
  where the diving is at a depth of 30 metres or more;
  where there is a special hazard, such as a current or the risk of a diver being trapped or his equipment entangled;

- the diving supervisor or members of a team required to attend any plant, equipment, etc. shall not be treated as stand-by or extra divers;
- the stand-by diver or any extra diver required may go to the assistance of any other diver in an emergency.

1.5.7.5 Diving rules

The diving rules issued by the diving contractor (Section 1.5.7.1) for each diving operation shall include measures for the health and safety of those engaged in the diving operation. In addition, a number of detailed matters which form Schedule 1 of the Regulations must be included. These fall under four headings which are given below with a summary of the matters covered.

- planning, to include considerations of weather, tides, proposed shipping movements, air and water temperatures, underwater hazards, depth and type of operation, suitability of plant and equipment, availability and qualifications of personnel, effect on diver of air pressure changes if he flies after diving and the particular activities of anyone who may be diving in connection with the operation;
- preparation, to include consultation with everyone involved in the safety of the operation, selection of breathing apparatus and mixtures, check on plant and equipment, allocation of personnel, fitness of divers, precautions against cold, signalling procedures and precautions against underwater hazards;
- procedures during diving, to include responsibilities of all involved, use of personal diving equipment, supply and pressure of gas and gas mixture, operations from the surface, operations related to a diving bell, work in different locations, use of equipment under water, limits on depth and time under water, descent, ascent and recovery of divers and of diving bell, diving tables for use in decompression procedures, control in changing conditions, time divers must remain near compression chamber and keeping of log books;
- emergency procedure, to include details on emergency signalling, emergency assistance (both under water and on the surface), therapeutic recompression and decompression and availability of chambers, first aid, medical assistance, calling assistance from emergency services, precautions in the event of evacuation of the work site, etc. and provision of emergency electrical supplies.

1.5.7.6 Qualifications of divers

The subjects in which a diver has to pass a test before being granted the necessary certificate of training are stated in Schedule 4 of the Regulations. Included are such subjects as the theory and practice of basic air diving, mixed gas or bell diving, air diving where no surface compression chamber is required on site and air diving with self-contained equipment where no surface compression chambers are required on site.

In the U.K. there are three government-recognized schools for divers offering training to two standards: (1) for basic air diving down to 50 m (164 ft) and (2) for mixed deep gas diving. These are:

- Fort Bovisand near Plymouth, Devon;
- Underwater Training Centre, Inverlochy, Fort William, Highlands; and
- Prodive Commercial Training Centre, Services Area, Falmouth Exploration Base, Falmouth Docks, Cornwall.

1.5.7.7 Certificate of medical fitness to dive

These certificates are issued only after thorough medical examination including x-rays, and are valid for 12 months only, after which the diver must be re-examined and re-certified.

The working life of divers, particularly in deep water, is quite short. Many have been found by the age of forty or less to be incapacitated from further diving by bone necrosis. The shock of this to a professional diver is well described by Webster, an experienced deep sea diver and diving supervisor, in 'The Danger Game'[29].

1.5.7.8 Plant and equipment

A brief outline of the main items required follows:

- a means of supplying a breathing mixture suitable in composition, quantity, pressure and temperature (including reserve supply for emergencies, including therapeutic recompression or decompression);
- a lifeline for each diver (with certain exceptions);
- means of communication between each diver and the diving supervisor;
- where reasonably practicable, a means of oral communication for the same purpose;
- plant and equipment needed for divers to enter and leave the water;
- a surface compression chamber (complying with Schedule 5) for all dives at depths in excess of 164 feet (50 metres) and for certain dives in excess of 33 feet (10 metres);
- a diving bell for diving operations in excess of 50 metres;
- means of heating and controlling the temperature of the diver's body in depths exceeding 50 metres;
- means of heating the diver's breathing mixture at depths in excess of 492 feet (150 metres);

116 High risk occupations

- a lamp attached to the diver to indicate his position on the surface at night;
- means of illuminating the surface diving station at night;
- depth measuring devices which can be monitored from the surface;
- a means of keeping the diving station if floating in a fixed position;
- all plant and equipment used in a diving operation should meet the standards required for that operation;
- every gas cylinder used in a diving operation should be marked with the name and formula of its contents.

1.5.7.9 Maintenance, examination and testing of plant and equipment

The requirements are again quite complicated, but in practice the most essential thing is that all plant and equipment used in diving operations shall have been inspected by a competent person. A certificate, which authorises its safe use within a specified period should have been issued by the competent person and be correct.

1.5.7.10 A tragic example[27]

Despite the fact that modern diving operates at the frontiers of technology, the causes of many fatal diving accidents have been found to be neglect on the part of the contractor and supervisor of regulations and the adoption of improper procedures. An example which happened in 1979 in the Thistle field in which two divers lost their lives is given in reference 24. The main features were as follows:

Work was being done in the Thistle field for BNOC. A diving contractor had subcontracted the work to another contractor, without the consent of BNOC (in violation of the contract).

Two divers in saturation were lowered in a bell to 490 feet—the bell being lowered over the side of the vessel instead of through the moonpool.

After the bell reached its designated position the pin broke or worked loose in the shackle attaching the bell to the main lifting wire.

There were still two methods of rescue which should have been available for use in an emergency:

(1) By the umbilical and the umbilical winch on the vessel. Neither were fit for the task.
- the umbilical, an assemblage of wires and tubes round a central cable should have been enclosed in a stocking of steel wire or nylon mesh. Instead it was merely taped together with adhesive tape at 15 inch intervals. This caused it to splay out and jam on the winch;
- the umbilical winch included inadequate components unsuitable for the weight, and had never been certified for this reserve function of rescue.

This method, therefore, failed.

(2) By releasing ballast weights slung on the outside of the bell from inside it.
- contrary to the requirements of the system the weights had been lashed onto the outside of the bell;

- even then, the rope might have been cut by one of the divers if he had been able to get to it;
- the door of the bell could not, however, be opened sufficiently, as the bell was on the bottom, due to the fact that the guide wire system had not been used when lowering the bell.

Both divers died in the bell.

1.5.8 Steeplejacks[30]

Most firms of steeplejacks in the U.K. belong to the National Federation of Master Steeplejacks and Lightning Conductor Engineers. Their work consists mainly of the building, repair and demolition of church spires, towers, chimney shafts and tall monuments, and the installation of lightning protection and earthing systems. In addition, they erect and maintain antennae, flagstaffs, and steel chimneys, clean and gild clock faces, maintain cooling towers, water tanks and tall factory buildings, rig and externally paint radomes. Much of their work is done from special ladders, bosun's chairs and steeplejack scaffolds known as flying stages. Their main hazard is that of falling and the main problem is that of providing safe access and working platforms for their work.

The traditional steeplejack scaffold does not have hand rails or toeboards and is only two boards wide. In the U.K. these features are exempted from the Construction (Working Places) Regulations, 1966 by Certificate F 2485, granted in modified form on 7 April 1976.

Steeplejack scaffolds differ from most others in use in that they are not built up from the ground, but rely for support on the structure (steeple, chimney, etc.) on which work is being done. They are usually attached to the chimney or other structure by battens fastened by long bolts which grip the chimney or structure.

Scaffold dogs which are driven into softwood plugs placed in holes drilled in the brickwork are also used to position and support scaffolds and ladders.

A booklet, 'Recommended Safe Working Methods for the Steeplejack Industry' by Max Beaumont, is published by the National Federation[30]. The methods fall under four parts: (1) Laddering, (2) Bosun's Chairs, (3) Steeplejack Scaffolds, and (4) General.

Since the profession is small in numbers and its members specially trained and experienced, it is not considered appropriate to detail these safe working methods here. The practices in this industry are dictated by the difficulties of providing safe access by means standard in other branches of construction. In consequence, they would not be considered safe for painters, bricklayers or even roof workers. To include them here might therefore be confusing.

The key to safety in the industry is that its members should all be 'qualified by training and experience to do such work'.

The hazards of some medium to low level work traditionallly done by steeplejacks can be reduced by the use of mobile aerial platforms. This is likely to increase. Its effect, however, is just as likely to be the reclassification of the job, thus allowing it to be done by other workers,

118   High risk occupations

than increasing the safety of the steeplejack's profession. Indeed, to the hazards he already faces, it adds the further one of redundancy.

References

Lists of relevant British Standards and HSE Guidance Notes follow these references.
1. Health and Safety Executive, *Safety in Steel Erection*, HMSO, London (1979).
2. Health and Safety Executive, *Construction, Health and Safety, 1979–1980*, HMSO, London.
3. Health and Safety Executive, *Health and Safety Statistics, 1979–80*, HMSO, London.
4. Mangrulkar, V.G. and D'Souza, L.T., 'Safety in Steel Erection Works', from *Proceedings of the National Seminar on Safety in Construction Industry*, Nov. 1975, Bombay. Central Labour Institute, Sion, Bombay.
5. Home Office Departmental Committee, *The Dangers Attendant in Building Operations*, HMSO, London (1907).
6. The Construction (Working Places) Regulations, 1966, Regulation 4, Interpretation (see Redgrave p. 864).
7. *Final Report of the Advisory Committee on Falsework* (The Bragg Report), Health and Safety Executive, HMSO, London (1976).
8. IYER, T.W., 'Safety in Centering Scaffolding and Formwork', same seminar as Ref. (4).
9. ILO, *Occupational Safety and Health Series 42, Building Work. A Compendium of Occupational Safety and Health Practice*, International Labour Office, Geneva (1979).
10. BRAND, R.E., *Falsework and Access Scaffolds in Tubular Steel*, McGraw Hill (1975).
11. Health and Safety Executive, *Safety in Scaffolding*, HMSO, London (1979).
12. Health and Safety Executive, *Fatal Accidents in Construction, 1977*, HMSO, London (1979).
13. Health and Safety Executive, *Guidance Note GS 10', Roof Work: Prevention of Falls*, HMSO, London.
14. Health and Safety Executive, *Safety in Demolition Work*, HMSO, London (1979). (Work carried out by a sub-committee set up by the Joint Advisory Committee on Safety and Health in the Construction Industries before the creation of the HSE.)
15. POWELL-SMITH, V., Secretary of the National Federation of Demolition Contractors, in Introduction to *A Complete Guide to Demolition* by D.M. Pledger, Construction Press (1977).
16. *Construction Safety, Section 8, Demolition*, National Federation of Building Trades Employers, London (December 1972 and later).
17. WHYTE, B.A.C., *Safety on the Site*, United Trade Press, London (1970).
18. ILO, *Code of Practice, Safety and Health in Building and Civil Engineering Work*, International Labour Office, Geneva (1972).
19. ILO, *Occupational Safety and Health Series 42, Building Work. A Compendium of Occupational Safety and Health Practice*, pp. 44–47, 'Demolition', International Labour Office, Geneva (1979).
20. ILO, *Occupational Safety and Health Series 45, Civil Engineering Work*, International Labour Office, Geneva, (1981).
21. National Joint Utilities Group, *Recommendations on the Avoidance of Danger from Underground Electricity Cables* (1979) (available from any U.K. Electricity, Gas or Water Board.)
22. National Joint Utilities Group, *Cable Locating Devices*, (1980) (available as Ref. 21).
23. IRVINE, D.J. and SMITH, R.J.H., *Trenching Practice, CIRIA Report 97*, Construction Industry Research & Information Association, London (1983).
24. BEAVER, P., *A History of Tunnels*, Peter Davies, London (1972).
25. YOUNG, D., *The Man in the Helmet*, Cassell, London (1963).
26. Department of Energy, *Development of the Oil and Gas Resources of the United Kingdom, 1979–1980*, HMSO, London.
27. CARSON, W.G., *The Other Price of Britain's Oil*, Martin Robinson, Oxford (1981).
28. VALLENTINE, R., *Divers and Diving*, Blandford Press, Poole, Dorset (1981).
29. WEBSTER, D.H.F., *The Danger Game*, Robert Hale, London (1978).
30. BEAUMONT, M., *Recommended Safe Working Methods for the Steeplejack Industry*, National Federation of Master Steeplejacks and Lightning Conductor Engineers, London (1982).

British Standards

BS 1139:1964	Metal Scaffolding, Materials, Workmanship Tests and Finish.
BS 5973:1981	Code of practice for access and working scaffolds and special scaffold structures in steel
BS 5974:1982	Code of practice for temporarily installed suspended scaffolds and access equipment.
BS 5975:1982	Code of practice for Falsework.
BS 6187:1982	Code of practice for Demolition.
BS 6164:1982	Code of practice for Safety in Tunnelling in the construction industries.
BS 6031:1981	Code of practice for Earthworks.

Health and Safety Executive Guidance Notes
(available from the HSE and HMSO, London)

GS 5	Entry into confined spaces
GS 6	Avoidance of danger from overhead electrical lines.
GS 7	Accidents to children on construction sites.
GS 10	Roofwork: prevention of falls.
GS 15	General access scaffolds.
CS 6	The storage and use of LPG on construction sites.
PM 3	Erection and dismantling of tower cranes.
PM 9	Access to tower cranes.
PM 14	Safety in the use of cartridge-operated tools.
PM 27	Construction hoists.

PART 2
HEALTH HAZARDS OF CONSTRUCTION WORKERS

Introduction

Neither of the authors has had medical training, although one is a trained scientist and engineer and the other is a safety specialist with extensive experience in the construction industry. When this book was first planned the main emphasis was intended to be placed on accidents to construction workers, with health problems occupying a secondary role. As work progressed, however, it became apparent that health problems occupied a serious place in the lives of construction workers, quite comparable to those of being injured by falls or other causes. Although a good deal of information on the subject had been published this was mostly scattered in journals and reports which were not readily available to the construction industry.

Whilst occupational health services have been steadily growing in scope and importance in manufacturing industry, apart from a few special exceptions such as diving, they were virtually non-existent in construction. We therefore felt there was a need to present a coherent picture of the many aspects of this complex subject—one derived from many different sources—which would help to put occupational health in construction on the same footing as in manufacturing industry. At the same time we realise that this is only a beginning and that the picture presented here is still far from complete.

Work on this book had already been started in the summer and early autumn of 1983, before the publication of the HSE's biennial report, *Construction, Health and Safety, 1981-82*.

It was hence an agreeable surprise to find when this report was published that the sentiments which prompted our expansion of this part of the book were reflected in the report. To quote from two paragraphs in the HSE report:

> "185. Although health surveillance is required for certain specific hazards such as lead, in general terms an occupational health service does not exist within the construction industry. As a result, medical records are inaccurate and proper studies of environmental health hazards remain poorly developed. Any early improvement in this situation seems unlikely."

"186. It is increasingly being recognised that a safe site can be efficient, productive and economic under the inter-disciplinary control of site manager, safety adviser and safety representative. Surely this framework could be expanded to include competent occupational health input and monitoring?"

An example of this lack of awareness of occupational health hazards quoted from the same report was of two experienced painters employed in painting new flats, who, after a fall, complained to their supervisor of tiredness and lack of coordination in the afternoons. On investigation it was found that the atmosphere at face height during painting of the walls contained grossly excessive levels of the vapour of white spirit. It would seem that this example is but the tip of an iceberg.

Chapter 2.1 considers the health of workers in various occupations in construction in relation to the rest of the working population. Chapter 2.2 discusses officially recognized occupational diseases in relation to construction. Chapter 2.3 deals with health problems in general terms, how various toxic materials enter and attack the human body and the roles that various health professionals play in dealing with them. Chapter 2.4 looks at chemical health hazards, how and where they arise, what harm they do, what levels are considered acceptable, and broad precautionary measures. Chapter 2.5 considers physical health hazards, primarily noise, vibration, various forms of electromagnetic radiation and electricity. Chapter 2.6 investigates other likely health hazards, i.e. biological, environmental and stress. Chapter 2.7 deals with the monitoring of various health hazards and, finally, Chapter 2.8 deals with protective clothing and personal devices available.

2.1
How healthy are construction workers?

Judged by a career leaflet for school leavers[1] some construction workers enjoy a healthy, manly, independent open-air life, with bulging muscles and bronzed torsos which make them the envy of pale-faced office workers (*Figure 2.1.1*).

> "We're out in all weathers. Its great when the sun is shining and you can strip your shirt off. We laugh about all those people paying a fortune to get an expensive tan on the Riviera and here we are getting paid for doing the same thing".

Since similar inferences are made on cigarette advertisements and army recruiting posters we may well suspect a little bias. Trade union sources, on the other hand, paint a more sombre picture of construction workers' health.

Figure 2.1.1 How healthy are construction workers?

According to a recent pamphlet[2] by the Union of Construction Allied Trades and Technicians (UCATT) there are about 1 in 6 'walking wounded' with some form of disability within the U.K. construction industry. UCATT points to six main hazards likely to affect construction workers:

Cold
Dust and fibres
Fumes and vapours
Vibration
Excessive noise

Popular American and British books[3,4] aimed at educating the workers emphasise the health hazards of work of almost every kind. Whilst advising workers on how to reduce the hazards of work they are more reticent on how to avoid work altogether. A hint on this was given by Russell[5]:

"What is work? Work is of two kinds: first, altering the position of matter at or near the earth's surface relative to other such matter, second, telling other people to do so. The first kind is unpleasant and ill-paid; the second is pleasant and highly paid. The second kind is capable of indefinite extension; there are not only those who give orders, but those who give advice on what orders should be given. Usually two opposite kinds of advice are given simultaneously by two organised bodies of men: this is called politics".

Why, we may wonder, should any work be unhealthy? Work is not war. There is no unavoidable need for casualties. Why should we accept the idea that disease, disablement and death are part of the inevitable costs of production? Man is remarkably adaptive and has, during the course of evolution, developed a series of natural barriers and protective measures against most hazards of nature. His body has remarkable powers of self-healing. He has a good natural defence system.

Evolution, however, is a slow business. The hazards man faces at the workplace are very recent in his history.

Perhaps in a few million years a new breed of man will evolve with natural inborn protection against such present-day health risks as asbestos, short-wave radiation and carbon monoxide. We have no means of knowing. All we do know is that man's present natural defence system is powerless against these and many others.

The health of construction workers is partly determined by home, social and economic circumstances, partly by age, partly by climate and partly by the hazards of work. Some of these, as we have seen already, cause sudden and rapid effects. These include bodily injuries through falls, explosions, contact with moving parts of machinery, suffocation through entry into confined spaces such as pits, tanks and sewers which are deficient in oxygen, drowning and carbon monoxide poisoning. In these cases the relation between the circumstances and their consequences is usually obvious, and the circumstances lend themselves to analysis, control and elimination. They are dealt with under the banner of 'Safety'.

Other hazards are much slower in producing significant effects on the health of the worker. These include toxic vapours and dusts in the working environment, excessive noise and vibration and various biological infections. Most cancers take many years to develop and diagnosis may come only years after the victim has moved from the environment or occupation where he was exposed to the carcinogen which caused it. The problem of relating cause and effect in such cases is more difficult and the connection becomes less certain.

To safeguard workers against such hazards the new professions of occupational medicine, occupational hygiene and industrial toxicology have emerged. So far their practice has been mainly confined to industrial production, where workers are exposed for long periods to a static working environment with the same recurring hazards. They have had relatively little direct impact on construction, except in cases such as diving and working in compressed air. In these, medical participation, e.g., through regular examination of workers, is a legal requirement.

Only the very largest construction companies in the U.K. employ trained occupational physicians. The materials used in the construction industry do, however, come under close control since they are nearly all industrial products.

Manufacturers are generally under a legal obligation to warn users of any health risks associated with their products, and a number previously used are now forbidden or much restricted (such as asbestos, lead paint, benzene and carbon tetrachloride).

Occupational health was defined by a joint Committee of the International Labour Office and the World Health Organisation in 1950 as follows: "Occupational health should aim at the promotion and maintenance of the highest degree of physical, mental and social well-being of workers in all occupations; the prevention among workers of departures from health caused by their working conditions; the protection of workers in their employment from risks resulting from factors adverse to health, the placing and maintenance of the worker in an occupational environment adapted to his physiological and psychological equipment". Most member states of the ILO signed a convention in 1961 which committed them to the provision of occupational health care for their workers.

2.1.1 Occupational mortality of construction workers in the U.K.

The idea that some occupations are associated with higher than average death rates from particular diseases is an old one. Once well established for a particular case, people, medical men especially, will look for a special factor or material found in that occupation as the cause of the disease. Until detailed causes of death were available from death certificates for the mass of the population, and until reasonably accurate statistics of the occupations of the working population were available, these associations were difficult to prove, except in the most clear-cut cases. An early example was the identification of soot as the cause of scrotal cancer among chimney sweeps by Percival Pott in 1775[6].

Today there is a great deal more data on occupations, diseases and mortality. Refined statistical tools and powerful computers are also available. Even so, associations between particular activities and substances with particular diseases seldom pass unchallenged when first reported. Vested interests often spend much money and time in trying to refute them. The linkages between smoking and a higher incidence of respiratory diseases or between asbestos and cancer are cases in point.

A new type of medical scientist, the epidemiologist has also appeared. He tries to diagnose the causes of outbreaks of diseases from the activities and environments of the population exposed to them.

Once in about ten years the data on all death certificates in the U.K. is surveyed and correlated with factors such as age, sex, social class, region, smoking habits and occupations. This is done by the Medical Statistics Division of the Office of Population Censuses and Surveys under the direction of the Registrar General. The most recent survey[7] covered the period from 1970 to 1972. The data was derived from the 1971 Census return and all death certificates over the period 1970–1972.

Two widely presented figures in the survey are standard mortality rates (SMRs) and relative age gradients (denoted here as RAGs). The SMR is a number which may be higher or lower than 100 and denotes a higher or lower chance of death than that of the population as a whole in the same age group. SMRs are given for a wide range of groups of people dependent on sex, region, occupation, age and social class. They are given also for all deaths (overall SMR) and for deaths arising from particular causes and diseases as recorded on the death certificates. For age, the working population (age 15 to 64) is split into five groups, each with a 10 year span (15 to 24, etc.).

Comparing SMRs of working men in different age groups with those for all men in the range 15–64, there is as expected a large and progressive increase in the SMR with age, as shown below:

Age	SMR	Age	SMR
15–19	102	40–44	318
20–24	112	45–49	463
25–29	144	50–54	676
30–34	156	55–59	1,043
35–39	202	60–64	2,083

Six social classes are recognised by the survey:

I	Professional, etc., occupations
II	Intermediate occupations
IIIN	Non-manual skilled occupations
IIIM	Manual skilled occupations
IV	Partly skilled occupations
V	Unskilled occupations

Those in the higher social classes tend to have higher incomes, more home amenities and material possessions and tend also to be most conscious of health hazards such as tobacco and poor diet. SMRs from

Occupational mortality of construction workers in the U.K.   129

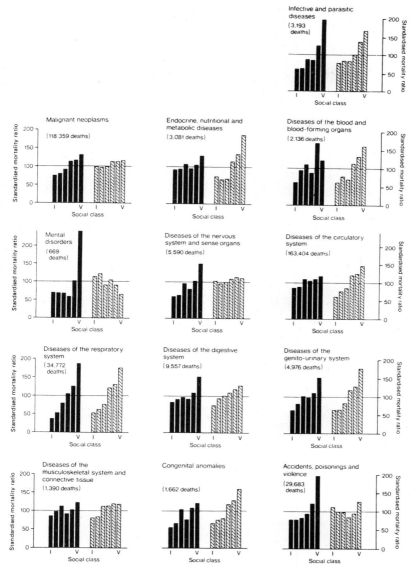

Figure 2.1.2 Mortality by social class and cause of death: standardized mortality ratios for men and married women (by husband's occupation) ages 15–64.

almost all causes therefore increase from the highest to the lowest social groups. This is shown in *Figure 2.1.2*.

Because of this strong dependence of mortality on social class, occupation SMRs are not merely quoted as relative to the whole population of the same sex and age group, but are also adjusted for social class. This gives a truer picture of whether the risks of any particular occupation are higher or lower than average.

TABLE 2.1.1. Standard Mortality Ratios (SMR) and Relative Age Gradients (RAG) for construction trades and occupations

Unit	Description	Overall SMR Unadjusted for social class	Overall SMR Social class adjusted	Relative age gradient	SMRs for selected causes of death All neoplasms	Circulatory diseases	Respiratory diseases	Accidents
026	Linesmen, cable jointers	102	96	12	92	105	112	139
027	Electricians	95	92	7	103	103	80	85
034	Steel erectors; riggers	164	158	4	181	145	165	246
036	Gas, electric welders, cutters, braziers	122	118	8	126	127	124	112
045	Plumbers, gas fitters, lead burners	101	96	14	117	99	91	89
046	Pipe fitters, heating engineers	114	110	27	136	114	114	101
055	Carpenters and joiners	98	92	14	111	97	82	96
057	Sawyers & wood working machinists	86	81	7	92	88	96	55
093	Bricklayers, tile setters	108	101	25	131	99	113	88
094	Masons, stone cutters, slate workers	118	110	21	122	109	125	115
095	Plasterers, cement finishers, terrazzo workers	112	106	35	124	103	137	93
096	Builders (so described), clerks of works	134	127	33	152	136	78	138
097	Bricklayers, etc. labourers nec	273	240	−96	255	239	419	403
098	Construction workers, nec	93	84	−18	104	83	90	143
100	Painters, decorators, nec	112	105	4	121	108	112	123
103	Crane & hoist operators, slingers	124	116	−8	127	120	138	139
104	Operators of earth moving and other construction machinery, nec	93	90	1	86	103	77	117
113	Building and contracting	80	57	−30	86	69	96	99
176	Managers in building and contracting	54	66	18	62	60	25	32
195	Civil, structural, municipal engineers	95	125	21	96	111	49	79
211	Surveyors	82	107	−6	88	89	38	86
212	Architects, town planners	74	96	15	70	79	47	79
218	Draughtsmen	80	82	4	87	86	56	66

Code	Occupation						
Other occupations with high SMRs							
001	Fishermen	171	151	31	183	137	235	253
008	Coalmine workers, above ground	160	143	-23	127	168	226	83
061	Shoemakers and shoe repairers	156	145	24	136	161	159	109
154	Publicans, inn-keepers	155	190	18	146	151	146	103
019	Rolling, tube mill operators, metal drawers	144	137	58	158	148	180	54
007	Coalmine workers, underground	141	132	45	119	132	252	156
Other occupations with low SMRs							
197	Electrical engineers	42	56	18	44	53	14	28
192	University teachers	49	64	-5	49	47	17	63
188	Physiotherapists	55	67	58	32	68	58	–
174	Local authority senior officers	57	70	6	51	70	29	31
141	Typists, shorthand writers, secretaries	59	59	15	59	63	45	30
173	Ministers of the Crown, MPs, nec, senior government officials	61	76	-6	70	66	29	35
190	Public health inspectors	64	79	21	63	71	49	20

It should be clear from this that SMRs have no absolute meaning, but are simply relative. If the whole population has a shortened life expectancy from a particular and perhaps unknown health hazard, this will not be apparent from the SMRs.

On the other hand, a table of SMRs of different occupations can alert us to serious work related health hazards, and stimulate us to investigate them further. Such investigations often lead to the identification of materials and factors which were responsible, a better understanding of how they operate and how the hazard can be eliminated.

The relative age gradients (RAGs) give the difference (+ or −) between the SMRs for the age group 45 to 64 and 15 to 44. A positive RAG for a particular occupation may indicate that the occupation has some hazard which has a cumulative or slow-acting effect on the body, such as the growth of cancers. A negative RAG suggests a hazard which greater experience helps the worker to avoid.

The 1970–72 survey contains a number of detailed tables and comments from which the following have been abstracted. *Table 2.1.1* gives the overall SMRs before and after adjustment for social class and also the SMRs for four selected causes of death, (1) neoplasms (tumours) (2) circulatory diseases, (3) respiratory diseases and (4) accidents. It also gives the RAGs. The data is given for all quoted occupations in construction as well as for a few other occupations with high and low overall SMRs. Not all of the occupations quoted under construction were in it exclusively. Some of the workers were employed in other industries. Thus crane and hoist operators are employed in transport and manufacturing as well as construction.

The figures warrant careful study. Most are plausible and confirm our general impressions of the hazards of particular occupations. Such anomalies as appear seem to be due to the fact that a person who claimed to follow one occupation on his Census return was recorded as following another on his death certificate.

The figures incidentally bear out Russell's cynical comment quoted earlier. Thus the manager of a building firm has an overall SMR of 54, a bricklayer an SMR of 108, while the bricklayer's labourer has an SMR of 273.

More detailed comments on particular causes of death of construction workers listed in *Table 2.1.1* follow.

The phrase 'than usual' means usual for all men of the same age group, usually before adjustment for social class. Some comments refer to more detailed data which could not be included in *Table 2.1.1*.

Unit 026—Linesmen, cable jointers
There were three times more accidental deaths and deaths from falls in this group than usual.

Unit 027—Electricians
No valid conclusions could be drawn due to the existence of six similar categories and to the resulting confusion in classification.

Unit 034—Steel erectors, riggers
The high SMR for this group was due to accidents, cancers and respiratory

diseases. There were 55 deaths from falls compared with 4 usual, and nearly twice as many deaths from cancers and respiratory diseases.

Unit 036—Gas, electric welders, cutters, braziers
These had a high SMR from pneumonia (144), although this had fallen since 1949–53 (207). The survey suggests that many of these cases were actually 'metal fume fever' from metals such as zinc, cadmium and vanadium. There were also more deaths from cancer than usual. The survey suggests asbestos as the cause.

Unit 045—Plumbers, gas fitters, lead burners
These had a slightly higher than usual number of deaths from cancers.

Unit 046—Pipe fitters, heating engineers
These again had a higher than usual number of deaths from cancers, particularly among the older workers. (This may well have been caused by asbestos lagging).

Unit 055—Carpenters and joiners
These had more deaths than usual from cancer of the lung, cancer of the bladder and accidental falls.

Unit 057—Sawyers and woodworking machinists
Deaths from all causes were lower than usual except for cancer of the nose. Deaths from this cause were nearly ten times higher than usual.

Unit 093—Bricklayers, tile setters
These had high death rates from cancer of the stomach, of the lung and of the lip. The last type is somewhat rare.

Unit 094—Masons, stone cutters, slate workers
These had an exceptionally high SMR (1328) from pneumoconioses (probably silicosis from fine silica dust in the air inhaled).

Unit 095—Plasterers, cement finishers, terrazzo workers
These had higher than usual death rates from cancer of the lung and from accidental falls. Deaths from bronchitis, emphysema (abnormal inflation of the lungs) and asthma were also higher than usual, although this was not statistically very significant.

Unit 097—Bricklayers, etc., labourers nec
Part of the reason for the very high overall SMR of this group lay in differences between the occupations given on the Census return and on the death certificates. It was clear, however, that this group had higher than usual numbers of deaths from cancer of the stomach, from bronchitis, emphysema and asthma, from accidental falls and from homicide. This group had a considerable proportion of migrant foreign workers.

Unit 098—Construction workers, nec
These had high death rates from cancer of the lung, from accidental falls and other accidents.

Unit 100—Painters, decorators, nec
These had more deaths than usual from cancer of the lung and bladder and from accidental falls.

Unit 103—Crane and hoist operators, slingers
These had a higher than usual death rate from accidents, particularly falls. Deaths from bronchitis, emphysema and asthma and from ischaemic heart disease (deficiency of blood flow due to constriction of blood vessel) were also somewhat higher than usual.

Unit 104—Operators of earth moving and other construction machinery, nec
Although the overall SMR was lower than usual, deaths from accidents and from chronic rheumatic heart disease were higher than usual.

Unit 113—Labourers in building and contracting
The lower overall SMR recorded was due to the same cause as the high SMR for Unit 097. Deaths from falls and other accidents were, however, nearly twice as high as usual.

Unit 176—Managers in building and contracting
These had fewer deaths from all causes than usual.

Unit 195—Civil, structural, municipal engineers
These had fewer deaths than usual for most causes except cancer of the large intestine.

Unit 211—Surveyors
These also had low death rates from most causes except cancer of the lung.

Unit 212—Architects, town planners
These showed a similar pattern to surveyors, with leukaemia replacing cancer of the lungs as a significant cause of death.

Unit 218—Draughtsmen
These had lower than usual death rates from most causes except cancer of the large intestine and cancer of the brain.

In conclusion, the survey provides a useful insight into the most common causes of death among construction workers. The high preponderance of accidents, particularly falls, among most construction workers was already clear from the accident statistics in Chapter 1.3. In addition, many construction occupations have high death rates from cancers and respiratory diseases. Some of the cancers, such as cancer of the lip and of the nose, are rather uncommon and seem to be related to quite specific occupations, such as bricklaying (lip) and woodworking (nose).

2.1.2 Health of construction workers in other countries

The International Occupational Safety and Health Information Centre (CIS) of the International Labour Office publishes abstracts eight times a year in seven languages of all important literature which appears internationally in this field. This covers of course the health problems of construction workers. Brief notes based on CIS abstracts about this subject follow in the next sub section.

The health problems of construction workers in the European Economic Community are also discussed in a German study commissioned by the EEC and published in 1982. Notes based on this study are given in Section 2.1.2.2.

2.1.2.1 Notes of papers covered by CIS Abstracts[8]

The following notes give a fleeting glimpse of work done over many years in many countries. After various attempts, the abstracts have been grouped under the body organ under attack or discussion. The following body organs are represented:

Skin
Lungs
Heart and circulatory system
Stomach
Brain
Central nervous system
Eyes
Ears
Spine

All cases where two or more organs are under attack are listed under 'whole body and general'.

The country of origin and the language of the original paper are prefixed by the letters C and L, respectively. Only the CIS abstract numbers are given. These must be referred to for the full reference. Occupations of workers involved are, of course, given with details of the diseases found.

The fact that a study was made in the first place generally means that there was a real occupational health problem. In many cases comparative health and mortality studies showed that it had a high statistical significance. Thus there was little doubt that the diseases or disorders referred to in the following notes were occupationally related. Unusual and bizarre problems such as snake bites, man-eating lions and tropical diseases have been omitted.

Skin
CIS 80-793. *C.* USA. *L.* English
Housepainters
Dermatitis
Note: Chloroacetamide and solvents used for skin cleaning are important causes.

CIS 82-202. *C.* FDR. *L.* German
Cement workers and plasterers
Skin lesions and dermatitis
Note: Compounds of chromium, cobalt and nickel in cement and plaster were considered to be important causes.

CIS 81-1347. *C.* Spain. *L.* Spanish
Cement workers and building workers in general
Dermatitis
Note: Cement and chromium compounds in it were considered to be the main cause.

CIS 77-489. *C.* Denmark. *L.* English
Concrete workers
Skin burns
Note: Workers suffered severe leg burns from contact with premixed concrete which spilled over the tops of their boots. The high alkalinity of the concrete was thought to be responsible.

LUNGS
CIS 76-1246. *C.* Switzerland. *L.* German
Brickwork and reinforced concrete workers
Silicosis
Note: Hazard arises from use of power-operated drilling tools. Protection required includes built-in dust removal device with tool, surfaces to be kept wet, wearing of dust masks.

CIS79-1142. *C.* Finland. *L.* Finnish
Building workers generally
Respiratory diseases
Note: Concrete dust containing free silica is main hazard. This is found in dry grinding of concrete, concrete cleaning operations using compressed air, floor surface grinding.

CIS 78-648. *C.* France. *L.* French
Tunnel workers, concrete finishers, others.
Silicosis
Note: Need for long term screening. Problems of prevention, treatment and compensation.

CIS 78-531. *C.* Switzerland. *L.* German
Brick and concrete workers
Silicosis
Note: Drilling holes, cutting brick or concrete walls and demolishing walls with power tools produce respirable dust containing free silica much in excess of TLV. (See Chapter 2.4). General or local exhaust ventilation and improved construction methods are needed to solve the problem.

CIS 81-61. *C.* France. *L.* French
Tunnel workers, concrete grinders
Silicosis
Note: Incidence of silicosis has increased in recent years due to use of new cements with pozzolana as additive and new techniques of concrete grinding.

CIS 80-344. *C.* France. *L.* French
Stone masons, boiler fitters, plasterers
Silicosis, asbestosis
Note: High incidence of silicosis and asbestosis in these workers increases with age.

CIS 80-13179. *C.* Sweden. *L.* English
Construction and metal workers generally
Bronchial carcinoma (cancer)
Note: Higher incidence than in most other industries.

BLOOD AND CIRCULATORY SYSTEM

CIS 78-534. *C.* Finland. *L.* English
Concrete workers
Vibration
Note: Older workers especially suffer from myocardial infarction (dead tissue in heart muscle caused by blocking of artery). See also CIS 75-1299 and CIS 76-1599 under 'Whole Body and General'.

CIS Unnumbered. *C.* Netherlands. *L.* Dutch
Compressed air workers, divers
Decompression sickness
Note: High incidence of decompression sickness was aggravated by heavy alcohol consumption and obesity.

CIS 76-234. *C.* France. *L.* French
Divers
Decompression sickness and other hazards
Note: Diving techniques, training and hazards including toxicity of oxygen under pressure are reviewed.

STOMACH

CIS 78-276 *C.* France. *L.* French
Construction workers in general
Dietary habits
Note: Construction workers had inadequate breakfast, lunchbox midday meal, bad eating conditions, unbalanced diet with too much fat and starch and not enough fruit and vegetables. Regulations on canteens, kitchen, etc., are adequate and far reaching but not fully observed. Worker education on diet needed.

BRAIN AND CENTRAL NERVOUS SYSTEM

CIS 77-191. *C.* Sweden. *L.* Swedish
Housepainters
Psychosomatic syndrome (a group of body symptoms caused by mental or emotional factors)
Note: Memory loss, personality changes, lassitude. Found in housepainters exposed for many years to solvents. Brain damage confirmed. No question of alcoholism in these cases.

CIS 77-2048. *C.* France. *L.* French
Workers at heights.
Balance and factors which affect it.
Note: Safety measures should be supplemented by medical screening of workers employed at heights, including clinical, neurological, audiometry tests, ECG and checks for disorders of the labyrinth (inner ear) etc.

CIS 78-1784. *C.* France. *L.* French
Construction workers generally
Aptitude and disability
Note: There is a need for construction workers to be screened for general and specific work with high hazards (e.g. scaffolders). This should be done by occupational physicians and tests discussed with workers.

CIS 81-537. *C.* France. *L.* French
Steel erectors and others.
Psychotechnical tests
Note: Workers, such as steel erectors, most likely to be injured by falls should be screened out by tests on I.Q., manual precision, visual-manual co-ordination and complex reactions.

EYES
CIS 77-692 and 77-696. *C.* France. *L.* French
Construction workers generally
Eye damage from lasers
Note: Ocular microlesions due to lasers are very difficult to detect. Eye protection or isolation from hazardous zone is needed.

CIS 79-1905. *C.* France. *L.* French
Machine and crane drivers, plumbers, welders, tile layers, and construction workers in general.
Eyesight.
Note: Importance of good eyesight discussed and need to correct visual defects with glasses or contact lenses.

CIS 80-1428. *C.* France. *L.* French
Construction workers generally.
Eye injuries.
Note: Eye injuries among construction workers discussed. Need for safety glasses and counselling of young workers by occupational physician.

EARS
CIS 77-496. *C.* France. *L.* French
Carpenters, bricklayers, site managers, operators of earth moving equipment.
Hearing damage.
Note: Few pre-employment audiometric examinations are now carried out. Protective measures are discussed.

CIS 77-982. *C.* FDR. *L.* German
All workers with machinery.
Study of noise problems.
Note: Machinery used on construction sites, compressors, power shovels, concrete mixers, truck mixers, concrete pumps are discussed as noise sources.

SPINE
CIS 78-1483. *C.* Finland. *L.* English
Concrete reinforcement workers.
Lumbar disc degeneration.
Note: Problem of slipped disc among these workers discussed.

CS 80-1174. *C.* France. *L.* French
Bricklayers and others.
Injuries to spinal column.
Note: Role of ergonomics in preventing this problem is discussed.

CIS 81-224. *C.* France. *L.* French
Kerbstone setters.
Back disorders.
Note: Problems of this occupation are discussed, including crushing of hands and feet.

WHOLE BODY AND GENERAL
CIS 74-1447. *C.* FDR. *L.* German
Painters.
Toxicity problems.
Note: Polyurethanes, tributyl tin and other biocides, and solvents used in paints are discussed from toxicity viewpoint. Statistics and diagrams given.

CIS 75-1836. *C.* FDR. *L.* German
Construction workers in general.
Hazards of electric current.
Note: Although injuries from electricity are rare, probability of fatality is 27 times higher than with other types of accident.

CIS 75-1299. *C.* USA. *L.* English
Heavy construction machine operators.
Whole body vibration.
Note: High incidences found of ischaemic heart disease (defective flow of blood caused by constricted blood vessel), and obesity of non-endocrine origin (i.e., not due to ductless glands).

CIS 76-1598. *C.* USA. *L.* English
Construction plant operators.
Whole body vibration.
Note: Heart disease, diseases of endocrine (ductless) glands and male genital organs found in these workers.

CIS 77-537. *C.* France. *L.* French
Demolition workers.
Multiple health hazards.
Note: Occupational diseases include osteoarticular disease (i.e., in joints between bones) gastro-intestinal disorders, chronic bronchitis and tetanus. Demolition in France has the worst safety record of any occupation in construction.

CIS 77-2049. *C.* France. *L.* French
Floor workers.
Multiple health hazards.
Note: Study of hazards of laying glued floor coverings—osteoarticular, traumatic and toxic lesions. Knee and low back disorders. Effects of glues and solvents discussed.

CIS 78-843. *C.* France. *L.* French
Plumbers, sanitary fitters and other workers in inaccessible work places.
Multiple hazards.
Note: Electricity, toxic vapours, cramped postures. Hazards and safety measures discussed.

TABLE 2.1.2. Diseases among construction workers attributable to occupational factors

Type of disease	Cause	Typical occupations affected	Names and variants of disease
Skin diseases	Cement Paint Varnish Other allergens Solvents Strong chemicals Abrasives	Concrete workers Bricklayers Floor tilers Building labourers Painters Road construction workers Insulation workers Flooring layers Joiners	Dermatitis, especially from cement cysticerosis
Hardness of hearing	Noise	Joiners & wood working machinists Crane drivers Carpenters Operators of earth-moving machinery Furnace bricklayers Machine operators Building labourers Concrete workers Locksmiths Formworkers	
Respiratory disease	Toxic dusts Vapours and gases	Bricklayers Concrete workers Quarry workers Painters Varnishers Civil engineering workers	Asbestosis Silicosis Bronchial asthma Tuberculosis Pulmonary allergens Pulmonary oedema

Muscular and bone diseases	High static stress Unnatural working posture	Floor tilers Floor layers Terrazzo layers Carpenters Bricklayers Reinforcing-rod benders	Lumbago Sciatica Spinal degeneration Epicondylitis Bursitis Meniscus degeneration Muscular strain
Cancer	Carcinogenic material	Painters Varnishers Insulation workers	Liver cancer Stomach cancer Lung cancer Throat cancer Cancer of testicles
Mental illness	Stress Toxic materials affecting brain and central nervous system	Bricklayers Formworkers (heavy shuttering) Concrete reinforcement workers Concreters Plant and machine operators Painters	Stress syndrome Myocardial infarction
Diseases caused by vibration	Vibration	Machinery operators Concrete workers	White finger disease Osteonecrosis Exostosis Osteophytis

CIS 78-1372. *C.* Sweden. *L.* Swedish
Carpenters, joiners, tile setters, painters and others.
Multiple hazards.
Note: Hazards of epoxy resins, glues, hardeners and reactive thinners discussed with advice on protection.

CIS 81-1731. *C.* France. *L.* French
Plasterers.
Premature aging.
Note: Plasterers suffer from early aging due to heavy work load in rendering and smoothing and unnatural working postures.

CIS 82-232. *C.* Finland. *L.* Finnish
Welders and braziers.
Respiratory irritation, headaches, malaise.
Note: Symptoms specially noted after welding painted and greasy parts. Regular medical examination of these workers recommended.

2.1.2.2 Notes from EEC Survey

The following notes are again highly condensed, being based on the Summary Report of the survey by Eisenbach[9]. The survey was carried out in 1981. It was based partly on answers to questionnaires sent to Government agencies, scientific institutions and associations in the member states, and partly on published reports such as those abstracted by the CIS and discussed in Section 2.1.2.1.

A number of diseases found in construction workers are attributable to occupational factors. In addition, the demanding nature of construction work produces both mental and physical stress. As a result of all these factors construction workers tend to degenerate prematurely in their working capacity. There is also a common desire among most construction workers to transfer to some healthier and less demanding industry or occupation.

A list of diseases among construction workers which can be attributed to occupation factors is given in *Table 2.1.2*. The types of disease are listed in order of frequency.

Skin diseases caused mainly by cement, and to a lesser extent by paints, varnishes etc., account for over half the occupational diseases in construction; hardness of hearing and deafness, which are commonest

TABLE 2.1.3. Six construction operations with highest frequency of occupational disease

Occupation	Most common disease	Second most common disease
Furnace bricklayer	Silicosis	Pulmonary tuberculosis
Refractory bricklayer	Silicosis	Pulmonary tuberculosis
Floor layers	Bursa† damage	Skin disease
Flooring layers	Skin diseases	Bursa† damage
Floor tilers	Skin diseases	Bursa† damage
Concrete workers	Skin diseases	Hearing damage Silicosis

†A bursa is sac containing fluid found between movable parts of the body, e.g. the knee joints, which facilitates smooth movement between them.

among joiners and wood working machinists account for about 25%; diseases of the joints caused by pressure account for about 10%; respiratory diseases also account for about 10%.

The six occupations with the greatest exposure to occupational disease are given in order in *Table 2.1.3*.

One of the main difficulties found in making this study within the EEC lay in the different data bases and legislation in the various member states. There are considerable differences in the occupational diseases among construction workers which are legally recognized and qualify for compensation under national insurance schemes. This partly affects the differences in the ratios between various occupational diseases as reported in the member states. The subject of recognized occupational diseases is dealt with in Chapter 2.2.

References

1. Manpower Services Commission, Careers and Occupational Information Centre, *Leaflet CU 67*, HMSO, London.
2. Union of Construction Allied Trades and Technicians, *Organise for Safety's Sake*, London (1981).
3. STELLMAN, J.M. and DAWN, S.M., *Work is Dangerous to your Health*, Random House Transatlantic Book Service (1980).
4. KINNERSLY, P. *The Hazards of Work and How to Fight Them*, Pluto Press, London (1980).
5. RUSSELL, B., *In Praise of Idleness*, Allen and Unwin, London (1932).
6. POTT, P. *Chirugical observations relative to the cataract, the polypus of the nose, the cancer of the scrotum, the different kind of ruptures and the mortification of the toes and feet*, London (1775).
7. Office of Population Censuses and Surveys, 1970–1972, *Occupational Mortality*, decennial supplement, HMSO, London (1978).
8. CIS, *Bibliography on Occupational Safety and Health in the Building Industry*, CIS-ILO, Geneva (1982).
9. EISENBACH, B. *Evaluation and Comparison of Existing Data and Criteria for the Definition of Working Conditions in the Construction Industry in the European Community*, (original in German), Bundesanstalt für Arbeitsschutz und unfallforschung, Dortmund (1981).

2.2

Occupational diseases and their causes

The term 'occupational disease' appears frequently in the foregoing chapter, but it is difficult to define. The concept is essentially a legal one. Lists of a few recognized occupational diseases first appeared in some countries in the 1920s. The legal importance of occupational diseases varies from country to country depending on its social security system. Broadly there are three groups:

- Countries with a general social security system which allows occupational diseases to be dealt with in a similar manner to disease or disablement due to any other cause.
- Countries which offer special benefits to victims of occupational diseases, in addition to a general system of social security.
- Countries in which a system of benefits exists only for victims of occupational accidents and diseases.

The question of a cause-effect relationship is most acute for the last group of countries.

A number of countries have quite detailed legislation on occupational diseases whereby certain diseases contracted in particular occupations are covered by social insurance. These occupational diseases are also of international interest. The International Labour Office has been concerned with them since its inception, and so more recently has the World Health Organisation (WHO).

The first international schedule of occupational diseases was embodied in the Workmen's Compensation (Occupational Diseases) Convention (Revised) 1934 (No. 42). This was superseded in 1964 by the Employment Injury Benefits Convention (No. 121), with a new and expanded list. This convention had by the beginning of 1983 been ratified by only 18 countries. Subsequent advances in medicine and epidemiology established the occupational origin of other diseases. The ILO list of occupational diseases was in consequence amended and extended as a result of a meeting of experts in Geneva in January 1980 in co-operation with the WHO.

The committee of experts could agree only on a broad definition of occupational diseases, as 'diseases having a cause-effect relation with work'. The ILO has had to leave it to member states to make their own lists and definitions of occupational diseases, which may differ from those of the ILO: 'Each member state should, under prescribed conditions, regard diseases known to arise out of the exposure to substances or dangerous conditions in processes, trades or occupations as occupational diseases'.

Legislation is complicated by the need to distinguish between diseases caused by work, aggravated by work and unrelated to work.

Those related to work can be divided into these two categories:

- Those due to a single cause
- Those caused by a variety of harmful factors at the workplace.

Diseases with a low incidence in the general population can more readily be linked to a particular occupation than those with a high incidence in the general population such as chronic bronchitis. For the last category, it is much more difficult to establish a cause-effect relation with an occupational activity. Only diseases whose relation to a particular occupation can be clearly established are recognized by the ILO committee as occupational. The incidence of a disease in a given occupational group compared with that in the non-exposed population is most important. Other important factors are work history, length of exposure, dose or nature of exposure and individual susceptibility.

Another distinction sometimes has to be made (especially with toxic substances) between an accident caused by acute poisoning or exposure at high concentrations for only a short period and a disease caused by chronic poisoning or exposure at lower concentrations over long periods. There is no general agreement among member states where the line should be drawn. Most occupational diseases are, however, contracted through prolonged exposure over many years. In many cases workers claiming compensation need to prove that they were exposed for a minimum number of years to the hazard to substantiate their claim.

2.2.1 The amended ILO list[1]

A list of occupational diseases, as amended by the 1980 meeting of ILO/WHO experts, is given in *Table 2.2.1*. This, in many ways, is a wide and general list. In most cases it is the substance or agent which is specified, not the disease itself. Thus the ILO/WHO list includes diseases caused by most materials on national lists (with exceptions such as nickel). National lists, however, vary considerably in the occupational diseases which they include.

2.2.2 Occupational diseases in the U.K.

Most employed workers in the U.K. who become partly or wholly disabled through a disease related to their work can make a claim against their employer and/or claim special benefit from the Department of Health and Social Security.

In the first case, the claim, whilst nominally against the employer, is usually in fact against his insurance company, since most employers are

TABLE 2.2.1. List of occupational diseases as amended by the ILO/WHO meeting of experts, January, 1980

Number	Occupational disease	Work involving exposure to risk	Nature of hazard
1	Pneumoconioses caused by sclerogenic mineral dust (silicosis, anthraco-silicosis, asbestosis) and silico-tuberculosis providing that silicosis is an essential factor in causing the resultant incapacity or death	All work involving exposure to the risk concerned	Chemical
2	Broncho pulmonary diseases caused by hard metal dust (e.g. tungsten carbide)	Ditto	Chemical
3	Broncho pulmonary diseases caused by cotton, flax, hemp or sisal dust (byssinosis)	Ditto	Chemical
4	Occupational asthma caused by recognised sensitising agents or irritants inherent to the work process	Ditto	Chemical or biological
5	Extrinsic allergic alveolitis caused by the inhalation of organic dusts as prescribed by national legislation	Ditto	Chemical
6	Diseases caused by beryllium or its toxic compounds	Ditto	Chemical
7	Diseases caused by cadmium and its toxic compounds	Ditto	Chemical
8	Diseases caused by phosphorus or its toxic compounds	Ditto	Chemical
9	Diseases caused by chromium or its toxic compounds	Ditto	Chemical
10	Diseases caused by manganese or its toxic compounds	Ditto	Chemical
11	Diseases caused by arsenic or its toxic compounds	Ditto	Chemical
12	Diseases caused by mercury or its toxic compounds	Ditto	Chemical
13	Diseases caused by lead or its toxic compounds	Ditto	Chemical
14	Diseases caused by fluorine or its toxic compounds	Ditto	Chemical
15	Diseases caused by carbon disulphide	Ditto	Chemical
16	Diseases caused by halogen derivatives of aliphatic or aromatic hydrocarbons	Ditto	Chemical
17	Diseases caused by benzene or its toxic homologues	Ditto	Chemical
18	Diseases caused by toxic nitro- and amino-derivatives of benzene or its homologues	Ditto	Chemical

19	Diseases caused by nitroglycerin or other nitric acid esters	Ditto	Chemical
20	Diseases caused by alcohols, glycols or ketones	Ditto	Chemical
21	Diseases caused by asphyxiants; carbon monoxide, hydrogen cyanide or its toxic derivatives, hydrogen sulphide	Ditto	Chemical
22	Hearing impairment caused by noise	Ditto	Physical
23	Diseases caused by vibration (disorders of muscles, tendons, bones, joints, peripheral blood vessels or peripheral nerves)	Ditto	Physical
24	Diseases caused by work in compressed air	Ditto	Physical
25	Diseases caused by ionising radiations	All work involving exposure to the risk of ionising radiations	Physical
26	Skin diseases caused by physical, chemical or biological agents not included under other items	All work involving exposure to the risk concerned	Chemical, physical or biological
27	Primary epitheliomatous cancer of the skin caused by tar pitch, bitumen, mineral oil, anthracene or the compounds, products or residues of these substances	Ditto	Chemical
28	Lung cancers or mesotheliomas caused by asbestos	Ditto	Chemical
29	Infectious or parasitic diseases contracted in an occupation where there is a particular risk of contamination	a) Health or laboratory work b) Veterinary work c) Work handling animals, animal carcases, parts of such carcases or merchandise which may have been contaminated by animals or animal carcases d) Other work carrying a particular risk of contamination	Biological

covered under the Employers' Liability (Compulsory Insurance) Act 1969. Since insurers do not fall over backwards to recognize such claims, the disabled employee (generally through his union) often has to initiate legal action against his employer. Most cases, however, are settled out of court at the eleventh hour to minimize legal costs. There are many delays and uncertainties and the final outcome is usually something of a gamble. If the claimant is also receiving or has received special DHSS benefit, one half of the amount so received in the five years after the injury (or contraction of occupational disease) will be deducted from the damages awarded for loss of earnings and profits (Law Reform (Personal Injuries) Act 1968).

In the second case, the victim, armed with a medical certificate and a satisfactory record as a national insurance contributor, may submit a claim for industrial disease benefit to the insurance officer at his local DHSS office. To qualify for such benefit, the victim has to show:

(1) that the disease from which he is suffering is on the official list of prescribed diseases; and
(2) that the nature of his occupation is one which qualifies for benefit.

The procedure for applying for industrial disease benefit is generally faster than taking legal action against the employer, and less financially burdensome on the victim. Unfortunately there are many instances where a construction worker who suffers a disabling disease at work is unable to claim DHSS benefit. A frequent problem is that the worker has not been employed for long enough in a particular occupation to qualify for this benefit. If his claim is otherwise a strong one, he may decide to proceed against his employer.

However, if the worker is not an employee, but a self-employed 'labour only subcontractor', which is very common in construction, he may find that he can neither take action against his 'employer' nor make a claim for industrial disease benefit from the DHSS. He can only cover himself for loss of earnings from a work-related disease by taking out a special personal insurance policy. He might, however, have a case as a member of the public against his employer, who is obliged to take out an insurance policy for public liability. (Note on author's memo pad: Check on the cost of insuring against writer's cramp!)

The law on health insurance, compensation and liability is too complicated to discuss further here, besides being subject to change at short notice. Several books written for the layman on the subject are available[2,3].

We are more concerned here with avoiding the hazards which give rise to occupational and industrial diseases than with legal aspects of compensation to the victims.

2.2.2.1 The official list of occupational diseases in the U.K.

The official list forms part of The Social Security (Industrial Injuries) (Prescribed Diseases) Regulations, 1980[4] which consolidate earlier Regulations made in 1975 under the Social Security Act, 1975. The list is given in *Table 2.2.2*. It includes additional diseases added by subsequent amendments resulting from more recent medical and epidemiological research. It excludes pneumoconiosis and byssinosis which are dealt with, together with occupational deafness, in parts II and VI of the Regulations. With the name and description of the disease, the list gives the nature of the qualifying occupation which allows the disease to be treated as occupational from the viewpoint of DHSS claims. Lastly, in *Table 2.2.2* we have made notes on:

- the type of hazard (chemical, physical or biological) responsible for each disease
- possible construction occupations where exposure to the hazard might reasonably occur,
- examples of the hazard in construction.

Some of the diseases are also 'notifiable' under the U.K. Factories Act 1961 and subsequent legislation. That is to say, a doctor visiting a patient whom he finds to be suffering from the disease is required to report it to the Health and Safety Executive. Notifiable diseases in *Table 2.2.2* are indicated with a (N) after their numbers.

Many of the diseases are specific and described precisely in medical terms. The more generalized descriptions on the ILO list (*Table 2.2.1*) may, in some cases, include two or more specific diseases from the U.K. list.

The following occupational diseases on the ILO list (*Table 2.2.1*) are not included in the U.K. list (under any description recognizable to the authors).

(2) Bronchopulmonary diseases caused by hard metal dust.
(5) Extrinsic allergic alveolitis (characterized by shortness of breath, tightness of chest, cough and fever) caused by the inhalation of organic dusts as prescribed by national legislation.
(9) Diseases caused by chromium or its toxic compounds.
(14) Diseases caused by fluorine or its toxic compounds.
(19) Diseases caused by nitroglycerin or other toxic nitric acid esters.
(21) Diseases caused by asphyxiants: carbon monoxide, hydrogen cyanide or its toxic derivatives, and hydrogen sulphide. (Author's note: The compounds listed here are not asphyxiants, but highly toxic ones. True asphyxiants are inert gases such as nitrogen, argon, and propane which, when present at high concentrations, deprive the breather of oxygen.)
(23) Diseases caused by vibration.

Occupational asthma (No. 4 on the ILO list) was added to the U.K. list by an amendment in 1982.

Thus the U.K. list has some important omissions. It seems anomalous for instance that a clerk or typist suffering from writer's cramp (No. 28 on the U.K. list) can claim benefit, whereas a pneumatic-drill operator in a road gang who becomes incapacitated through 'banana fingers' cannot.

There are, however, many diseases on the U.K. list which are not included on the ILO list.

TABLE 2.2.2. List of prescribed diseases with the occupations for which they are prescribed (except pneumoconiosis and byssinosis) from 'The Social Security (Industrial Injuries) (Prescribed Diseases) Regulations, 1980

Number N = Notifiable	Description of disease or injury	Nature of occupation	Type of hazard	Possible construction occupations exposed	Examples of hazard
					Notes by authors
1N	Poisoning by: Lead or a compound of lead	Any occupation involving: The use or handling of, or exposure to the fumes, dust or vapour of, lead or a compound of lead, or a substance containing lead.	Chemical	Painters Demolition gas cutters Plumbers, jointers Welders (mechanics)	Use of lead primers, burning off old paint Cutting old painted ironwork Soldering, lead burning
2N	Manganese or a compound of manganese	The use or handling of, or exposure to the fumes, dust or vapour of manganese or a compound of manganese, or a substance containing manganese	Chemical		Welding with rods containing manganese
3N	Phosphorus or phosphine or poisoning due to the anticholinesterase action of organic phosphorus compounds	The use or handling of, or exposure to the fumes, dust or vapour of phosphorus or a compound of phosphorus or a substance containing phosphorus	Chemical	(Ground workers, labourers)	Contact with pesticide such as parathion, or ingestion of rat poison
4N	Arsenic or a compound of arsenic	The use or handling of, or exposure to the fumes, dust or vapour of arsenic or a compound of arsenic or a substance containing arsenic	Chemical	Woodworkers (groundworkers, labourers)	Working with wood treated with arsenic preservative. Working in arsenic treated soil
5N	Mercury or a compound of mercury	The use or handling of, or exposure to the fumes, dust or vapour of mercury or a compound of mercury or a substance containing mercury	Chemical	Electricians	Exposure to mercury vapour from contact breaker or other apparatus containing mercury
6N	Carbon bisulphide	The use or handling of, or vapour of carbon bisulphide or a compound of carbon bisulphide or a substance containing carbon bisulphide	Chemical	Unlikely	

No.	Agent	Description	Type	Occupation	Notes
		the fumes of or vapour containing benzene or any of its derivatives		labourers, carpenters	as solvent but less toxic homologues–styrene, xylenes– widely used in paints
8N	A nitro or amino or chloro derivative of benzene or of a homologue of benzene or poisoning by nitro-chloro benzene	The use or handling of, or exposure to the fumes of or vapour containing a nitro or amino or chloro-derivative of benzene or a homologue of benzene or nitrobenzene	Chemical	Unlikely	
9	Dinitrophenol or a homologue or by substituted dinitrophenol or by salts of such substances	The use or handling of, or exposure to the fumes of or vapour containing dinitrophenol or a homologue or substituted dinitrophenols or the salts of such substances	Chemical	Unlikely	These compounds are used almost entirely as intermediates in the chemical industry and are unlikely to be found as such in construction
10N	Tetrachloroethane	The use or handling of, or exposure to the fumes of or vapour containing tetrachloroethane	Chemical	Unlikely	
11	Tricresyl phosphate	The use or handling of, or exposure to the fumes of or vapour containing tricresyl phosphate	Chemical	(Mechanics, pipe fitters)	Use as fire resistant hydraulic fluids
12	Triphenyl phosphate	The use or handling of, or exposure to the fumes of or vapour containing triphenyl phosphate	Chemical		
13	Diethylene dioxide (dioxan)	The use or handling of, or exposure to the fumes of or vapour containing diethylene dioxide (dioxan)	Chemical	(Painters, varnishers)	Once used as solvent, now practically banned because of toxicity
14	Methyl bromide	The use or handling of, or exposure to the fumes of or vapour containing methyl bromide	Chemical	(Mechanics (refrigeration maintenance))	Once common as refrigerant and fire extinguisher, now rarely used
15	Chlorinated naphthalene	The use or handling of, or exposure to the fumes of dust or vapour containing chlorinated naphthalene	Chemical	(Painters, joiners)	Monochloronaphthalene once used as wood preservative. Chlorinated naphthalenes can cause dermatitis and liver damage
16	Nickel carbonyl	Exposure to nickel carbonyl gas	Chemical	Unlikely	Only found in manufacture of nickel
17	Nitrous fumes	The use or handling of nitric acid or exposure to nitrous fumes	Chemical	Tunnel workers, blasters	Produced by use of explosives

TABLE 2.2. Continued

Number N = Notifiable	Description of disease or injury	Nature of occupation	Type of hazard	Possible construction occupations exposed	Examples of hazard
				Notes by authors	
18	Gonioma kamassi (African boxwood)	The manipulation of gonioma kamassi or any process in, or incidental to, the manufacture of articles therefrom	Chemical	Unlikely (joiners)	Hardwood formally used in making shuttles
19	Anthrax	The handling of wool, hair, bristles, hides or skins or other animal products or residues or contact with animals infected with anthrax	Biological	Unlikely (painters)	Use of paint brushes made from infected animal hair
20	Glanders	Contact with equine animals or their carcases	Biological	Unlikely	
21	Infection by leptospira	a. Work in places which are, or are liable to be infested with rats, fieldmice or voles	Biological	Ground workers, demolition workers, labourers	Eating sandwiches contaminated with rats' urine
		b. Work at dog kennels or in the care of handling dogs	Biological		
		c. Contact with bovine animals or their meat products or pigs and their meat products	Biological		
22*	Ankylostomiasis	Work in or about a mine	Biological	Ground & tunnel workers	Caused by hookworm entering host through feet
23	a. Dystrophy of the cornea (inc. ulceration of the corneal surface) of the eye. b. Localised new growth of the skin, papillomatous or keratotic c. Squamous-celled carcinoma of the skin due in any case to arsenic, tar pitch, mineral oil (including paraffin), soot or any compound product (inc. quinone or hydroquinone) or residues of any	The use of or handling of, or exposure to arsenic, tar, pitch, bitumen, mineral oil (inc. paraffin), soot or any compound, product (inc. quinone or hydroquinone), or residue of any of these substances	Chemical	Road workers, asphalters, tar sprayers, roof workers, floor workers, mechanics	Clear from nature of occupation

152

24	—			
25	Inflammation, ulceration or malignant disease of the skin or subcutaneous tissues or of the bones or blood dyscrasia, or cataract, due to electromagnetic radiations (other than radiant heat) or to ionising particles	Exposure to electro-megnatic radiations other than radiant heat or to ionising particles	Physical	Electric welders radiographers, welding inspectors Surveyors	Effect of UV radiation on skin Exposure to radioactive sources Exposure to laser beams Use of lasers
26	Heat cataract	Frequent or prolonged exposure to rays from molten or red hot material	Physical	Welders	Clear from nature of occupation
27N	Decompression sickness	Subject to compressed or rarified air	Physical	Divers, tunnel and caisson workers	Clear from nature of occupation
28	Cramp of the hand or forearm due to repetitive movements	Prolonged periods of handwriting, typing or other repetitive movements of the fingers, hand or arm	Physical		
29	—				
30	—				
31	Subcutaneous cellulitis of the hand (beat hand)	Manual labour causing severe or prolonged friction or pressure on the hand	Physical	Reinforcing bar benders, ground workers & others	Clear from nature of occupation
32	Bursitis or subcutaneous cellulitis of the knee (beat knee)	Manual labour causing severe or prolonged external friction or pressure at or about the knee	Physical	Floor workers of all types, roofing workers & others	Clear from nature of occupation
33	Bursitis or subcutaneous cellulitis arising at or about the elbow due to severe or prolonged external friction or pressure at or about the elbow	Manual labour causing severe or prolonged external friction or pressure at or about the elbow	Physical	(Carpenters)	Clear from nature of occupation
34	Traumatic inflammation of the tendons of the arm or forearm or of the associated tendon sheaths	Manual labour or frequent or repeated movements of the hand or wrist	Physical	Plasterers, painters, decorators, bricklayers, tile setters, labourers	Clear from nature of occupation
35*	Miner's nystagmus	Work in or about a mine	Physical	(Tunnel workers)	Exact cause not clear

TABLE 2.2.2. Continued

Number N = Notifiable	Description of disease or injury	Nature of occupation	Type of hazard	Possible construction occupations exposed	Examples of hazard
				Notes by authors	
36	Poisoning by beryllium or a compound of beryllium	The use or handling or, or exposure to the fumes, dust or vapour of beryllium or a compound of beryllium or a substance containing beryllium	Chemical	Unlikely (plumber)	Brazing beryllium copper tubes or fittings
37	a. Carcinoma of the mucous membrane of the nose or associated air sinuses b. Primary carcinoma of a bronchus or a lung	Work in a factory where nickel is produced by decomposition of a gaseous nickel compound which necessitates working in or about a building where that process or other industrial process ancillary or incidental thereto is carried on	Chemical	Unlikely	Excluded by nature of occupation
38*	Tuberculosis	Close and frequent contact with a source or sources of tuberculosis infection by reason of employment– a) in the medical treatment or nursing of a person or persons suffering from tuberculosis, or in a service ancillary to such treatment or nursing b) in attendance upon a person or persons suffering from tuberculosis, where the need for such attendance arises by reason of physical or mental infirmity c) as a research worker engaged in research in connection with tuberculosis d) as a laboratory worker, pathologist or person taking part in or assisting at post mortem examinations of human remains where the occupation involves working with material which is a	Biological	Unlikely	Excluded by nature of occupation

No.	Disease	Agent	Occupation
	lining of the urinary bladder (papilloma of the bladder) or of the renal pelvis or of the ureter or of the urethra	the following substances is produced for commercial purposes: i) Alpha naphthylamine or beta naphthylamine ii) Diphenyl substituted by at least one nitro or primary amino group or by at least one nitro and primary amino group iii) Any of the above substances mentioned in sub-paragraph (ii) above if further ring substituted by halogeno methyl or methoxy groups, but not by other groups iv) The salts of any of the substances mentioned in sub-paragraphs (i) to (iii) above v) Auramine or magenta b) The use or handling of any of the substances mentioned in sub-paragraphs (i) to (iv) of paragraph (a) or work in a process of which any such substance is used or handled or is liberated c) The maintenance or cleaning of any plant or machinery used in any such process as is mentioned in paragraph (b), or the cleaning of clothing used in any such building as is mentioned in paragraph (a) if such clothing is cleaned within the works of which the building forms a part or in a laundry maintained and used solely in connection with such works	dyestuffs factory or dye works (maintenance and demolition workers in premises described)
40	Poisoning by cadmium	Exposure to cadmium fumes	Welders
41	Inflammation or ulceration of the mucous membrane of the upper respiratory passages or mouth produced by dust, liquid or vapour	Exposure to dust, liquid or vapour Chemical Chemical or physical	Cement workers, bricklayers, demolition workers, joiners, welders, others Welding cadmium-coated steel Clear from nature of occupation particularly in case of dust

TABLE 2.2.2. Continued

Number N = Notifiable	Description of disease or injury	Nature of occupation	Type of hazard	Notes by authors: Possible construction occupations exposed	Notes by authors: Examples of hazard
42N	Non infective dermatitis of external origin (inc. chrome ulceration of the skin but excluding dermatitis due to ionising particles or electro-magnetic radiations other than radiant heat)	Exposure to dust, liquid or vapour or any other external agent capable of irritating the skin (inc. friction or heat but excluding ionising particles or electro-magnetic radiations other than radiant heat)	Chemical or biological	Cement workers, bricklayers, demolition workers, painters and many others	Contact with cement dust, epoxy resin, hardener
43*	Pulmonary disease due to the inhalation of the dust of mouldy hay or of other mouldy vegetable products, and characterized by symptoms and signs attributable to a reaction in the peripheral part of the broncho pulmonary system and giving rise to a defect in gas exchange (Farmers lung)	Exposure to the dust of mouldy hay or other mouldy vegetable produce by reason of employment: a) in agriculture, horticulture of forestry, or b) loading or unloading in storage such hay or other vegetable produce, or c) handling bagasse	Biological	Thatchers (Demolition workers)	Renewing thatched roofs Demolition of farm buildings, vegetable warehouses
44	Primary malignant neoplasm of the mesothelium (diffuse mesothelioma) of the pleura or of the peritoneum	a) the working or handling of asbestos or any mixture of asbestos, b) the manufacture or repair of asbestos textiles or other articles containing or composed of asbestos c) the cleaning of any machinery or plant used in any of the foregoing operations and of any chambers, fixtures and appliances for the collection of asbestos dust, d) substantial exposure to the dust arising from any of the foregoing operations	Chemical	Demolition workers Insulators Heating and ventilating engineers Painters Joiners, roof workers	Demolition of old power stations containing asbestos insulation

45*	Adeno-carcinoma of the nasal cavity or associated air sinuses	Attendance for work in or about a building where wooden furniture is manufactured	Chemical	Woodworking machinists Joiners in close contact with such work	Inhalation of air containing suspended sawdust
46*	Infection by brucella abortus	Contact with bovine animals affected by brucella abortus, their carcases or parts thereof or their untreated products, or with laboratory specimens or vaccines of or containing brucella abortus, by reason of employment: a) as a farm worker b) as a veterinary worker c) as a slaughterhouse worker d) in any other work relating to the care, treatment examination or handling of such animals, carcases or parts thereof	Biological	Unlikely (demolition workers)	Demolition of abattoir
47	Poisoning by acrylamide monomer	The use or handling of, or exposure to acrylamide monomer	Chemical	Unlikely (painters) Tunnel workers	Heat stripping old acrylamide paint Use of acrylamide for water proofing
48*	Substantial permanent sensorineural hearing loss due to occupational noise amounting to at least 50 dB in each ear, being due in the case of at least one ear to occupational noise, and being the average of pure tone losses measured by audiometry over the 1,2 and 3 kHz frequencies (occupational deafness)	a) The use, or supervision of, or assistance in the use of pneumatic, percussive tools or the use of high speed grinding tools in the cleaning, dressing or finishing of cast metal or of ingots, billets or blooms; or b) The use or supervision of, or assistance in the use of pneumatic, percussive tools on metal in the shipbuilding or ship repairing industries; or c) the use or supervision of, or assistance in the use of pneumatic, percussive tools on metal or for drilling rock in quarries or underground or in coal mining for at least an average of one hour per working day; or	Physical	Woodworking machinists, joiners	

TABLE 2.2.2. Continued

Number N = Notifi- able	Description of disease or injury	Nature of occupation	Type of hazard	Notes by authors	
				Possible construction occupations exposed	Examples of hazard
		d) working wholly or mainly on the immediate vicinity of drop forging plant (inc. plant for drop stamping or drop hammering or forging press plant engaged in the shaping of hot metal; or e) work wholly or mainly in rooms or sheds where there are machines engaged in weaving man-made and natural (inc. mineral) fibres or in the bulking up of fibres in textile manufacture; or f) the use of machines which cut, shape or clean metal nails; or g) the use of plasma spray guns for the deposition of metal			
49*	Viral hepatitis	Any occupation involving: a) Close and frequent contact with human blood or human blood products; or b) close and frequent contact with a source of viral hepatitis infection by reason of employment in the medical treatment or nursing of a person or persons suffering from viral hepatitis, or in a service ancillary to such treatment or nursing	Biological	Unlikely	
50*	a) Angiosarcoma of the liver; and b) Osteolysis of the terminal phalanges of the fingers	Any occupation involving work in or about machinery or apparatus used for the polymerization of vinyl chloride monomer, a process which for the purposes of this provision, comprises all	Chemical	Unlikely (plastic welders)	Welding PVC pipe may cause some VC monomer to be liberated

		of the slurry produced by the polymerization and the packaging of the dried product; or work in a building or structure in which any part of the aforementioned process takes place		
51*	Carcinoma of the nasal cavity or associated air sinuses (nasal carcinoma)	a) Attendance for work in a building used for the manufacture of footwear or components of footwear made wholly or partly of leather or fibreboard; or b) Attendance for work at a place used wholly or partly of leather or fibreboard	Chemical	Unlikely	See No. 45
52	Occupational vitiligo	The use or handling of, or exposure to, para tertiary butylphenol, para tertiary butyl catechol, para amyl phenol, hydroquinone or the mono benzyl or mono butyl ether of hydroquinone	Chemical	Unlikely	
53	Asthma, which is due to exposure to any of the following agents: a) isocyanates b) platinum salts c) fumes or dusts arising from the manufacture, transport or use of hardening agents (inc. epoxy resin curing agents) based on phthalic anhydride, tetrachlorophthalic anhydride trimellitic anhydride triethylamine tetramine d) fumes arising from the use of resin as a soldering flux e) proteolytic enzymes f) animals or insects used for the purposes of research or education or in laboratories g) dusts arising from the sowing, cultivation, harvesting, drying, handling, milling, transport or storage of barley, oats, rye, wheat or maize or the handling, milling, transport etc of flour made therefrom	Any occupation involving exposure to any of the agents set out in column 1 of this paragraph	a Chemical b Chemical c Chemical d Chemical e Chemical/biological f Biological g Biological	(a) and (c) painters, particularly in spray painting, and joiners, tiler layers	One pack and two pack surface coatings and adhesives containing materials listed. Clear from description of disease

The descriptions of the occupations listed in *Table 2.2.2* are of two types:

- Those which apparently include any occupation where the material in question could be used or handled, or the hazard in question encountered;
- Those which include only quite specific occupations or places of work which would exclude construction workers in all but exceptional circumstances.

Those of the second type have been marked with an asterisk against the number of the disease in the first column of *Table 2.2.2*. They comprise those numbered 22,35,38,39,43,45,46,48,49,50 and 51. At least three of these have been and probably still are caused by particular construction occupations, and they deserve comment here since the logic of excluding disabled construction workers from claiming consequential benefit is unclear. They are:

- 22 *Ankylostomiasis*—restricted to 'work in or about a mine'. This disease is caused by a hookworm infestation, acquired either by eating infested food or by hookworm penetration through the skin, generally the soles of the feet. The first symptoms include skin eruptions and intense itching. These are followed by anaemia, shortage of breath and bowel upsets. It was once common in Europe among tunnel workers and diggers of trenches, as well as miners.
- 45 *Adeno-carcinoma*—(cancer) of the nasal cavity or associated air sinuses—restricted to workers in factories making wooden furniture. This disease appears to be caused by inhaling air containing fine sawdust in suspension, and to affect other woodworkers, e.g. joiners and woodworking machinists, as well as those making furniture.
- 48 *Occupational hearing loss caused by noise*—restricted to seven clearly-stated occupations. Although the number of occupations has been extended since the last list was published in 1975, the only construction occupation included is that of 'drilling in tunnels with pneumatic percussive tools'. This is surprising when one considers the number of very noisy occupations in construction and the frequently reported loss of hearing among construction workers. The problems of noise and noise-induced hearing loss is considered later in Chapter 2.5.

Of the occupations of the first type, described merely by the material used or handled, or the hazard exposed, some are commonly found among certain construction workers, some only occasionally and some hardly ever. In the notes given in *Table 2.2.2* the word 'unlikely' is used for the last type, and those occupations which might only rarely be involved are shown in brackets.

Of all construction occupations those with the greatest assortment of occasional health hazards are demolition and ground work. Old buildings of all sorts which may contain largely hidden deposits of toxic dusts and infectious germs from previous use and occupancy are constantly having to be demolished. They are often veritable Pandora's boxes. The same applies, perhaps to a lesser extent, to excavations in any but virgin ground.

Occupational diseases in the U.K. 161

The remaining occupational diseases and hazards listed in *Table 2.2.2* against which disabled construction employees appear eligible to claim benefits are discussed in more detail in subsequent chapters under the type of hazard involved—chemical, physical or biological.

2.2.2.2 Pneumoconiosis and byssinosis and occupational deafness

The first two of these diseases are not included in *Table 2.2.2* and all three are dealt with separately in the Regulations[4].

Occupations for which pneumoconiosis is prescribed are covered in Part II of Schedule 1 of the Regulations. Only extracts which may apply to certain construction workers are quoted here:

1. Any occupation involving
    (a) the mining, quarrying or working of silica rock or the working of dried quartzose (quartz) sand or any dry deposit or dry residue of silica or any dry admixture containing such materials (including any occupation in which any of the aforesaid operations are carried out incidentally to the mining or quarrying of other minerals) or to the manufacture of articles containing crushed or ground silica rock.
    (b) the handling of any of the materials specified in the foregoing paragraph in or incidental to any of the operations mentioned therein, or substantial exposure to the dust arising from such operations.

3. Any occupation involving sand blasting by means of compressed air with the use of quartzose sand or crushed silica rock or flint, or substantial exposure to the dust of such sand blasting.

7. Any occupation involving the dressing of granite or any igneous rock by masons or the crushing of such materials or substantial exposure to the dust arising from such operations.

9. Any occupation involving
    (a) the working or handling of asbestos or any admixture of asbestos;
    (c) the cleaning of any machinery or plant used in any of the foregoing operations and of any chambers, fixtures and appliances for the collection of asbestos dust.
    (d) substantial exposure to the dust arising from any of the foregoing operations.

10. Any occupation involving:
    (d) the sawing, splitting or dressing of slate, or any operation incidental thereto.

It would appear from the above that bricklayers (including furnace bricklayers who are particularly prone to silicosis) should be covered under occupations (1)(a) and (1)(b), while stone masons should be covered under occupation (7). Demolition workers might be covered under occupations (1) or (9).

Byssinosis is prescribed only for certain employees in textile factories involved in the spinning or manipulation of raw or waste cotton or of flax. It seems unlikely to affect construction workers.

Those who may claim disability for occupational deafness must have been employed in one of the occupations listed against it (No. 48, *Table 2.2.2*) for a period or periods amounting in the aggregate to not less than 20 years.

There are more occupations for which occupational deafness is prescribed in the 1980 Regulations compared with those of 1975, which included only parts of occupations (a), (b) and (d) of the 1980 list.

The probable reason for the rather restrictive treatment of benefits for occupational deafness in the U.K. is the high incidence of some degree of occupational deafness in the working population and fears that the DHSS would be swamped with claimants if the flood gates were opened wider.

References

1. ILO, *Amendment of the List of Occupational Diseases Appended to the Employment Injury Benefits Convention, 1964, (No. 121)*, International Labour Office, Geneva (1980).
2. DEWIS, M. *The Law on Health and Safety at Work*, MacDonald and Evans (1978).
3. HOLLIMAN, D. *Croner's Health and Safety at Work*, Croner Publications, New Malden, Surrey (1981, looseleaf, updated by subscription).
4. Department of Health and Social Security, *The Social Security (Industrial Injuries) (Prescribed Diseases) Regulations, 1980*, and subsequent Amendments, HMSO, London.

2.3

Health hazards, general

The illnesses suffered by construction workers in the U.K. which result from their work far exceed the number of cases which qualify for DHSS benefit as occupational diseases. This should be clear from the two preceding chapters and from the limited occupational disease statistics quoted in Chapter 1.3.

Enlightened employers are well aware of the situation and try to take a broader view of their workers' health than the mere avoidance of DHSS benefit claims. There are several good reasons why they should do so. First, the number of recognized occupational diseases is steadily rising, and new ones are being added at the rate of one every year or two. Secondly there is an increased public awareness that this or that familiar job-linked illness need not and should not be taken for granted. To change the situation for the better frequently needs new methods and different materials.

One can either try to recognize the situation in advance and plan the changes in an orderly way; alternatively, one can turn a blind eye to it and only make the changes at the last moment under pressure. The first course generally saves money, since changes are invariably less expensive and more effective when anticipated and introduced in an orderly way. It also creates much better relations between the employer and his work-force. This goodwill, while difficult to quantify, usually pays off in a thousand and one ways.

Health hazards have been classified in preceding chapters as chemical, physical and biological. This is by no means the end of the list. Weather, physical and mental stresses are additional health hazards. Chemical health hazards are discussed in Chapter 2.4, man-made physical health hazards in Chapter 2.5, and others (including weather, biological and environmental hazards) in Chapter 2.6.

The borderline between difficult classes of health hazard is not always clear cut. Is rapid decompression following work in a compressed air atmosphere, a physical or a chemical hazard? Should fine sawdust in the air, which can cause cancer of the nose or sinuses in furniture makers and wood workers, be classed as a chemical or a biological hazard?

Different health hazards may also have synergistic effects, so that their combined effect is more serious than the sum of the individual effects. As an example, a mixture of benzene and carbon tetrachloride vapour is far more harmful than either separately, since carbon tetrachloride reduces the capacity of the liver to convert benzene into compounds which are more readily excreted.

Some flexibility in the way health hazards are classified is therefore needed. We should not be surprised to see some hazard classified one day as chemical and another day as biological.

While health hazards, almost by definition, act slowly over a period of time to produce diseases, the same hazard if present at high levels for a short time will produce an acute and immediate effect. This is usually regarded as an accidental injury rather than a disease. A loud explosion close to a man could cause immediate damage to his hearing organs and permanent hearing loss. Exposure of someone to a high local concentration of carbon monoxide gas may cause rapid loss of consciousness and death. Both of these would be treated as accidental injuries rather than occupational diseases.

Physical hazards include such things as noise, laser beams, radiation from radioactive sources, vibration and electric shock. Physical stresses are also caused by exposure to heat and cold, cramped and unnatural working conditions, over-exertion and lack of a smooth working rhythm. Most arise from the man-made working environment, but some arise from weather and the natural environment.

Chemical health hazards arise almost entirely from the materials used or encountered by the worker as 'part of the job'. The material does not have to be classified as a chemical to constitute a chemical health hazard. Cement contains constituents (especially compounds of chromium) which can cause dermatitis.

Most chemical constituents of paint solvents, e.g. xylene, are potential respiratory hazards above a certain concentration in air.

Some chemical health hazards may lurk hidden in the natural environment. Pits and wells have been known to contain lethal concentrations of hydrogen sulphide and carbon dioxide which have seeped in from the ground.

Biological health hazards are living organisms or substances derived from them. The living organisms include micro-organisms, bacteria, fungi, protozoa and viruses, as well as insect and worm parasites and poisonous snakes. They originate in the natural environment. Most of them, such as measles, cholera, fleas and scabies, affect the general population.

A few diseases caused by biological hazards are regarded as occupational. Examples are Weil's disease, which is transmitted by rats and found among sewer workers, and farmer's lung which is caused by the spores of fungi growing on mouldy hay and straw. The first is a particular hazard of tunnel workers and the second of thatchers.

2.3.1 Occupational health professionals

With the growing specialization of the health services and the growing number of new materials made and used, new professions have arisen which deal partly or exclusively with occupational health and diseases.

Some are employed in government service and some within industry. Those serving industry are employed mainly by large manufacturing firms where particular health hazards arise from repetitive or continuous operations and processes.

Very few are employed by the U.K. construction industry. This is partly because of the small size of the average construction firm. There is, however, both scope and need for occupational health professionals within the industry. Other countries, such as France, appear to be in advance of the U.K. in this, as the abstracts quoted in Chapter 2.1 indicate.

The professionals listed below have their own special fields of responsibility but, due to their limited numbers, there is a good deal of overlap in their work. Thus, an occupational physician besides being concerned with the causes, symptoms and cure of the cases he handles will also take a keen interest in controlling the conditions which caused them. The range and scope of work of the different occupational health professionals also varies considerably from country to country.

The two types of professional most concerned with day-to-day problems of the workers and the work place are:

- the occupational health physician
- the occupational hygienist

The former is directly concerned with the health of the worker whereas the latter is primarily concerned with controlling health hazards at the work place.

In addition, there are a number of other professionals involved, including:

the epidemiologist
the dermatologist
the ergonomist
the industrial psychologist
the occupational nurse

All too often the small construction firm and its work force has to manage without the services of any of these professionals. In the medium sized firm the safety officer and welfare officer do their best to cover the field by their own studies and efforts.

Despite the present scarcity of occupational health professionals in the U.K. construction industry, a short description of each of their roles follows. This is partly in the hope of improving understanding of their work and the role they can play in the industry. It is also partly to help managers and safety professionals get some appreciation of what they can do themselves in this field in the absence of full-time professionals.

2.3.1.1 The occupational health physician[1]

Doctors in company employment have suffered many conflicts of interest and conscience, when the interests of the company and the health of their workers seemed to be diametrically opposed. There is little doubt that in the first half of this century when a few companies appointed doctors, some did it mainly to protect the company against claims at common law from industrial diseases. The theme is tense and dramatic and has inspired several books, plays and films.

The picture in the U.K. was improving by 1935 when the Association of Industrial Medical Officers was founded by some doctors anxious to improve the standards and image of the profession. Far sighted employers at the same time realized that it was in their own interests as well as their workers' to employ doctors to study and master the toxicological and other health hazards in their factories. The contribution that health services could make to production was appreciated in the 1939–45 war, when the appointment of doctors was made obligatory in factories over a certain size.

By the 1960s most major U.K. companies were employing full-time doctors whose impartial position and high ethical standards were recognized. Medium sized companies appointed part-time doctors, and group schemes for small companies were started with assistance from the Nuffield Foundation.

Prior to 1973, there were, in the U.K., some 1300 appointed Factory doctors, most of whom were general practitioners with no special occupational health training who carried out various statutory duties such as:

- Examination of young people first entering employment
- Regular examination and in some cases certification and medical supervision of workers in processes covered by Regulations.
- Investigation of gassing accidents and industrial diseases.

In 1973 the Employment Medical Advisory Service (EMAS) was founded under the Department of Employment and staffed by doctors with appropriate qualifications and experience to replace the system of appointed Factory Doctors. Its services are provided free to all interested parties, except for statutory medical examinations for which the employer must pay a fee. It has powers to investigate and advise on any occupational health problem at the request of:

Employers or managers
Employees
Trade Unions
Factory or other Inspectors
Disablement Resettlement Officers
Careers Officers
Teachers or parents

or on its own initiative.

However, in the case of doctors employed within industry, the onus rests with the employer to engage the doctor and supporting staff.

The terms of reference for the post of occupational physician should ensure that:

1. He is responsible for the health of the whole enterprise.
2. He concerns himself with the working environment and its health hazards as well as those working in it.
3. He is accessible to workers for individual consultations and investigates work-related health hazards arising from them.

4. He is a good communicator and teacher and advises and educates management and workers on work-related health problems, and takes a special interest in new processes and equipment, their design and their possible health hazards.
5. In a comprehensive health team, he should be the leader and co-ordinator, whilst recognizing the mastery of other specialists such as occupational hygienists and nurses in their own fields.

The occupational physician is to a certain extent involved with treatment and to a major extent with medical examination of workers.

Regarding treatment, the occupational physician should not duplicate or attempt to take over the role of the general practitioner. Many minor ailments and injuries can, however, best be diagnosed and treated by medical and nursing staff at work, especially when there is a well-equipped first aid room and medical department, with x-ray, physiotherapy and casualty facilities. Fractures following an injury may be speedily diagnosed or excluded if an x-ray machine is available. Minor septic wounds can be treated and dressed and most skin diseases can be best treated in the works medical department.

Medical examinations, whether pre-employment or periodic, must be based on practical needs as well as statutory requirements.

The main statutory medical examinations which apply to construction are for workers in compressed air atmospheres (The Work in Compressed Air Special Regulations 1958, Regulation 14) and in diving (The Diving Operations at Work Regulations 1981, Regulation 7). Most other statutory medical examinations in the U.K. apply to special factory processes which are not normally found in construction, though workers exposed to lead (painters, and welders in special cases) qualify for statutory examination under the Control of Lead at Work Regulations, 1980, as do those exposed to ionizing radiations (radiographers employed in welding inspection) under the Ionizing Radiations (Sealed Sources) Regulations, 1969.

Not all construction firms employing divers and tunnel workers employ an occupational physician, even on a part-time basis. Where they do, he may not have the equipment or training to carry out the full examination required. Where there is no occupational physician or facilities are lacking, the company must make special arrangements for their divers and compressed air workers to be regularly examined and proper records kept.

Examinations required on practical considerations fall into two groups, pre-employment examinations and periodic monitoring tests.

Pre-employment Examination
Priority should be given to examining those workers upon whose fitness and alertness the safety of themselves and other workers depend, i.e., truck and crane drivers, steel erectors, scaffolders, roof workers, steeplejacks and demolition workers. Workers in jobs calling for physical strength and full function of limbs and certain senses also require pre-employment examination. Plasterers, bricklayers, painters and decorators might be included in this category. The tests should be related to the man's work. For example, painters and decorators might well be tested for colour vision. (Surprisingly, some very talented artists have been blind to certain tone differences.)

Routine Examinations
These are most needed for workers in jobs carrying a specific health hazard. The examination should concentrate on monitoring the possible effects of the hazard on the worker: chest x-ray and lung function tests where there is a pneumoconiosis hazard, an audiogram for workers in noisy jobs, a blood count for workers exposed to ionizing radiations and blood and/or urine tests for lead and other metals in painters, welders, etc. In the case of ionizing radiation and lead exposure, the requirements are statutory.

Needless to say, all workers in jobs requiring routine examinations should be examined before employment as well. The initial examination provides a base line against which subsequent developments can be checked.

There are some concerns where a medical examination of every new employee is expected. Unfortunately such examinations have little predictive value on the person's subsequent work performance and absenteeism through sickness. These are determined more by social and psychological factors which will not be revealed by a medical examination. Thus one investigator found that the best attendance records were held by registered disabled persons!

It follows from the foregoing that there should be a clear purpose behind all medical examinations of workers, both pre-employment and routine monitoring. The examinations should be appropriate to the jobs performed and to their possible health hazards.

2.3.1.2 *Occupational (or Industrial) Hygienists*[2,3,4]

The profession of industrial hygiene (as it is called in the U.S.), received great impetus in 1970 with the passing of the Occupational Safety and Health Act. Many of our ideas and particularly our standards on industrial hygiene in the U.K. are 'technology imports' from the U.S.A. The following definitions, approved by the American Industrial Hygiene Association, give a clear appreciation of the subject and the profession:

> "Industrial Hygiene is that science and art devoted to the recognition, evaluation and control of those environmental factors or stresses, arising in or from the work place, which may cause sickness, impaired health and well-being, or significant discomfort and inefficiency among workers or among the citizens of the community."

> "An Industrial Hygienist is a person having a college or university degree or degrees in engineering, chemistry, physics or medicine or related biological sciences who, by virtue of special studies and training has acquired competence in industrial hygiene. Such special studies and training must have been sufficient in all of the above cognate sciences to provide the abilities:
>
> (a) to recognise the environmental factors and stresses associated with work and work operations and to understand their effect on man and his well-being;

(b) to evaluate, on the basis of experience and with the aid of quantitative measurement techniques, the magnitude of these stresses in terms of ability to impair man's health and well-being, and
(c) to prescribe methods to eliminate, control or reduce such stresses where necessary to alleviate their effects.

The categories of stresses which interest the industrial hygienist are basically all the health hazards considered in this and subsequent chapters,

(a) chemical—liquid, dust, fume, mist, vapour or gas
(b) physical—electromagnetic and ionising radiations, noise, vibration and extremes of temperature and pressure;
(c) biological—insects, mites, moulds, yeasts, fungi, bacteria and viruses; and
(d) ergonomic—body position in relation to the task, monotony, boredom, repetitive motion, worry, work pressure and fatigue.

'The industrial hygienist thus links the manufacturing or construction operations of a company to its medical department and helps the physician to correlate the worker's condition and complaints with the health hazards of his work."[2]

There are probably very few occupational (or industrial) hygienists employed by U.K. construction companies. This gap is partly filled by company doctors and safety professionals. Where a particular occupational health problem is suspected which the firm cannot solve unaided, it should seek advice from EMAS or an occupational hygiene consultant. A survey is usually required before detailed recommendations can be given.

A list of U.K. organizations offering industrial hygiene services on a fee-paying basis is given in *Table 2.3.1*.[3] These are mostly associated with universities. They will undertake surveys of worker exposure to a wide range of chemical and physical hazards. The Table does not include consultants who offer specialized services in one or two fields only (e.g., noise, vibration, asbestos or welding fumes).

To evaluate health hazards in the working environment first requires detailed standards on tolerable levels of such hazards and, secondly, means of measuring the levels at which they are present.

Most of our standards of tolerable levels are based on the work and publications of the American Conference of Governmental Industrial Hygienists, P.O. Box 1937, Cincinnati, Ohio. This organization sets and annually reviews limits for the intensities of physical agents and concentrations of chemical substances (so called) in the working environment. Up to 1980, these values were reprinted by the U.K. Health and Safety Executive and published in a series of Guidance Notes[5]. From 1983, the HSE is publishing its own values under the title *Occupational Exposure Limits*. Lists are also published by the International Labour Office and by a number of countries.

There is considerable variation in the limits considered acceptable in different countries. Most limits have hitherto been quoted as 'Threshold Limit Values', known as TLVs, for continuous exposure during an 8 hour

TABLE 2.3.1. List of hygiene consultants

The following are organisations offering industrial hygiene services on a fee-for-service basis. They will undertake measurements of worker exposure to chemical contaminants such as dust, fumes, vapour, gases, mists (oil etc.) and also to physical stresses such as noise, light, heat, etc. Source: Jones, A.L., Hutcheson, D.M., Ward Dymott, S.M., *Occupational Hygiene, An Introductory Guide*, Croom Helm, London (1981).

(1) North of England Industrial Health Service,
    20 Claremont Place,
    Newcastle upon Tyne.
    (In association with the University of Newcastle upon Tyne)
(2) Environmental Health Service,
    Level 5,
    Medical School,
    Ninewells,
    Dundee,
    Scotland.
    (In association with the University of Dundee)
(3) National Occupational Hygiene Service Ltd,
    12 Brook Road,
    Fallowfield,
    Manchester 14.
    (Formerly in association with the University of Manchester)
(4) TUC Centenary Institute of Occupational Health,
    London School of Hygiene and Tropical Medicine,
    Keppel Street (Gower Street),
    London WC1.
    (In association with the University of London)
(5) Occupational Health and Safety Group,
    Department of Epidemiology and Community Medicine,
    Welsh National School of Medicine,
    Heath Park, Cardiff CF4 4XN.
    (In association with the University of Wales)

working day for 5 days a week. Ceiling values which should never be exceeded, and short-term exposure limits for 5, 15 or 30 minutes have also been published for a number of substances[5]. The new 'Occupational Exposure Limits' (OEL) which replace TLVs for airborne chemical substances in the U.K. are discussed in Chapter 2.4. Limits for physical agents to which construction workers are liable to be exposed are given in Chapter 2.5.

OELs or TLVs are not legally defined values but represent levels for good working practice which would be enforceable under section (2) of the Health and Safety at Work, etc. Act of 1974. This states:

'It shall be the duty of every employer to ensure so far as is reasonably practicable, the health, safety and welfare at work of all his employees'.

TLVs or OELs have so far been set for only about 5% of all chemicals manufactured. Many have been revised drastically since they were first set, usually downwards, as new toxicological information has become available. Thus the TLV for benzene was reduced progressively from 100 to 5 ppm when its effects on the blood and bone marrow and its possible carcinogenity were fully appreciated. However, unless the intensities of physical agents (e.g., sound) and the concentrations of chemical substances in the working environment are actually measured and monitored, discussion of TLVs (or OELs) tends to be unrealistic and academic. The means are available and within the reach of most construction firms.

For sampling the air and measuring concentrations of airborne contaminants, several firms offer compact apparatus which can be worn by a worker without interfering with his job. These generally consist of three parts, a sampling pump, a sampling head and a detector tube. The sampling head is worn on the lapel of the worker's overall and air is drawn by the pump through the detector tube from the worker's breathing zone (see Chapter 2.7).

Industrial hygienists are specially trained in the use of such equipment, as well as apparatus for measuring intensities of noise, electromagnetic radiation of harmful wavelengths, heat stress and other physical and biological health hazards. These skills can also be learnt by safety professionals with a basic knowledge of chemistry and physics.

Tests such as these are mainly used in routine manufacturing operations which vary little from day to day. They are more difficult to apply and interpret in construction operations, where the work and weather conditions may change frequently. Nevertheless, they are valuable in construction when a particular health hazard is suspected.

Another difficulty with construction is that many of the materials used—paints, glues, chemical cleaning agents, etc.—are known only under trade names. Nevertheless, manufacturers of these materials are generally prepared to give data sheets with chemical compositions to bona fide users if requested to do so on health grounds. The obligations of manufacturers of substances for use at work are covered in Section 6 of the Health and Safety at Work etc. Act 1974.

The work of the occupational hygienist in monitoring health hazards in the working environment complements that of the occupational physician and nurse in monitoring the health of the worker and his intake of hazardous materials such as lead, discussed under Section 2.3.1.1.

As well as monitoring the working environment, the occupational hygienist is trained to recommend and initiate remedies. He should have sufficient knowledge of the subject matter to discuss it on equal terms with those whose job is to apply it. Remedies are basically of three types and are applied in the order given below:

- find a satisfactory, non-hazardous substitute for the hazardous material or physical agent. Thus a water-based emulsion paint might be substituted for one with a toxic solvent, or fibre glass insulation might be substituted for asbestos.
- employ suitable means for reducing the hazard intensity in the working area. Thus sanding and grinding tools may be provided with local exhaust ventilation, and sound insulating mountings and barriers used to reduce exposure to noisy machinery.
- provide personal protective devices, such as ear plugs, dust masks, respirators and ensure that these are used by the worker.

2.3.1.3 Nurses, first-aiders and other voluntary workers

Few construction companies have any full-time health workers on construction sites. Only on sites employing a large number of men would a full-time first-aider or even a first-aid room be expected. The treatment given on most sites is therefore limited to first-aid treatment by trained and certified part-time first-aid workers. Under the Health and Safety (First-Aid) Regulations 1981, employers in all industries including construction are obliged to have, in specified circumstances, workers with first-aid training and to provide and maintain properly equipped first-aid boxes at every site where workers are employed.

An Approved Code of Practice and Guidance Notes for these Regulations are available and should be consulted for detailed requirements. The first-aider is more concerned with treatment of accidental injuries than those disabled by occupational diseases. He should, however, appreciate hazardous conditions likely to injure health, such as excessive noise, high dust concentrations and the use of toxic and corrosive chemicals and solvents. He should take what steps lie in his power to ensure that exposure is minimized and his fellow workers adequately protected. The fact that he is a certificated first-aider should give his statements and advice a certain standing both among management and his fellow workers.

The first-aider should check, and where possible arrange that along with first-aid boxes, proper washing facilities, including means for irrigating eyes in case of accidental entry of particles, paints, etc., are available and always ready. First-aiders in the construction industry should have basic knowledge of the special hygiene needs of the industry, including skin care, for preventing allergies and dermatitis.

Special stretchers which enable injured workers to be secured and lowered from heights or raised from excavations should also be available, each with at least two blankets, on every construction site. Contingency plans for dealing with accidents should be made before work commences. Supervisors and part-time first-aid workers should take the lead in this. Certain men should be briefed and trained in advance to act as stretcher bearers, so that they know what to do immediately an accident requiring the use of a stretcher occurs.

On sites where there is a risk of asphyxia, e.g., from gassing or electric shock, appropriate means of resuscitation should be kept available, for instant use. The equipment must be maintained regularly and persons must be trained to use it.

Trained nurses when employed by construction companies, can take on many of the duties discussed under 'Occupational Physicians', both in medical examination of employees (e.g., of hearing and vision), in treatment of injuries and some diseases, especially of the skin, and in relating illnesses to particular health hazards at work.

Reports should be made and records kept of all accidents and illnesses which appear to be related to work. These reports should include sections to be written by the supervisor and by the injured or disabled person. Section 140 of the Factories Act 1961 requires a register to be kept for every factory where various details including those of industrial diseases have to be recorded. Further duties to report accidents and dangerous occurrences are set out in The Notification of Accidents and Dangerous Occurrences Regulations 1980. When nurses are available to treat the injured or disabled person they are usually in the best position to handle these reports and to ensure that all relevant details provided by the injured person, especially in regard to causation, are recorded. The nurse will normally keep one copy of the injury report on her own file, distribute other copies according to a pre-arranged system, and act as a clearing house for the information.

2.3.1.4 Other occupational health professions

On the medical side, a number of specialists may be brought into play to treat and rehabilitate victims as well as in a preventative capacity. Only those engaged in the latter capacity are considered here.

Epidemiologist
The work of the epidemiologist was discussed briefly in Chapter 2.1 under 'Occupational Mortality'. Every occupational physician and most occupational nurses have an epidemiologist's hat which they don from time to

Figure 2.3.1 '... the dermatologist has to be a bit of an epidemiologist and a bit of a psychologist as well'.

time. Full-time epidemiologists are more concerned with the frequency and distribution of diseases in defined populations than with day to day medical practice. The main uses of epidemiology are:

1. to quantify known hazards, e.g., the relationship between dermatitis among form-workers and the chromium content of cement.
2. to search for causes, e.g., relationship between cancer of the nasal cavity and exposure to sawdusts.

Dermatologist
The dermatologist[6] is primarily concerned with diagnosing, treating and preventing skin diseases, many of which are caused by allergic reactions to particular substances. Many skin disorders are also caused or aggravated by psychological causes. Thus the dermatologist has to be a bit of an epidemiologist and a bit of a psychologist as well.

In diagnosing allergens in the working environment that bring out skin complaints among particular workers, he can be greatly assisted by observations made by the workers themselves, supervisors, safety professionals and company nurses and physicians.

Ergonomist
The ergonomist[7] is concerned with the relationship between man and his working environment. There are few full-time ergonomists, since the subject is an inter-disciplinary one which involves anatomists, physiologists, psychologists, occupational physicians, industrial hygienists, design engineers, work study engineers, architects, illuminating engineers and others. The Ergonomics Research Society, founded in the U.K. in 1949, brought together workers in these various fields who were concerned with different aspects of the subject. Ergonomics has been increasingly applied over the past forty years to improve performance in industry, reduce human stresses and reduce or eliminate hazards at the man-machine interface.

The hard core of ergonomics is concerned with the matching of tools, seats, working positions, controls and instrument panels to the worker. In this it has many applications in construction where so much work is done, apparently because no better way has been found, in cramped, confined and uncomfortable working positions.

In its broader aspects, ergonomics is concerned with the thermal environment, heating, air conditioning and ventilation, with vision and illumination, with automation, with hearing, noise and vibration, with hours of work and shift work. In the U.S.A. ergonomics is generally known as 'Human Engineering'. The subject in its widest sense is a very broad one which overlaps many of those discussed in this book.

Industrial Psychologist
The industrial psychologist[8] is concerned with human relationships in industry and with mental stresses in individual workers. Some of the basic tasks in which psychologists are employed are:

- staff selection, starting with the application of psychological tests for new applicants.
- job analysis, to ensure that the mental demands of the job fit the characteristics of the person selected.
- staff training, to supplement the academic training of staff and help them to adjust to the rough and tumble of their work.
- motivation, to study the factors in a job which positively motivate workers, e.g., achievement, recognition, the work itself, responsibility and advancement, and those which cause dissatisfaction, e.g., company policy and administration, supervision, salary, inter-personal relationships and working conditions.
- job enrichment, often through quite simple means, such as involving workers in making suggestions and decisions.
- intervention in personal crises, caused by emotional reactions to hazardous situations and changes. This usually takes the form of skilled counselling, taking the subject's personality into account.
- stresses in management and employees arising partly from the physical and mental conditions of their work. Most people work best at a certain, fairly low, stress level, but suffer mentally and physically

when exposed to excessive stress. Whole managements can suffer in this way and behave irrationally, leading to serious accidents. Mental stresses can induce or aggravate illness such as heart disease, dermatitis and asthma. The industrial psychologist by his training can sometimes analyse the causes of mental stresses and recommend remedies.
- management of mental ill health and employment and rehabilitation of the mentally ill.
- management of alcoholics, drug addicts and epileptics.

2.3.1.5 Information sources and training in occupational health (see Reference 12)

The following sources of information on occupational health and health training in the U.K. should be useful. Additional sources were given earlier in *Table 2.3.1*.

1. The Society of Apothecaries
   Apothecaries Hall
   Blackfriars Lane
   London EC4

grants diplomas in Industrial Health and can advise students on suitable full and part time courses.

2. The Medical Adviser to the TUC
   Congress House
   Great Russell Street
   London WC1
   Tel: 01-636 4031

advises unions on occupational health problems.

3. The Health & Safety Executive and
   The Employment Medical Advisory Service
   Baynards House
   1–13 Chepstow Place
   Westbourne Grove
   London W2 4TF
   Tel: 01 229 3456

will advise on all aspects including legislation.

4. The Institute of Occupational Medicine
   Roxburgh Place
   Edinburgh
   EH8 9SU
   Tel: 031 667 5131

5. The Society of Occupational Medicine
   Royal College of Physicians
   11 St Andrews Place
   London NW1
   Tel: 01-486 2641

6. The British Occupational Hygiene Society
 Hon Secretary Mr T Anderson
 Esso Research Station
 Abingdon, Berks
 Tel: 0235 21600

7. The Institute of Occupational Safety and Health
 222 Uppingham Road
 Leicester LE5 0Q6
 Tel: 0533 768424

2.3.2 Modes of entry and attack[1,3]. Target organs, allergies

Only the barest outline of the way health hazards affect the human body can be given here. For a clear and fuller account in layman's terms, Stellman and Daum's *Work is Dangerous to your Health*[9] can hardly be beaten.

Chemical and biological hazards enter the body in one of three main ways, in the air breathed (the 'inhal' route), through the skin (the 'dermal' route) and through the mouth (the 'oral' route). They sometimes attack the body at the point of entry, but many pass into the bloodstream or alimentary canal and concentrate their attack on other organs, in particular the lungs, the liver, the kidneys, the blood, the nervous system and the heart.

Physical hazards tend to attack their target organs directly. Some affect the whole body. Excessive noise damages the hair cells (otiliths) of the inner ear, vibration of low frequency affects the organ of balance (semi-circular canals); x-rays and ionizing radiation can cause damaging chemical changes in most parts of the body; UV light affects the skin and eyes, and lasers as used industrially are primarily a hazard to the eyes.

An allergy is a hyper-sensitive condition in one or more parts of the body, mainly the skin and respiratory system, to minute traces of specific types of foreign matter which produce reactions out of all proportion to their dosage. They are specific both to the individual and to the particular foreign matter or allergen, so that one person may suffer acutely, whilst another is quite unaffected.

The entry routes and target organs are discussed below with only brief reference to some of the chemical and biological hazards involved. These are discussed in more detail in Chapters 2.4 and 2.6.

2.3.2.1 *The inhal route and respiratory hazards*[10]

The respiratory system enables air to come into close contact with the blood. It allows oxygen to pass from the air into the blood and carbon dioxide to pass back from the blood to the air. The oxygen combines with haemoglobin in the blood and is conveyed to various parts of the body where it reacts on demand with 'fuels' in the body to supply heat and energy and to form carbon dioxide.

The human system is shown diagrammatically in *Figure 2.3.2* and consists of an upper and a lower part. The upper part consists of the nose,

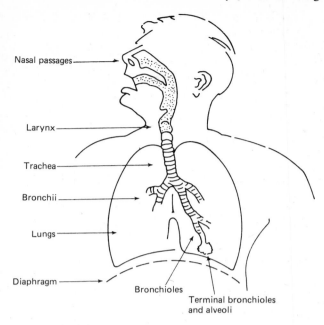

Figure 2.3.2 The human respiratory system.

mouth and larynx, and the lower part consists of the trachea, bronchus, bronchioles and alveoli, the last two of which are the main components of the lungs. These have a light porous and spongy elastic texture. The bronchioles which are air passageways, divide into progressively smaller tubes until they reach a diameter of about 1 mm. From then on, they lead into a vast number of blind cup like pouches termed air cells or alveoli. These have a diameter of about 0.3 mm and there are about 700 million of them in the human body. Their walls which are very thin and elastic have an enormous surface, and they are permeable to gases and vapours which pass easily through them into and out of the blood stream. The air expired contains 16 to 17% by volume of oxygen and 4 to 5% by volume of carbon dioxide.

The human respiratory system has a wide range of working capacity, from about 5 litres/minute for a person at rest to about 50 for someone engaged in heavy bodily exertion. It is impossible to breathe entirely through the nose at rates much in excess of 12 litres per minute.

The respiratory system presents the easiest route for air contaminants to enter the human body and airborne dusts are readily trapped in it, in many cases causing temporary or permanent damage. Pollens and other particles to which a person is allergic (see Section 2.3.4) may cause the bronchioles to contract, thus restricting breathing (hay fever and asthma), while in bronchitis, the tubes are obstructed by mucus. Generally these particle contaminants dissolve in the protective fluids in the lung and are rendered harmless. However, some fine particles such as crystalline silica and asbestos fibres which enter the alveoli are practically insoluble in these fluids. They thus remain and can react with the cell walls causing them to become fibrotic, harden and lose their flexibility.

The respiratory system has various forms of natural protection against harmful air contaminants, including a wet filter formed by the fine hairs (cilia) of the nose and trachea. These are irrigated by mucus which enables trapped particles to be eliminated by sneezing, blowing the nose, spitting or swallowing. This natural protection is quite effective for most coarse visible particles at low respiration rates when a person is breathing through the nose. It is far less effective though for very fine invisible particles, and it is ineffective even for coarser particles when a person is engaged in heavy exertion and breathing through the mouth.

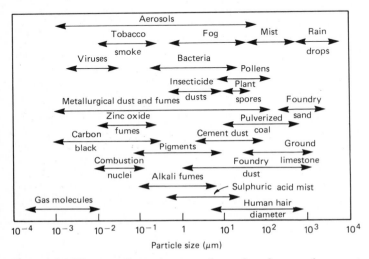

Figure 2.3.3 The approximate size range of a number of commonly encountered airborne materials (from *Occupational Hygiene*, edited by H.A. Waldron and J.M. Harrington, Blackwell Scientific, 1980)

The approximate size ranges of a number of common airborne materials are shown in *Figure 2.3.3* in microns. One micron (symbol μ or μm) is a thousandth of a millimetre. By way of example, a human hair has a diameter of about 80 microns, while the smallest particle usually seen by the naked eye has a diameter of about 40 microns. 'Coarse' particles with diameters above 10 microns are mostly filtered out in the nose and trachea. Smaller particles which reach and become trapped in the alveoli have a size range of 0.2 to 7 microns while particles smaller than 0.2 microns tend to remain in the air and are exhaled.

Fibrous airborne particles, such as asbestos, are less effectively removed in the upper respiratory system than compact particles and are more likely to reach the alveoli.

Some air contaminants have their main effects on the respiratory system, whilst others can pass through the very thin walls of the alveoli to dissolve in the bloodstream. The latter include most gases and vapours, and finely divided solid compounds of many metals (e.g., lead, mercury and manganese). Once there, they or their metabolites (i.e., compounds produced from them by metabolism) are readily transported to other target organs, their effects on which are discussed later.

Contaminants which attack the respiratory system directly fall into four groups:

- reactive gases, such as ozone, chlorine and formaldehyde which cause intense irritation of the upper respiratory system and swell the walls of the airways;
- allergens, such as isocyanates, chromates and some wood dusts, which cause bronchial restriction in sensitive persons and also asthma;
- most inhaled mineral dusts are deposited in the lungs. Some, such as iron oxide, merely tend to block the passageways and reduce breathing capacity. Others cause harmful changes in the lung structure, resulting in silicosis, asbestosis, etc.
- some airborne contaminants lead eventually to the growth of malignant and usually fatal tumours in the respiratory system. Blue asbestos (crocodolite), compounds of chromium and arsenic and some complex hydrocarbons found in tars have this effect.

While considering the respiratory system, we should not forget the olfactory organ of the nose which is connected by special nerves to the brain and enables us to smell, thus warning us of the presence of many gases and vapours. Unfortunately, this organ is readily deactivated by prolonged exposure to toxic gases and vapours such as hydrogen sulphide, so that it ceases to warn when there is a dangerous concentration. This deactivation is rarely permanent, and the sense of smell recovers after a period in clean air.

2.3.2.2 The dermal route and skin hazards

Many toxic liquids and vapours pass readily through the skin and thereby enter the bloodstream. They include benzene, toluene, nicotine, trichloroethylene, tetraethyl lead and several insecticides and herbicides and paint solvents.

Once in the bloodstream they attack the same target organs and have much the same effects as if they had entered via the 'inhal route'. The target organs and the way in which they are affected by contaminants reaching them through the bloodstream are discussed in subsequent sections.

The skin itself is also very vulnerable to damage or irritation from external agents, encountered at work. In 1977, 65% of all industrial injury benefits paid out in the U.K. were for non-infective dermatitis. The skin is also vulnerable to disorders originating elsewhere in the body, as well as to infestations (lice, fleas, scabies, etc) and psychosomatic disturbances which are beyond the control of the conscious mind (e.g., blushing). These may also intensify the symptoms of some existing skin disorder whether occupational or not. A typical case of psychosomatic disturbance quoted in a medical book[6] was a man whose eczema always got worse whenever he went to Leeds. His mother-in-law lived there!

Three key questions therefore have to be asked regarding skin disorders which may have an occupational cause:

1. Is it dermatitis or not? (i.e., an inflamed skin which may be red, swollen, blistered, oozing, scaled, cracked or thickened).
2. If dermatitic, is it primarily caused by external factors in contact with the skin or not?
3. If an external factor is involved, is it a skin allergen (sensitizer) or a skin irritant?

The diagnosis of specific allergens often involves patch tests, which are beyond the scope of the safety specialist. He should, however, be in a position to know most of the substances with which a worker has come in contact. This can be of great help to medical people in diagnosing the irritant or allergen which is causing the worker's dermatitis.

Strong sunlight and ultraviolet light (from welding) also cause skin reactions in many people. (These can often be controlled by a UV barrier cream). Others, however, develop allergies to certain chemicals only in the presence of UV or strong sunlight.

Construction supervisors and safety specialists have an important part to play in preventing chemicals and other substances which are irritants and allergens from coming into contact with workers' skins. Solvents for paints, sealants, etc., remove protective fats and create dry and cracked skins which then become more vulnerable to allergens and other irritants. Specific skin irritants and allergens encountered in construction are discussed in subsequent chapters.

2.3.2.3 The oral route

The oral route has the best natural protection (intestines, liver, kidneys) against harmful substances, and it is also the easiest to control and protect during work exposure. This is a simple matter of hygiene. Substances and articles encountered during construction should not be eaten or drunk or put in the mouth. Food and drink should be kept well separate from such substances and not allowed to be contaminated by them in any form. Food and drink should always be kept well covered in vermin-proof boxes or containers. A half-eaten sandwich left lying about is an invitation to rats and can lead to diseases such as Weil's disease. Hands and faces should be washed well before eating or drinking.

2.3.3 Other target organs of the body

The respiratory system and the skin have been discussed both as modes of entry and as target organs for harmful substances. Other principal target organs for attack are:

the liver
the kidneys
the blood itself
the nervous system
the eyes

Other target organs of the body

These are discussed in the following sections as well as other organs such as the heart and ears which are less readily affected.

2.3.3.1 The liver

The liver is the major detoxifying organ in the body. Since it has a large supply of blood from the lungs, most chemicals entering the body eventually land up in the liver. There they are converted into forms which can be removed from the body in the urine or faeces.

Whilst the liver is relatively resistant to toxins, it can be overloaded by massive doses and become inflamed (hepatitis) and/or diseased causing jaundice. It is also seriously affected by chlorinated hydrocarbons, including insecticides such as DDT. Vinyl chloride monomer, produced when PVC is heated to the point of decomposition can cause a rare malignant liver tumour known as angiosarcoma.

Many compounds entering the body by one of the routes described earlier are converted or metabolized to other compounds in the liver. These may be more or less toxic than the original compound, but they are usually more soluble and hence more readily secreted in the urine. As an example benzene is converted to phenol and other water soluble compounds in the liver.

2.3.3.2 The kidneys

The kidneys are delicate organs which maintain the levels of water and salt in the body as well as the pH (acidity/alkalinity) of the blood. They are damaged by compounds of mercury, cadmium and lead and some organic solvents, particularly carbon tetrachloride, and cease to function properly. The result may be inability to urinate, swelling of the body tissues, anaemia and in severe cases coma and death.

2.3.3.3 The blood

The blood with the heart, veins and arteries is the main transport system of the whole body. Its components are:

- *red blood cells* contain haemoglobin which carries oxygen to the tissues.
- *white blood cells* make antibodies and fight infections.
- *platelets* initiate blood clotting after a wound.
- *plasma* contains numerous proteins containing fats, minerals and vitamins and transports fuel and building materials to all parts of the body.

The red and white blood cells and the platelets are produced in the bone marrow cavities. The red blood cells and the bone marrow are very vulnerable to carbon monoxide, lead compounds, benzene and other toxic compounds.

Carbon monoxide combines with haemoglobin in the blood thereby preventing it from carrying oxygen. Lead compounds, benzene and other chemicals cause red blood cells to burst and release haemoglobin molecules into the blood, which causes malfunction of the kidneys.

Benzene also affects the bone marrow, interfering with the production of blood cells. It sometimes causes cancerous blood cells to be formed, resulting in leukaemia.

2.3.3.4 The nervous system

This consists of the central and peripheral systems over which we have some control, and the autonomic system which controls heartbeat and other reflex actions.

The vapours of many paint solvents are mild narcotics. After passing via the lungs into the blood stream, they reach and affect the central nervous system, dulling the senses.

The incidence of glue sniffing among deprived children and adolescents, sometimes with fatal consequences, testifies to their danger. Other air contaminants, mercury vapour and compounds of lead and manganese, as well, of course, as alcohol, can produce hallucinations, loss of memory and erratic and irrational behaviour.

2.3.3.5 The eyes

The eyes are easily injured through a variety of causes. Such injuries fall more in the category of accidents than diseases. According to Guidance Note MS11, accidents at work are the main cause of eye injuries; 75% of these are caused by particles from grinding, turning, drilling, etc; 15% are caused by burns from welding flashes, acids and chemicals and hot materials; 10% are caused by abrasions and blows.

Irritating gases, such as chlorine, at a concentration of 3 ppm by volume in air, formaldehyde (20 ppm), ozone (1 ppm) and ammonia (140 ppm) painfully affect the cornea and conjunctiva. The vapours of most volatile paint and glue solvents and chemicals used in construction have the same effect at concentrations from 50 to 500 ppm by volume in air. Some of these in order of decreasing effect are cyclohexanol, turpentine, styrene, xylene, amyl butyl and ethyl acetates, methyl ethyl ketone and acetone.

Corneal distrophy (fatty degeneration of the cornea) is caused by exposure to the fumes of coal tar pitch.

Optic neuritis (inflammation of the optic nerve in or behind the eyeball) is caused by the vapour of some chemicals which may affect the eye directly, or enter the body through one of the channels described previously. These include alkyl tin compounds (used as fungicides in paints, etc), benzene and tricresyl phosphate (a plasticizer sometimes found in PVC and other plastics).

Nystagmus (involuntary rolling of the eyes) an occupational disease of miners, is caused mainly by poor illumination, though it can also be induced by carbon monoxide and some chemicals.

2.3.3.6 Other organs

The *cardiovascular system* (heart and blood vessels) is fairly immune to direct attack, although chemical damage to other organs may cause hypertension and hence coronary disease.

The *male genitals* are also very sensitive to most of the gases and vapours which irritate the eyes and upper respiratory system. Impotence may be caused by heavy exposure to nitroglycerine (used in explosives) and other chemicals, although this does not appear to be a serious occupational disease of construction workers.

The *ears (including the inner ear)* are at risk from excessive sound but are not usually considered as special target organs for chemical hazards.

2.3.4 Allergies[6,11]

Allergies were briefly mentioned earlier under Section 2.3.2. They appear to be caused by a malfunction of the body's natural defence system due to a number of foreign materials, known as allergens. These cause the body to produce and release locally excessive quantities of a chemical, histamine, which has a marked physiological effect causing contraction of the muscles and other reactions.

Allergic reactions are manifested in various ways—in skin disorders such as dermatitis, eczema and nettlerash, in respiratory disorders such as hay fever, asthma, chronic bronchitis, sinusitis, and others. Some gastro-intestinal disorders, arising after eating a particular food (e.g., a cheese or strawberries) or taking a particular drug (e.g., aspirin) are allergic reactions, as is migraine.

Allergic reactions range from mild to fatal. Many allergens are natural and unrelated to occupation, but a number of man-made substances used in construction are allergens. The individuals in whom these produce reactions have usually been sensitized by exposure to them a number of times. A graphic example is quoted below:[6]

> "He had been born in Iceland and had spent most of his 54 years fishing in the North Sea, where Dogger Bank Itch is well recognised amongst sailors. It is due to an allergy to the 'sea chervil' and occurs upon contact with the seaweed that is caught up in the nets. The dermatitis quickly clears on shore but each successive attack is worse. There is weeping and swelling of the exposed areas of the hands and forearms and puffiness of the skin around the eyes; he tells a tale of having to prop his eyes open with matchsticks in order to see the compass! . . The true degree of his sensitivity was shown by his story of waking one night in his captain's cabin with irritation and swelling of the skin around the eyes from the minute amount of the allergen in the air; on looking at his chart he found he had been crossing the North Dogger."

Once a person has become sensitized to an allergen, the condition is often permanent and it is usually then advisable to avoid similar exposure in the future, even if this means change of occupation or employment.

Several chemicals and other materials encountered in construction are allergens. These include:

Toluene di-isocyanate (component in two packs polyurethane kits)
Cement dust (believed due to chromium compounds)
Coal Tar products
Some detergents

Some dyestuffs
Some disinfectants, including iodine, lysol and formalin
Mineral oils, especially motor oil
Some resins and glues, particularly fish glues
Dusts from some tropical woods

Many construction occupations have higher-than-average exposure to allergens. They include:

Carpenters, joiners, formworkers
Demolition workers
Painters
Roof workers
Tilers
Flooring layers
Concreters
Asphalters
Labourers

Two phenomena connected with sensitization should be mentioned, cross sensitization and photosensitization.

Cross sensitization means that a person who has become sensitized to a particular chemical by exposure to it may become automatically sensitized to other closely related chemical compounds.

Photosensitization means that a particular allergen only causes a reaction when the person is exposed to sunlight. Creosote is an example. Thus a sensitized person who gets a splash of creosote on his skin may feel no ill effects until he goes out in the sun when severe inflammation and dermatitis result.

Many people's skin becomes sensitized to strong ultra violet light alone, without any apparent chemical agent.

Some cases of 'Monday morning feeling' are the result of allergies. This is particularly found in byssinosis victims among cotton workers.

References

1. TYRER, F.H. and LEE, K., *A Synopsis of Occupational Medicine*, John Wright, Bristol (1979).
2. Chapter 37, 'Industrial Hygiene' in *Accident Prevention Manual for Industrial Operations*, National Safety Council, 444 North Michigan Avenue, Chicago, Illinois 60611, USA (periodically updated).
3. JONES, A.L., HUTCHESON, D.M.W. and DYMOTT, S.M., *Occupational Hygiene, an Introductory Guide*, Croom Helm, London (1981).
4. WALDON, W.A. and HARRINGTON, J.M. (editors), *Occupational Hygiene*, Blackwell Scientific, London (1980).
5. HEALTH AND SAFETY EXECUTIVE, *Guidance Note EG 15/8 Threshold Limited Values* (1980) and *Guidance Note EG 40/83, Occupational Exposure Limits* 1983), HMSO, London.
6. SEVILLE, R., *Dermatological Nursing and Therapy*, Blackwell Scientific, London (1981).
7. MURRELL, K.F.H., *Ergonomics*, Chapman and Hall, London (1979).
8. TIFFIN, J. and MCCORMICK, E.S., *Industrial Psychology*, 7th ed., Allen & Unwin, London (1981).
9. STELLMAN, J.M. and DAUM, S.M., *Work is Dangerous to your Health*, Random House, Transatlantic Book Service, (1980).
10. SCHRATER, R.C. and LEVER, M.J., *The Deposition of Materials in the Respiratory Tract*, Chapter 4 in Ref. 4.

11. GRAHAM-BONNALIE, *Allergies*, David and Charles, Newton Abbott (1970).
12. GAUVAIN, S., *Occupational Health: A Guide to Sources of Information*, Heinemann Medical, London (1974).

See also the following *Guidance Notes* published by the Health and Safety Executive:

MS 18 *Health surveillance by routine procedures*
MS 20 *Pre-employment health screening*
EH 18 *Toxic substances: a precautionary policy*
EH 22 *Occupational skin diseases; health and safety precautions*

2.4
Chemical health hazards

The effects of toxic substances on the health of construction workers were seen in Chapter 2.1 and some occupational diseases caused by them were discussed in Chapter 2.2. The modes by which they enter the human body and the target organs they attack were discussed in Chapter 2.3. Here we focus attention on the substances which can damage health by their chemical action on one or more organs of the body.

Whilst some substances have acute chemical or toxic effects and are known as 'toxins', it must be realized that all substances can harm the human body if taken in excess. This was put aptly by Paracelsus 450 years ago: "All things are poisons, for there is nothing without poisonous qualities. It is only the dose which makes a thing a poison". At the same time, many substances which are toxic in quite small doses, particularly metals such as copper and manganese, are necessary to health as trace elements in even smaller doses. (These traces are generally present in the diet.) Hence we take a broad view of substances which may harm the body chemically.

All matter is made up of chemical elements, whether in the elemental state like oxygen, nitrogen or lead, or in combination with other elements as compounds, such as titanium dioxide or calcium carbonate. Most substances encountered in construction are complex mixtures of chemical compounds. The principal toxic effects are often due to small concentrations of some toxic element or compound in a mixture of compounds of low toxicity. Thus most of the harmful effects of cement dust on the skin and respiratory system are due to a few parts per million of chromium compounds in it.

Some of the substances discussed here as chemical health hazards, such as acetone, oxygen and organic peroxides (used as resin catalysts) are also fire and explosion hazards. This aspect is discussed later in Chapter 3.2.

A growing number of substances referred to broadly as 'chemicals' are today employed in construction. Most of these are synthetic and some have been found to have acute toxic effects a few years after first being introduced. These 'chemicals' are, therefore, given special attention here.

The toxic substances which may present health hazards to construction workers are classified for the purposes of this chapter into the following groups:

1. gases;
2. organic solvents (used in paints, varnishes, glues, resins, etc.);
3. water-soluble acids, alkalis, salts, etc. (used in stone cleaning fluids, rust inhibitors and cements, etc.);
4. metal fumes and metal pigment dusts (from welding, brazing, cutting and paint spraying, etc.);
5. other non-solvent organic compounds excluding plastics and their components (fungicides, insecticides, weedkillers, concrete additives, water-proofing compounds);
6. components used in the site preparation of resins (used in adhesives, paints, grouts and insulating foams);
7. other airborne dusts and fibres including both very harmful ones such as silica and asbestos, and less harmful or nuisance dusts such as limestone and iron oxide.

Radioactive materials are not included here since the principal harm they cause is due to their radiation. This is considered as a physical hazard in Chapter 2.5.

It is clear from the list that we have to consider a wide variety of substances and chemical health hazards, some of which may be met by all construction workers whilst others are only met by workers in specific occupations.

Lists of chemicals used in construction have been compiled by the Construction Industry Research and Information Association (CIRIA). These lists are based on their application, and give an outline of the hazards and the precautions needed.[1]

Most of the substances affect construction workers either as a result of breathing air contaminated by them (the inhal route) or by skin contact (the dermal route). In planning protection for workers liable to be exposed to toxic airborne substances, a survey is required to provide information on:

- the toxic materials liable to be inhaled by individual workers, and the circumstances under which this happens;
- the toxicity of individual substances and the maximum concentrations which can safely be permitted in the breathing zone of workers;
- the concentrations of toxic substances actually present in the air inhaled by workers exposed to them;
- the health of exposed workers;
- data on workers' up-take of toxic substances as determined by analysis of blood, urine, body tissue and air breathed out.

Whilst a great deal is known about the toxicity and lethal dosage of individual substances, this information cannot directly tell us whether a particular environment is safe or not. For a number of years, however, occupational hygiene experts in various countries have considered and published data on what they believe to be the maximum concentrations in the air of a number of substances to which a worker may be exposed.

At the same time, a number of firms now produce kits which enable the air breathed by workers to be sampled and analysed. Such air monitoring techniques were first employed for factory workers who were liable to be exposed to particularly hazardous substances—radioactive materials, compounds of lead, chromium and beryllium, and compounds intended for riot control and chemical warfare, nerve gases, etc. As yet they have been little applied to the health problems of construction, but there is no reason why they should not be used, at least to survey processes and occupations where there is clear evidence of a health problem. The monitoring of the working atmosphere to which construction workers may be exposed is discussed in Chapter 2.7.

2.4.1 Occupational exposure limits for airborne substances

Guidelines for the maximum concentrations of airborne substances which should be allowed in the breathing zone of workers have, in the U.K. up to 1980, been based on figures known as 'Threshold Limit Values'. These are published and revised annually by the American Conference of Governmental Industrial Hygienists. They have been adopted with minor modifications by the HSE and published by them as a 'Guidance Note' in the U.K. The figures were reached by a consensus of experts after studying all relevant information. Corresponding but by no means identical 'threshold limit values' are published in several other countries.

From 1983 the figures adopted in the U.K., which form part of the criteria used by the HSE in assessing compliance with the Health and Safety at Work, etc. Act 1974 (HASWA) and other regulations are being published[2] under the name 'Occupational Exposure Limits' (OELs). These differ only slightly from the previously adopted 'threshold limit values'. They will, however, be revised annually by the HSE and will be national U.K. limits, harmonizing with those of E.E.C. countries. U.K. Occupational Exposure Limits have legal significance under HASWA (Sections 2, 3 and 6), as well as the Factories Act 1961, the Mines and Quarries Act 1954, the Agriculture (Poisonous Substances) Act 1952, and any regulations made under these Acts.

So far as construction is concerned, the law as set out in Regulation 20 of the Construction (General Provisions) Regulations, 1961 is quite clear:

> "20. *Inhalation of dust and fumes to be prevented.* Where, in connection with any grinding, cleaning, spraying or manipulation of any material, there is given off any dust or fume of such a character and to such extent as to be likely to be injurious to the health of persons employed, all reasonably practical measures shall be taken either by securing adequate ventilation or by the provision and use of suitable respirators or otherwise to prevent the inhalation of such dust or fume."

Published official OELs are, of course, of little use in themselves unless the air breathed by workers is regularly monitored by appropriate sampling and analysis. Such monitoring is a legal requirement in the U.K. for specific air contaminants under the Chromium Plating Regulations, 1931 (SR and O No. 455 as amended by SI 1973 No. 9) and under the Control of Lead at Work Regulations, 1980 (SI 1980 No. 1248).

Table 2.4.1 gives the 1983 limits for a number of substances liable to be encountered in construction, together with notes on their use or occurrence, their main health hazards, and suggestions for personal protection. The substances are arranged in the groups previously discussed, except for group 2, 'Organic Solvents' and Group 7, 'Other Airborne Dusts and Fibres'. Limits for organic solvents, of which there are a large number, are given separately in *Table 2.4.3* with their boiling points and other details. Limits for other airborne dusts and fibres are given in *Table 2.4.5* in Section 2.4.8.

The limits in *Table 2.4.1* are given as concentrations in parts per million (ppm) by volume in air for gases and vapours, as well as in milligrams per cubic metre of air (mgm^{-3}) at a temperature of 25 °C. Concentrations of airborne particles (fumes, dusts, etc.) are given as mgm^{-3}. Two limits are given for each substance, one for long-term and the other for short-term exposure. The long-term exposure limits are for a period of eight hours in any working day, while the short-term limits are for a period of only 10 minutes. Both limits are given as time weighted concentrations (TWA), which are concentrations (in air) averaged over the period in question. The period of 10 minutes was chosen for short-term limits as the shortest practicable time over which samples can be taken for analysis.

Limits for a few substances are contained in Regulations, Approved Codes of Practice and E.E.C. Directives agreed by the HSC. These are referred to as 'Control Limits' (CL) and indicated by these letters in *Table 2.4.1*. Failure to comply with Control Limits may result in enforcement action by the HSE. Limits for the other substances shown in *Table 2.4.1* are referred to as 'Recommended Limits'. These are intended to be used as criteria for the control of exposure, equipment design, and, where necessary, for the selection and use of personal protective clothing and equipment. They do not have quite the same legal status as Control Limits.

Substances liable to enter the body through the skin or to injure the skin directly are designated by the letters Sk in *Tables 2.4.1* and *2.4.3*.

OELs are based on the judgement of experts and there is no correlation between them and the probability of an industrial disease. As there is no sharp dividing line between 'safe' and 'dangerous' concentrations, exposure to all substances, whether listed or not, should be kept as low as reasonably practicable.

In the sections which follow, the health hazards of a large number of chemical substances used in construction are discussed, including a number for which OELs are not at present published. The absence of a published OEL must not be taken to imply that the material is non-toxic; it may be that its toxicity in an airborne condition has not been evaluated and it may be that it presents a more serious hazard in some other way, e.g. by skin contact.

The substances and their groups as defined in Section 2.4 are discussed next.

2.4.2 Gases[3,4,5,6]

Besides toxicity as given by the OEL, other important characteristics affect the hazardousness of any gas. The concentration at which it can be readily

TABLE 2.4.1. Occupational exposure limits of airborne materials encountered in construction

Type or where encountered	Substance	Long term OEL 8 hour ppm	Long term OEL 8 hour mg/m³	Short term OEL 10 min ppm	Short term OEL 10 min mg/m³	Notes on use or occurrence	Main health hazard	Suggested personal protection Eye protection	Rubber or PVC gloves	Rubber boots	PVC overalls	Respiratory protection (min.)
Gases	Acetylene					Welding & cutting gas	Asphyxiant. May contain toxic impurities	—	—	—	—	—
	Ammonia	25	18	35	27	In toxic washes for brickwork & concrete	Eye & nose irritant. Eye damage	/	—	—	—	—
	Carbon dioxide	5000	9000	15000	27000	Dry ice. Released from some mineral waters in wells & springs	Simple asphyxiant. Heavier than air	—	—	—	—	—
	Carbon monoxide	50	55	400	440	Formed by incomplete combustion of fuels in welding	Poisoning through use of burners in enclosed spaces with inadequate ventilation	—	—	—	—	—
	Chlorodifluoromethane	1000	3500	1250	4375	Refrigerant. Aerosol propellant	Simple asphyxiant. Heavier than air	—	—	—	—	—
	Chlorine	1	3	3	9	Water treatment. Disinfectant	Strong irritant to eyes & respiratory system	/	—	—	—	—
	Dichlorodifluoromethane	1000	4950	1250	6200	Refrigerant. Aerosol propellant	Simple asphyxiant. Heavier than air	—	—	—	—	—
	Formaldehyde (resin constituent)	2	3	2	3	Component of foamed insulation	Dermatitis; irritant to respiratory system	/	/	—	—	—
	Hydrogen chloride	5	7	5	7	As solution in water for cleaning & descaling masonry	Eye & skin burns from aqueous acid	/	/	—	—	—
	Hydrogen cyanide (sk)	10	10	10	10	Burning polyurethanes. Fumigant	Rapid poison. Prevents body's use of oxygen	/	/	/	—	—
	Hydrogen fluoride	3	2.5	6	5	As solution in water for cleaning masonry	Deep skin burns	/	/	/	—	—

Substance					Source/Use	Health effects										
sulphide					...found in sewers & some pits. Produced in some soils	Foul-smelling & nose irritant. Rapid poison causing complete arrest of respiration & death	—	—	—	—	—	—	—	—	—	—
LPG's	1000	1800	1250	2250	Paint stripping. Drying. Space heating. Brazing & soldering	Asphyxiant. Frostbite from liquid contact	—	—	—	—	—	—	—	—	—	—
Nitric oxide	25	30	35	45	In welding fumes & engine exhausts	Respiratory and eye irritant. Damage to lungs										
Nitrogen					As liquid for freezing pipes	Asphyxiant										
Ozone	0.1	0.2	0.3	0.6	In welding fumes. Produced by UV light & some office copiers	Respiratory and eye irritant. Damage to lungs	—	—	—	—	—	—	—	—	—	—
Phosgene	0.1	0.4	—	—	Produced from chlorinated hydrocarbon vapours in contact with hot metals	Acute lung irritant causing permanent lung damage	—	—	—	—	—	—	—	—	—	—
Phosphine	0.3	0.4	1	1	Impurity in commercial acetylene. Repulsive smell	Lung irritant; nausea; vomiting; paralyses CNS	—	—	—	—	—	—	—	—	—	—
Sulphur dioxide	2	5	5	13	From sulphur candles (fumigant). Coke, coal and oil fires	Lung & eye irritant. High exposure harmful to lungs	—	—	—	—	—	—	—	—	—	—
Vinyl chloride (CL)	(7)		8 (1 hour)		Burning & overheated PVC	Liver cancer on prolonged exposure	—	—	—	—	—	—	—	—	—	—
Calcium oxide		2			In producing slaked lime	Deep skin burns. Fine dust a respiratory hazard	/					/		/		/
Calcium hydroxide (slaked lime)		5			In mortars											
Water soluble acids, alkalis, salts, etc. used in stone cleaning fluids, rust inhibitors and cements, etc.																
Hydrogen peroxide	1	1.5	2	3	Cleaning & removing stains	Eye, nose & skin irritant; danger of accidental drinking						/		/		/
Nitric acid	2	5	4	10	Cleaning & descaling masonry	Eye & skin burns; eye damage from splashes; nitric oxide from action on metals						/		/		/

191

Type or where encountered	Substance	Long term OEL 8 hour ppm	Long term OEL 8 hour mg/m³	Short term OEL 10 min ppm	Short term OEL 10 min mg/m³	Notes on use or occurrence	Main health hazard	Suggested personal protection (min.) Eye protection	Rubber or PVC gloves	Rubber boots	PVC overalls	Respiratory protection
	Phosphoric acid	1		3		Rust remover & inhibitor	Eye & skin contact	✓	✓	—	—	—
	Sodium chlorate	2				Weed killer	Burns eyes & skin tissue; acute poison if ingested	✓	✓	—	—	—
	Sodium hydroxide	2		2		Limestone cleaner & ingredient of poly phenol grouts	Burns eyes & skin tissue	✓	✓	—	—	—
	Sodium hypochlorite	1				General disinfectant & cleaner	Eye, skin & respiratory irritant; possible eye injury from splashes	✓	✓	—	—	—
	Sulphuric acid	1				May be used for cleaning masonry	Skin irritant & burns; eye damage from splashes; burns holes in clothing	✓	✓	—	—	—
Metal fumes & metal dusts from welding, brazing, cutting & paint spraying, etc.	Aluminium metal & oxide		10		20							
	Arsenic & compounds except arsine and lead arsenate		0.2			In some timber preservatives	Inhaling dust from treated timber; liver damage; cancers	—	—	—	—	—
	Barium sulphate		2			Paint pigment	From paint stripping; slight danger from inhaling dust	—	—	—	—	—
	Beryllium		0.002			Alloying element in copper & steel alloys (electrical switchgear)	Inhalation of dust and brazing fumes; acute lung damage	—	—	—	—	—
	Cadmium oxide fume		0.05		0.05	Constituent of some brazing rods		—	—	—	—	—

Substance	Value	Use/Source	Effects					
Chromates (as Cr)	0.05	Paint pigments. Constituent of alloy steels	Inhalation of dust from sanding old paintwork. Paint spraying. Welding fumes. Bronchial asthma	/	—	/	—	/
Cobalt and compounds (as Co)	0.1	Occasional constituent of alloy steels	Inhalation of welding fumes; bronchial fever	—	—	—	—	/
Copper fume, dusts & mists	2	Compounds used in timber & masonry preservatives. Brazing	Metal fume fever from brazing & spray & dust from timber treatment, etc.	—	—	—	—	—
Iron oxide fume (as Fe)	10	Welding fumes. Grinding	Mild effect on lungs from inhalation of welding fumes & dust	—	—	—	—	—
Lead compounds inorganic (as Pb) (CL)	0.15	Use & stripping of lead-based paints; demolition. Plumbing & soldering	Cumulative poison; interferes with blood cell production. Lethargy, headache	—	—	—	—	/
Manganese fume (as Mn)	1	Constituent of many steels. Welding fumes	Irritant to respiratory tract. Heavy exposure affects CNS					
Mercury & inorganic compounds (as Hg)	0.05	From old electrical switchgear & mercury vapour lamps	Inhalation of vapour. Attacks CNS. Tremors. Liver & kidney damage					
Molybdenum compounds, insoluble (as Mb)	10	Constituent of some steel alloys	Fairly harmless					
Nickel compounds, insoluble (as Ni)	1	Constituent of stainless steels	Fumes from welding stainless steel. Skin sensitiser. Dermatitis. Asthma					/
Tin compounds, inorganic	2	Fungicides used in treatment of timber and masonry	Skin irritant & sensitiser. Dermatitis. Liver & kidney damage					

194

Type or where encountered	Substance	Long term OEL 8 hour ppm mg/m³		Short term OEL 10 min ppm mg/m³		Notes on use or occurrence	Main health hazard	Suggested personal protection (min.) Eye protection	Rubber or PVC gloves	Rubber boots	PVC overalls	Respiratory protection
	Vanadium pentoxide fume		0.05		0.05	Constituent of steel alloys. Dust from cleaning oil fired boilers	Skin & respiratory irritant. Heavy exposure affects CNS. Tremors, etc.	/				
	'Welding fume'		5		10	As described. OEL's refer to plain steels not alloys	Flu-like symptoms, difficulty in breathing					
	Zinc oxide fume		5		10	Cutting galvanised steel. Paint spraying	Flu-like symptoms					

Type or where encountered	Substance	Long term OEL 8 hour ppm mg/m³		Short term OEL 10 min ppm mg/m³		Boiling point °C	Notes on use or occurrence	Main health hazard	Suggested personal protection (min.) Eye protection	Rubber or PVC gloves	Rubber boots	PVC overalls
Other non-solvent organic compounds excluding those used as resin components	Calcium formate (as formic acid)		9				Concrete accelerator	Dust & ingestion	/			
	Chlorinated biphenyls		1		2		Electrical insulation waxes & lacquers	Skin sensitisers. Toxic vapour on heating attacks liver & kidneys.				
	Cresols (Sk)	5	22				As creosote in wood preservatives	Skin burns & skin sensitiser. Attacks respiratory & nervous system	/	/		

Substance						Use	Effects	
1,4 Dichlorobenzene	75	450	110	675	173	Insect repellant	Vapour through inhal route attacks liver	
DDT				1	3	Insecticide	Inhalation of spray & vapour & absorption through skin affects on CNS—tremors, convulsions & on liver	/
Dieldrin (Sk)		0.25			0.75	Insecticide used in wood treatment		/
Ethanolamine	3	8	6	15	172	In soaps used in emulsion paints	Vapour through inhal route may cause liver & kidney damage	
Ethyl silicate	10	85	30	255	165	Waterproofing treatment for brick & cement	Narcotic. Liver damage on long exposure. Nose & skin irritant	/
Organo tin compounds		0.1			0.2	Pesticides used in wood preservatives	Skin & lung irritant & sensitiser. Liver damage on long exposure	/
Nitroglycerine + (Sk)	0.2	2	0.2	2	explodes	Explosive as dynamite	Absorbed through skin. Dilates blood vessels. Headaches. Tremors	/
Oil mist +		5	5		10	Lubrication of pneumatic tools	Dermatitis, lung inflammation. Some oils cause cancer	
Paraquat		0.1				Weedkiller	Skin & nose irritant. Very toxic. Delayed effects on liver, kidney, CNS.	/
Pentachlorophenol (Sk)		0.5			1.5	Mould growth inhibitor for masonry. Timber preservative	Absorbed through skin. Very toxic. Severe dermatitis. Attacks CNS	/
Phenol (Sk)	5	19	10	38	182	Disinfectant. Toxic wash. Resin component	Absorbed through skin. Affects CNS & respiratory muscles	/
Pitch, coal tar, volatiles		0.2				Damp-proof membranes for floors	Skin & lung irritant & sensitiser. Cancer risk from vapours	/

Type or where encountered	Substance	Long term OEL 8 hour ppm	Long term OEL 8 hour mg/m³	Short term OEL 10 min ppm	Short term OEL 10 min mg/m³	Boiling point °C	Notes on use or occurrence	Main health hazard	Eye protection	Rubber or PVC gloves	Rubber boots	PVC overalls
	Pyridine	5	15	10	30	115	Denaturant in methylated spirits	Foul smelling. Local irritant. Narcotic. Dermatitis. Attacks eyes & CNS				
	Acetic acid	10	25	15	37	118	In silicone sealants	Eye & skin irritant	/	/		
	Barium peroxide (as Ba)		0.5				Curing agent for polysulphide sealants	Dust strong eye & skin irritant	/	/		
Components of resins produced on site	Benzoyl peroxide		5			decomposes	Catalyst for styrene-polyester resins	Skin irritant & sensitiser. Dermatitis	/	/		
	Bisphenol A						Component of epoxy resins	Nuisance dust causing eye & skin irritation, possible dermatitis		/		
	Diethylene triamine	1	4			207	Curing agent for epoxy resins	Skin irritant & sensitiser	/	/		
	Diglycidyl ether	0.5	3	0.5	3		Uncured epoxy resin	Skin irritant. Vapour from hot material affects nose, eyes & skin	/	/		
	Dipenyl methane di-isocyanate (as N = C = O) (CL)		0.02		0.07		Rigid polyurethane foam component (insulation)	Irritant dust—lung sensitiser. May cause cancer	/	/		
	Epichlorhydrin	2	8	5	20	117	Component of epoxy resin	Skin & lung irritant. Skin sensitiser. Can damage kidneys				
	Methyl methacrylate	100	410	125	510	101	Component of acrylic polyester resins	Vapours irritate eyes, nose & throat & mildly narcotic				
	Methyl styrenes	100	480	150	720	165	Curing agent for polyester resins	Damage to liver & CNS after long exposure				

Resorcinol	10	45	20	90 270	Component of grouting resin (with formaldehyde)	Absorbed through skin. Affects CNS
Styrene	100	420	125	525 145	Curing agent for polyester resins	Vapour irritates eyes, nose & throat. Mildly narcotic. Liver and CNS affected
Toluene di-isocyanate (as N=C=O) CL		0.02		0.07 / /	Flexible polyurethane foam components	Vapour irritates eyes, skin & respiratory tract. May cause cancer

detected by smell or other effects (e.g. eye irritation) is obviously important, especially when this is related to its OEL. Its specific gravity relative to air is very important in relation to its tendency to remain in pits and confined spaces or to disperse easily. Finally, it is important to know if it is formed naturally or as an incidental result of certain constructional activities, or whether it is unlikely to be present unless deliberately brought to the construction site.

All gases other than oxygen and those listed in *Table 2.4.1* which have toxic effects, are simple asphyxiants. Their presence in appreciable concentrations in the breathing zone of a worker reduces the oxygen content of the air breathed and inhibits normal respiration. Such asphyxiant gases which are liable to be encountered in construction are listed in *Table 2.4.2*, with notes on their flammability, density relative to air and occurrence.

TABLE 2.4.2. Asphyxiant gases liable to be found in construction

Gas		Flammable	Density relative to air	Occurrence
Acetylene		Yes	0.9	Used in welding
Argon	*	No	1.38	Shielding gas used in special electric welding. No smell
Butane	*	Yes	1.93	Used in LPG for heating
Carbon dioxide	*	No	1.52	Combustion product. Danger in pits, etc.
Chloro difluoro methane	*	No	3.0	Refrigerants and aerosol propellants
Difluoro dichloro methane ('Freons')	*	No	4.2	
Hydrogen		Yes	0.07	Used in welding aluminium, lead
Methane		Yes	0.55	Town gas, for heating, etc.
Nitrogen		No	0.97	Atmosphere and for inerting vessels, etc. prior to hot work on them. Liquid used for freezing pipes, etc.
Propane	*	Yes	1.52	Used in LPG for heating

The OELs of asphyxiant gases which are non-flammable and have no other physiological effects are taken as 5000 ppm (5% by vol.) for long-term exposure, and 15 000 ppm (15% by vol.) for short-term exposure.

The most dangerous of these asphyxiants are those heavier than air gases marked with an asterisk (*) in *Table 2.4.2*. Of them, only commercial butane and propane have any smell (due to the addition of a 'stenching agent' before distribution). Argon, carbon dioxide and the Freons are odourless. There is always a risk when people enter an unventilated pit, cellar, tank, etc. of being overcome by one of these gases (especially carbon dioxide). Several casualties, usually fatal, are recorded every year when men enter confined spaces where such gases have accumulated[7]. Often there are multiple fatalities as colleagues rush in to help without proper breathing apparatus and become unwitting victims themselves. Casualties have also occurred through men entering tanks filled with nitrogen for inerting purposes before it has been properly displaced with air.

Gases with toxic effects, given in *Table 2.4.1* (which kill at much lower concentrations than are required to asphyxiate) are next discussed.

Acetylene, when pure, has a faint ethereal smell and is a mild narcotic as well as asphyxiant. Commercial acetylene is, however, often contaminated with extremely toxic impurities such as phosphine[8], hydrogen sulphide, arsine and carbon disulphide.

Phosphine is the commonest impurity in commercial acetylene and imparts a garlic-like smell to the acetylene. It irritates the eyes, nose, skin and lungs, causes kidney damage and paralyses the central nervous system (CNS). The other named impurities in commercial acetylene have very similar effects.

Ammonia is commonly used as an aqueous solution for cleaning and as a component of toxic washes for brickwork and concrete. It may also be used to neutralize acids in descalants, rust removers, etc. Its odour is universally familiar and allows ammonia to be detected at a concentration a little lower than its OEL. Thus, if there is a strong persistent smell of ammonia, the OEL, long or short term, is probably being exceeded. Ammonia is intensely irritating to the eyes and can damage the cornea.

Inhalation of ammonia gas is so irritating to the air passages that people generally leave the area promptly. Large doses burn the air tubes and cells of the lung and cause the lungs to fill with fluid (pulmonary oedema), which may be fatal. Long-term exposure at levels above the OEL results in chronic bronchitis.

Workers using ammonia solutions for cleaning should at least have eye protection, and eye wash bottles filled with clean water or weak boracic solution on hand to deal with accidental splashes.

Carbon dioxide, whose main hazard is as an asphyxiant, is produced in a pure state by fermentation, the reaction of chalk and other carbonates with acidified water, or heating solid carbonates used in dry powder extinguishers or by evaporation of dry ice. It sometimes occurs naturally in mineral waters which enter springs and wells.

Besides acting as a simple asphyxiant, carbon dioxide is mildly toxic. At low concentrations of 5% volume at atmospheric pressure it stimulates respiration but at higher concentrations it has the opposite effect. It is a well-known hazard of deep sea divers.

Because of the possible presence of carbon dioxide in pits, vats, etc., such spaces should be well ventilated before being entered, and the air inside should be sampled (at the lowest level) by a probe and tested by a recognized portable test kit, details of which are given in Chapter 2.7.

Carbon monoxide is not detectable by odour and is slightly lighter than air. It results from the incomplete combustion of fuels. The exhaust gases from internal combustion engines contain lethal concentrations of carbon monoxide which make the use of such engines quite hazardous in buildings and confined spaces, e.g. fork lift trucks fuelled with LPG used in store buildings, and cars and trucks in tunnels. Any fuel-burning heater or burner which is starved of oxygen is a source of carbon monoxide; it is also present in tobacco smoke. Several nightwatchmen who have lit a heater in their huts and closed all the doors, windows and vents have been poisoned by carbon monoxide and found dead the next morning.

The current long-term OEL for carbon monoxide is 50 ppm. An hour's exposure to a concentration of 5000 ppm in the air may be fatal.

The first symptom of exposure to carbon monoxide is a headache. Further exposure causes breathlessness, cherry-red lips, dizziness, dimness of vision and nausea, and finally coma, suffocation and death.

Casualties of carbon monoxide poisoning rarely recognize the symptoms in themselves in time to escape. However, when such symptoms are recognized or suspected, casualties should be removed to fresh air and made to rest. If the casualty is not breathing, artificial respiration should be given, and cardiac massage should be applied if the heart stops. The presence of carbon monoxide should be confirmed by sampling and analysis of the atmosphere.

Chlorine is used only incidentally on construction sites for water and sewage treatment and disinfection, and is supplied for these purposes as a liquefied gas under pressure in cylinders. The main danger is a sudden escape from an open valve or broken pipe which overwhelms people before they can escape.

Chlorine is detectable by odour at a concentration of 1 ppm which is the present long-term OEL. It is very irritating to the eyes, nose, sinuses, throat, larynx and larger air tubes. Inhalation can cause inflammation of the sinuses and hoarseness, and heavy concentrations may cause pulmonary oedema if it reaches the lungs, as in the case of ammonia. The gas is so irritating that workers will usually escape before serious injury occurs.

Chloride of lime, or bleach, releases chlorine and hypochlorous acid which has a characteristic smell and behaves similarly to chlorine. Chlorine acts on urine, ammonia and other nitrogenous matter to form nitrogen trichloride, which is an even stronger eye irritant than chlorine itself.

Formaldehyde[9] is a gas with the same density as air, but it is generally supplied as a 37% solution in water. Such solutions affect the skin and fingernails making them scaly and brittle (a form of dermatitis). The principal use of formaldehyde in construction is in the production 'in situ' of urea-formaldehyde foams, used for thermal insulation, especially of cavity walls. This operation is normally done by specialist contractors who provide their operatives with protective clothes and respirators. If this process is not carried out properly the foam may contain unreacted formaldehyde which gradually escapes to the atmosphere. This can create a hazard and nuisance to finishing craftsmen (carpenters, tile layers, decorators) and subsequent occupants of the building.

Like ammonia and chlorine, it is extremely irritating to the respiratory system and readily detectable at low concentrations. Although it can cause nasal congestion, severe bronchial inflammation and pneumonia, people exposed to it usually have enough warning to escape.

Hydrogen chloride and *hydrogen fluoride* are gases which are very soluble in water when they are known as hydrochloric and hydrofluoric acids. They are considered under these names as 'water soluble acids, etc.'

Hydrogen cyanide is occasionally used as a fumigant, but is also produced when polyurethane materials burn and when cyanides come into contact with acids. It is odourless, has a low OEL and its solution in water is rapidly absorbed through the skin.

Hydrogen cyanide is a rapidly-acting poison and interferes with the body's use of oxygen for energy production. This results in the heart and all other body processes stopping.

The first symptoms of hydrogen cyanide poisoning are rather similar to those of carbon monoxide. The treatment is also similar, but in addition the casualty should inhale amyl nitrite vapour to help relieve the symptoms. Where there is a risk of hydrogen cyanide poisoning ampoules of amyl nitrite should be kept with the first aid kit.

Great care must be taken with bonfires on building sites to avoid worker exposure to the smoke and fumes, especially when polyurethanes and other nitrogenous materials are burnt.

Hydrogen sulphide is not used in construction, but it is produced during the decomposition of proteins and from sulphates in the soil by reducing bacteria. It is often found in sewers and underground workings.

The gas is usually detected at concentrations below the OEL of 10 ppm by someone freshly exposed to it. There is a characteristic rotten egg smell, but one of its effects is to obliterate the sense of smell so it is possible for a person to be exposed to a gradually-increasing concentration which reaches a serious or fatal level without him being aware of its presence.

The gas is slightly denser than air and can hang about in pits. These can be a deadly trap to a person entering who quickly falls unconscious and rarely survives.

This emphasizes the need to test the atmosphere in such spaces for hydrogen sulphide, carbon dioxide and oxygen content, and to ensure that they are positively and continuously ventilated with clean air before anyone is allowed to enter. If the atmosphere cannot be guaranteed, as in a sewer where there is continuous formation of hydrogen sulphide, it should only be entered by a trained person wearing an air-line respirator, attached to a lifeline, and with trained support and rescue personnel ready and in attendance.

The early symptoms of hydrogen sulphide poisoning are similar to those of carbon monoxide poisoning and casualties should be given similar first aid treatment.

Liquefied Petroleum Gases (LPG's) which consist of butane and propane and similar hydrocarbons are dealt with extensively as fire and explosion hazards in Chapter 3.1. They present the same asphyxiation hazard in pits and cellars, etc. as carbon dioxide. However, their presence can normally be readily detected by the smell imparted to them by a 'stenching agent' which is added before they are bottled.

Strict rules, as detailed in Chapter 3.1, relating to their storage and use should prevent the accidental presence of these gases in pits, cellars, etc.

The gases can vary considerably in their chemical composition. They may contain variable quantities of unsaturated hydrocarbons which are mild narcotics, as well as saturated hydrocarbons which are mainly simple asphyxiants.

Nitric oxide and ozone are formed from air in low concentrations by welding operations and electric discharges. Nitric oxide is also present in the exhaust gases of internal combustion engines. Both are highly irritating as well as toxic.

Ozone can be detected by its smell at very low concentrations before any irritation is felt, but nitric oxide is chiefly apparent from its irritating effects. Massive exposure causes pulmonary oedema, and chronic lung damage.

Phosgene is found when chlorinated hydrocarbons, such as methylene chloride, used as a paint stripper, come into contact with hot metal surfaces. It is an acute lung irritant but the discomfort caused is usually far less than the harm done to the lungs. Hence care must be taken to avoid all electric fires and naked lights where chlorinated hydrocarbons are used.

Sulphur dioxide is a pungent irritating gas which is heavier than air. Although the pure gas is used in industry, it is not, so far as is known, used in construction, and its presence on construction sites will be mainly in the flue gases of coal, coke, oil and rubbish fires. Whilst quite toxic, its pungent properties usually give adequate warning of dangerous concentrations.

Vinyl chloride[10], from which the commonly used plastic PVC is made, is most likely to be found on building sites in the smoke and fumes from bonfires of waste plastic materials. Since it is now recognized as the cause of a rare liver cancer, the need to keep such fires well away from and on the down-wind side of construction workers and other people will be clear.

2.4.3 Organic solvents[3,4,5,6]

Volatile organic solvents are widely used for applying surface coatings, polishes, adhesives, grouts and sealants, as well as for stripping old paintwork. Those using them are generally obliged to inhale air contaminated by their vapours, and frequently receive splashes or spills on their skin. Most of these solvents can be grouped into the following classes:

- hydrocarbons, which may be aliphatic, cyclic or aromatic;
- alcohols (mostly monohydric, that is, containing a single free hydroxy-group);
- ketones, both aliphatic and cyclic;
- esters, in the main simple esters of the lower alcohols and carboxylic acids;
- ethers, especially cyclic ethers such as dioxane and tetrahydro furane and complex ethers which are both ether and alcohol;
- chlorinated hydrocarbons, such as methylene chloride, trichloro ethylene;
- nitroparaffins, such as 2-nitropropane.

The hydrocarbons, chlorinated hydrocarbons and nitroparaffins are completely soluble in each other but practically insoluble in water. However, the lower boiling alcohols and ketones (methanol, ethanol, acetone) are completely soluble in water, but their solubility decreases on moving up the series (increasing boiling point and number of carbon atoms per molecule). Simple ethers and esters occupy an intermediate position,

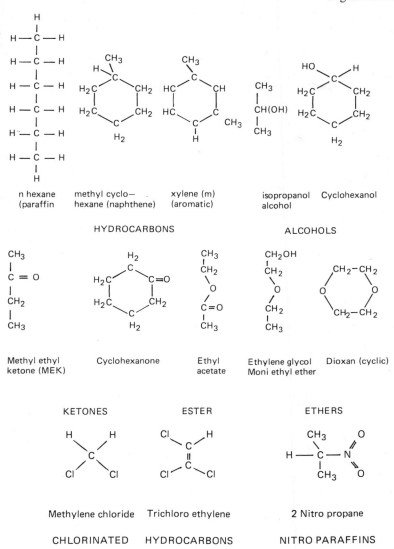

Figure 2.4.1 Structures of common constituents of organic solvents

and even the lowest boiling ones are only partly soluble in water. Most organic solvents are completely soluble in most hydrocarbons with the exception of the lowest boiling alcohols.

The chemical structures of some members of the classes listed are shown in *Figure 2.4.1*.

Most organic solvents for which OELs have been published in the U.K., plus a few others (shown in brackets) for which TLVs were published earlier, are given in *Table 2.4.3*. These have been arranged into the same

TABLE 2.4.3. Occupational exposure limits for vapours of organic solvents used for paints, lacquers, glues, resins, grouts, sealants, varnishes and strippers. Grouping by chemical type

Chemical type	Compound	OEL mg/m³ Long term 8 hours	OEL mg/m³ Short term 10 mins	Boiling point °C at atmospheric pressure	Main solvent usage and notes	Main health hazard	Rubber or PVC gloves
Hydrocarbon							
paraffinic	n hexane	360	450	69	Components of petroleum naphtha used as paint thinner	narcotic, leading to dizziness, intoxication, nausea and loss of appetite	
	n heptane	1600	2000	98			
	n octane	1450	1800	125			
naphthenic	cyclohexane	1050	1300	81			
	turpentine	560	840	155–165	Paint solvent	Irritant and narcotic	
aromatic	benzene	30	—	80	Former rubber adhesive solvent	Narcotic, leukaemia and damage to bone marrow	/
	toluene	375	560	111	Components of coal tar naphtha, rubber adhesive, paints	Narcotic and dermatitis (removes natural fat from skin); swells liver; affects judgement	/
	xylenes, mixed (sk)	435	650	139–145			/
	ethyl benzene	435	535	136		As above	/
mixture	Petroleum naphtha (low aromatic naphtha)	(500)		80–130	Paint solvent or thinner	As above	
	Coal tar naphtha (high aromatic petroleum naphtha)	(100)		80–130	Rubber solvent		/
Alcohols, monohydric							
aliphatic	Methanol (sk)	80	120	65	Denaturant for ethanol used as solvent	Eye pain; impairment of vision; headache; nausea	
	Ethanol	1900	—	78	Natural resin solvent, e.g. Shellac	Headache; intoxication; liver damage	
	isopropanol	980	1225	83		Narcotic; mild irritant	
	1 butanol (sk)	150	150	117	Lacquers; plastics; rubbers	Irritant; narcotic; may cause eye damage	/
	2 butanol	450	—	99			
cyclic	cyclohexanol	200	200	161		Eye, nose and throat irritant; affects most body organs	
complex	di-acetone alcohol	240	360	168	Anti-blush agent in cellulose paints	Eye and nose irritant	
Ketones							
aliphatic	acetone	2400	3000	57	General solvent, especially	Mild narcotics causing	

Category	Substance				Use	Health effects	
Esters	methyl formate	250	375	32			
	methyl acetate	610	760	58			
	ethyl formate	300	450	54			
	ethyl acetate	400	—	77	Lacquer solvents	Irritant to eyes, nose and mucous membrane, leading to headache, dizziness and tightness of chest; possible liver damage; little effect on skin	/
	isopropyl acetate	950	1185	89			
	n-butyl acetate	710	950	147			
	sec & tert butyl acetate	950	1190	118			
	sec amyl acetate	650	800	142			
	di-ethyl carbonate			126			
Ethers, linear & cyclic	ethylene glycol mono-ethyl ether (sk)	370	560	135	Cellulose paint solvent	Irritant on mucous membranes; produces inflammation of kidneys	/
	ethylene glycol mono butyl ether (sk)	240	720	171	Wax, paint and lacquer	Affects central nervous system, liver and kidneys	/
	1.4 dioxan (sk)	180	—	101			
Chlorinated hydrocarbons	tetrahydro furane	590	735	65	Resin and lacquer	Nausea; dizziness; headache	
	dichloromethane (methylene chloride)	700	870	41	Paint stripper	See text	
	chloroform	350	225	61	Adhesive for plastics (Perspex)	Anaesthetic; headache; vomiting	
	carbon tetrachloride (sk)	65	130	76	Rubber adhesive	Affect liver and kidneys	/
	1.2 dichloro ethane	40	60	84			
	1.2 dichloro ethylene	790	1000	60			
	1.1.2 trichloroethane (sk)	45	90	114	Paint stripper; waxes; lacquers; de-greaser	Affect central nervous system and liver; de-grease skin and lead to dermatitis	
	1.1.2 trichloroethylene CL (sk)	535	802	87		Affects nervous system	/
	1.1.2 tetrachloroethane	(30)		146		Affects livers and kidneys	
	1.1.2 tetrachloroethane	670		121	Cellulose solvent; de-greaser		
	chlorobenzene	350		132	Paint stripper	Skin and eye irritant; liver and kidney damage	/
Nitro-paraffins	nitro methane			101		Nausea and headache; affect liver, kidneys and central nervous system	
	Nitro ethane			115			
	2 nitro propane	90	90	120	Lacquer solvent and paint stripper		

groups as before. The OELs for organic solvents are given here only in mg m^{-3} for both 8 hour and 10 minute exposures (where published) since it is easier to relate to these units than to ppm by volume. Also given are their boiling points at atmospheric pressure and notes on their use and health hazards. As will be seen, their largest use in construction is in paints and lacquers.

The word paint in its broadest sense is taken to mean all pigmented materials. A familiar example of paint is an 'oil paint' used in house decorating. This is based on a solution of a drying oil and a synthetic resin in a volatile solvent, with a fine pigment powder and perhaps other solids to give weather resistance, intimately dispersed in it. The solvent is usually a mixture of hydrocarbons with occasionally a few percent of an oxygen containing compound such as n-butanol. Paint dries partly as a result of the evaporation of the solvent and partly by the chemical reaction of the drying oil with oxygen in the air to form a tough solid.

Chlorinated rubber paints and bituminous paints which also generally contain hydrocarbon solvents are used where higher chemical or water resistance is required. Bituminous paints are generally black, the bitumen itself providing the pigment.

Many adhesives, grouts and sealants used in construction are also based on chlorinated rubber or bitumen mixed with sufficient solvent to render them workable or tacky.

Non-pigmented paints and lacquers are often termed varnishes. Lacquers are usually defined as surface coatings which dry purely by solvent evaporation, without chemical action. They are sometimes pigmented, as in the nitrocellulose lacquers used in repairing car bodies, and sometimes unpigmented, as in clear vanishes containing a synthetic polyurethane or natural shellac resin and used to protect wood block flooring or wooden furniture. Because no chemical change takes place when a lacquer dries, the coating will dissolve when the same solvent is reapplied. It is therefore usual to choose a resin which is insoluble in the commoner organic liquids such as kerosene, petrol or 'meths' which are liable to be spilt on the coated article. Examples are nitrocellulose, vinyl resins (usually co-polymers of vinyl chloride and vinyl acetate) and non-reactive acrylic resins. These generally require a specific oxygen-containing solvent, such as acetone, ethyl acetate or nitropropane. But although these resins are insoluble in hydrocarbons alone, some hydrocarbon is often needed in the solvent mixture to give the required combination of solubility, viscosity and rate of drying.

Most paint and lacquer solvents are thus not single compounds but complex mixtures blended to give the best combination of properties. Usually a certain proportion of a low boiling volatile compound is needed so that the viscosity of the paint film rises quickly once it has been applied to prevent it from running (thixotropic paints achieve the same result in other ways). Less volatile compounds are, however, also needed to prevent the film drying too fast and leaving brush marks. Crazing and blushing are other faults of paints which are largely caused by injudicious choice of solvent mixture.

The 8 hour exposure OELs for the solvent compounds listed in *Table 2.4.3* span a wide range from 30 mg m^{-3} for benzene (now largely banned), through 560 mg m^{-3} for turpentine to 1600 mg m^{-3} for n heptane and 2400 mg m^{-3} for acetone. The following example is given to show what these figures mean in practice:

A painter spends two hours in applying 1 litre of paint containing 500 g of a low boiling solvent to the woodwork of a medium-sized room measuring 5 m × 4 m × 3 m, a total volume of 60 m^3. Assuming that the air in the room mixes (e.g. as a result of convection from radiators, etc.) calculate the number of air changes per hour needed to keep the average solvent vapour concentration in the room below various levels as set by the OEL.

First, if all doors and windows were closed and the solvent vapour simply accumulated in the air of the room, the evaporation of 500 g of solvent in a volume of 60 m^3 would produce an average concentration of 8300 mg m^{-3} of solvent vapour in the air over a period of two hours. This is well above the OEL of the safest solvent on the list, acetone. If the OEL for the solvent in question were 1000 mg m^{-3}, 4 air changes per hour would be needed, whilst if it were only 100, 40 air changes per hour would be needed.

If all the air had to pass through a window opening of 0.5 m^2 in area, an average air velocity of 1.2 ms^{-1} through the window would be required to keep the vapour concentration below 100 mg m^{-3}. This might be achieved without artificial ventilation on a windy day, but on a still day a ventilation fan in the window opening would be needed.

This example gives some quantitative appreciation of the problem, and the frequent need for ventilation when using paints, varnishes, etc. containing volatile solvents inside buildings.

The different types of solvent compounds are next considered.

2.4.3.1 Hydrocarbons

Hydrocarbons are the most widely used solvents in construction. The solvent power, odour and toxicity generally increase in passing from paraffinic through naphthenic (cyclic) to aromatic compounds. Mixtures comprised of paraffin (saturated straight chain) compounds only generally have insufficient solvent power for paint resins, and a certain percentage of naphthenes (cyclo paraffins) and aromatics are required. For some rubbers, chlorinated hydrocarbon solvents are needed.

Where possible, the paint manufacturer chooses a standard grade of a special boiling spirit as solvent, the properties of some of which are listed in Table 2.4.4. The solvents contain a number of different paraffin

TABLE 2.4.4. Special boiling spirits—standard U.K. grades

Description	Specific gravity at 15°C	Distillation range °C	Aromatics content % vol.
SBP No. 1	0.68	45–105	1
SBP No. 2	0.70	70–95	Below 1
SBP No. 3	0.73	100–120	0.1–10
SBP No. 4	0.71	50–150	7–8
SBP No. 5	0.72	90–105	0.1–8
SBP No. 6	0.77	140–166	12–18

hydrocarbons, heptanes, octanes and nonanes with branched chains (iso compounds) as well as straight chains (n-compounds). Only OELs for the straight chain compounds have been published, though it is understood that the toxicities of the iso compounds are very similar to the normal (n-) compounds, i.e. generally mild except perhaps for n-hexane.

The wide variations possible in the aromatic contents of SBPs 3 and 5 are to a large extent compensated by the naphthene contents, so that the solvent power and specific gravity remain practically the same.

All hydrocarbon vapours have some narcotic effect, leading in high concentration and prolonged exposure to dizziness, intoxication, nausea and loss of appetite. Of the straight chain paraffin hydrocarbons, n-hexane is considerably more toxic than n-heptane and n-octane, and also more toxic than cyclohexane.

Of the aromatics, benzene, which was once widely used as a solvent for polystyrene, methyl methacrylate and in many cellulose esters and ethers, as well as rubbers, is now recognized as a highly toxic compound[11] particularly depressing the capacity of the bone marrow to produce blood cells and also causing leucaemia. It has now been largely replaced by toluene and xylene which have also long been used as solvents for lacquers and adhesives. These are less dangerous than benzene, but they cause more irritation to the eyes and upper respiratory tracts than the paraffinic hydrocarbons, hexane, heptane and octane. In addition, they have more effect on the skin leading to a dry, scaly, cracked dermatitis. They are also more likely to cause liver damage[12].

The vapour of natural turpentine, the traditional paint solvent, now used mainly by artists, produced by distillation of resin which is exuded from fir and pine trees, is highly irritating to the eyes.

2.4.3.2 Alcohols

Of the alcohols, methanol is found in some paint stripping and adhesive formulations and small percentages are used as a denaturant for industrial ethanol. The latter is used as a solvent for a few natural resins, such as shellac (as in French polish). Methanol vapour is the most toxic of the lower alcohols and has a particular effect on vision and the optic nerves.

Iso and normal propanol are also used as solvents for natural gums, resins and waxes, e.g. shellac and copal. They are mixed with ester solvents and hydrocarbons for nitrocellulose and are somewhat more toxic than ethanol.

N-butanol is used as a solvent for some waxes and synthetic paint resins (e.g. alkyds), but it has some of the toxic effects on the optic system as methanol, as well as producing skin dermatitis. Isobutanol has much the same solvent power but is reported to be less toxic.

2.4.3.3 Ketones

The ketones are mild narcotics. Acetone, one of the safest organic solvents, is a good solvent for nitrocellulose, cellulose esters and many resins. It is used mainly for cellulose lacquers and cements.

Butanone (methyl ethyl ketone) is a very good solvent for resins and nitrocellulose, but is unfortunately more toxic than acetone.

Cyclohexanone, with a considerably higher boiling point, is also used as a component of lacquer solvents.

2.4.3.4 Esters

A number of esters of the lower alcohols and carboxylic acids are also used as solvents. These are diluted with isopropanol and hydrocarbons for resins and nitrocellulose in lacquers, although they are generally more expensive than the ketones acetone and butanone.

Esters have strong characteristic fruity odours and some are used in artificial fruit essences. Iso amyl acetate smells of pears and ethyl butyrate of pineapples.

They tend to irritate the eyes, nose and mucous membrane, and are mild narcotics leading to headaches, dizziness and tightness of the chest. Though they have little effect on the skin they may cause liver damage at high concentrations.

2.4.3.5 Ethers

The lower boiling ethers, whilst good solvents for many resins, are little used in paints because of the ease with which they form explosive peroxides in contact with air. However, the ethyl and butyl ethers of ethylene glycol are used as higher boiling components in the formulation of cellulose paint solvents to prevent blushing. Their vapours are somewhat irritating to the mucous membrane and may produce kidney inflammation. They are readily absorbed through the skin into the bloodstream.

1,4-dioxan is an excellent solvent for waxes and special paints and lacquers. It is, however, very toxic and affects the central nervous system, liver and kidneys, and is also readily absorbed through the skin. Its use should, if possible, be avoided.

Tetrahydrofuran is also used as a solvent for some resins and lacquers. It is mildly narcotic and exposure to its vapour can lead to nausea, dizziness and headache.

2.4.3.6 Chlorinated hydrocarbons

Most of the chlorinated hydrocarbons considered are non-flammable and they tend to be more powerful solvents than the hydrocarbons. The vapours, however, of the fully saturated compounds carbon tetrachloride, 1,2-dichloro-ethane, 1,1,2-trichloro-ethane and tetrachloro-ethane are very toxic, poisoning the liver and kidneys and sometimes causing toxic jaundice, a notifiable industrial disease. They have never been used to any extent as paint solvents, but carbon tetrachloride has been used as a rubber solvent. These four solvents should not be used in construction.

The other less toxic chlorinated hydrocarbons produce mainly narcotic effects.

Methylene chloride, the lowest boiling of the chlorinated solvents is used mainly as a paint stripper, mixed with about 15% of industrial ethanol. It was formerly regarded as one of the safest chlorinated hydrocarbons, but it has been found recently that it is broken down in the body to form carbon monoxide which reacts with haemoglobin in the blood and interferes with oxygen transfer. Its present relatively high OEL is thus open to question[3].

Chloroform is a solvent for various gums and cellulose esters, but is not much used in construction except as a solvent for Perspex (polymethyl methacrylate). It is well known as a powerful anaesthetic and can cause headache and vomiting.

1,1-dichloro- and trichloro-ethylene[13,14] are used as solvents in adhesives, particularly tile adhesives. The latter has a sweet seductive smell and it appears to be addictive since this is the solvent mainly preferred by 'glue sniffers'. It is one of the few solvents whose quoted OEL is a statutory Control Limit. The other chlorinated hydrocarbons do not appear to be much used as solvents in construction, although some are widely used for dry-cleaning and industrial degreasing.

Chlorobenzene, believed to be used to a limited extent as a paint stripper is a skin and eye irritant and high exposure can cause liver and kidney damage.

2.4.3.7 Nitro-paraffins

Nitromethane, nitro-ethane and both 1- and 2-nitro-propane are used in nitrocellulose ester and vinyl resin lacquers as well as in strippers. Their vapours are mild irritants to the lungs, upper respiratory tract and central nervous system and severe exposure can cause headache, dizziness, nausea, vomiting and diarrhoea.

2.4.3.8 Conclusions and precautions for solvents

Painters and others handling organic solvents in construction need to be better informed about the materials whose vapours they are inhaling and which frequently come into contact with their skin. More information is also needed on the actual concentration of solvent vapours in the breathing zones of those using them. This can only be achieved by a programme of monitoring employing highly compact sampling and analytical kits which can be worn by or attached to the worker's clothing (see Chapter 2.7).

A number of highly toxic compounds could be present among the paints, lacquers and stripping solvents used, particularly in developing countries which do not place the same legal obligations on suppliers as those contained in the Health and Safety at Work, etc. Act, 1974. A definite policy of excluding the use of solvents containing highly toxic compounds such as benzene, dioxan, carbon tetrachloride, dichloro-ethane, should be followed. A careful check also needs to be kept on the health of painters and others, such as floor tilers, who are constantly exposed to solvent vapours.

Where possible, the use of solvent free (e.g. emulsion) paints should be preferred to paints with solvents, although even here a check is needed to ensure that these do not contain other toxic and volatile compounds (such as ethanolamine).

Suitable gloves and eye protection should be worn by painters wherever solvents which cause dermatitis or may damage the eyes are used.

Those engaged in paint spraying[24] operations need special protection, generally impervious clothing and air line respirators supplied with clean air.

Those using solvents should be trained in their hazards and in measures needed to minimize them, such as keeping all solvents in closed containers. In addition, the need for good ventilation should be explained, especially when working indoors where solvent vapours tend to accumulate near floor level.

2.4.4 Water soluble inorganic compounds used in solution[3,4,5,6]

This large group of materials used in construction fall naturally into three groups, acids, alkalis and 'salts'. Their uses include cleaning and treating stone, brick and concrete, rust proofing metal, treating timber to control rot, fungi and insects, accelerating the curing of concrete, and killing weeds. However it should be noted that not all cleaners, wood preservatives or weedkillers fall into this category and others will be found in subsequent sections.

With odd exceptions, these compounds exhibit low vapour pressure when dissolved in water at the concentrations at which they are used. As a result they do not pose the same respiratory hazards as the organic solvents discussed in Section 2.4.3.

The published OELs for hydrogen chloride, hydrogen fluoride and nitric acid apply to the gases and vapours which will rarely be found in construction operations. Those for sulphuric and phosphoric acids and sodium hydroxide apply to airborne mists and dusts, which will only rarely be present. Most of the solutions, however, irritate and burn the skin and eyes by liquid contact, although they are not readily absorbed through the skin into the bloodstream.

2.4.4.1 *Inorganic acids*

Acids react violently with alkalis and with some salts, so that these materials have to be kept and used separately. Sodium chlorate presents a fire and explosion hazard (see Chapter 3.2) as well as one of toxicity.

The main general precautions with all these materials are to prevent skin and eye contact, to wash any affected parts of the body well with clean water, and to keep the materials in proper containers in a safe and secure place when not in use.

A possible hazard to be noted with most acids is that they react with cement, limestone and other carbonate-containing materials as well as many metals, effervescing strongly and giving off gases. The effervescence produces a fine spray of small droplets of acid which are carried by the air and may damage eyes, skin and clothing. The gases given off depend on the material which reacts with the acid. They range from the relatively harmless hydrogen (from metals) and carbon dioxide (from limestone), to highly toxic ones such as hydrogen cyanide, sulphide, arsenide or phosphide. Good ventilation is essential whenever acids are used, and artificial ventilation should be employed where natural ventilation is lacking.

Hydrochloric acid is a solution of hydrogen chloride in water, and is used in construction. Hydrogen chloride in the pure, water-free state is a toxic gas which irritates and burns the eyes, skin, air passages and lungs (see Section 2.4.2.1). It is also extremely soluble in water.

Although a 37% (weight) solution fumes strongly and gives off hydrogen chloride gas, at concentrations below 20% (weight) it no longer fumes and its main properties are those of a strong acid. Aqueous solutions of from 5 to 20% (weight) are used in construction to remove and descale concrete and mortar and to remove rust stains from stone, cement and similar surfaces.

The main hazard from its use is eye injury through accidental splashes, which cause painful burns and sometimes permanent damage. The acid is also a strong skin irritant and particularly affects cuts and abrasions. The liquid rots clothing, particularly cotton or rayon.

As previously mentioned, the acid effervesces strongly with carbonates and some metals liberating gas and forming a fine mist of acid droplets. These may reach the face, eyes and other parts of the body, causing severe irritation. Good ventilation is needed when hydrochloric acid is used because the gases evolved may be toxic, asphyxiating or flammable. Rubber or PVC gloves and eyeshields should always be used when handling hydrochloric acid and impervious PVC jackets and overalls are advisable if merely to protect the normal clothing. Acid splashes in the eyes or on the skin should be well rinsed with clean water and contaminated clothing should be well washed as soon as possible.

Hydrofluoric acid or hydrogen fluoride is a very low boiling liquid or gas in its pure state. Like hydrogen chloride it is very soluble in water. The pure water-free compound is usually called hydrogen fluoride and its solution in water is termed hydrofluoric acid. Solutions of over 30% concentration give off fumes of hydrogen fluoride. The solutions used in construction may contain from 5 to 20% hydrogen fluoride and seldom fume on their own.

Hydrofluoric acid has the unusual property of attacking silica and silicates such as glass and cement which contain combined silica. Thus it cannot be stored in glass or metal containers. Its main use in construction is for cleaning and etching sandstone, unpolished granite and slate. Hydrofluoric acid reacts with carbonates and metals in the same way as hydrochloric acid, liberating gases and forming persistent mists.

Hydrofluoric acid is far more damaging to the skin and dangerous than hydrochloric acid. Even the dilute acid can cause severe and deep skin burns which are slow to heal, although no warning sensation is felt when the first contact is made. The effects of hydrogen fluoride gas or acid spray on the lungs and eyes are also more serious than those caused by hydrogen chloride. Hydrogen fluoride and other fluorides (originating from fluoride fluxes in welding rods) which enter the body through various ways have long-term toxic effects (except in very low concentrations), causing softening of the bones.

Skin areas affected by hydrofluoric acid, whether as splashes or through acid mist, should be flushed with large quantities of water, even though no immediate damage is apparent. Anyone suffering a hydrogen fluoride skin burn or thought to have absorbed fluorides into his body (e.g. welders) should consult his doctor. Those handling hydrofluoric acid should take the same or more stringent precautions (gloves, goggles, etc.) as those handling hydrochloric acid.

Gloves of natural and GRS rubber become brittle in time in contact with hydrofluoric acid. Neoprene or butyl rubber are not affected in this way.

Nitric acid is sometimes used as a dilute solution (5 to 20%) in the same way as hydrochloric acid for removing or descaling concrete and mortar. At these concentrations its hazards are very similar to those of hydrochloric acid and the same precautions should be taken. Its reactions with metals and many organic and inorganic chemicals are, however, different to those of hydrochloric acid, since nitric acid is both an acid and an oxidizing agent. Many of these reactions (e.g. with copper or lead) form nitric oxide, a colourless toxic gas which reacts spontaneously with the oxygen in the air to give a brown gas, nitrogen dioxide (see Section 2.4.2 for hazards of nitric oxide).

Sulphuric acid may occasionally be used on construction sites in the same way as hydrochloric and nitric acids. The concentrated acid is very dangerous since it has a very strong affinity for water with which it reacts violently, generating heat and steam. Acid used on building sites is unlikely to exceed about 20% (weight) concentration in water, when the same precautions should be taken as when handling hydrochloric acid. Sulphuric acid is, however, more damaging to cotton, and nylon clothing than hydrochloric acid, and any drops of dilute acid on a cotton handkerchief, shirt, vest or overall can be certain to produce burn holes which slowly spread unless washed well soon after contact.

Phosphoric acid is used both as a rust stain remover for stone, etc. and a rust proofer for iron and steel. It is not as strong an acid as hydrochloric or sulphuric, but its effects are similar and the same precautions apply.

Sulphamic acid is used as a rust and scale remover for steel and other metals. Again its effects are a little less severe than those of hydrochloric and sulphuric, but the same precautions should be taken. No OEL has been stipulated for this acid.

Hydrogen peroxide is grouped with the acids since it is neither an alkali nor a salt, but an exceedingly weak acid although its properties are quite different from the acids previously discussed.

Hydrogen peroxide is sometimes used as a 5 to 20% solution in water as a cleaner, bleach and stain remover. Most people have no doubt used it in the home and are familiar with its properties, the most important of which is that it decomposes naturally, giving off several times its own volume of oxygen. To prevent this happening during storage, stabilisers are added to the solution.

When hydrogen peroxide solutions are heated, however, or come into contact with certain finely divided solids, such as manganese dioxide, they effervesce vigorously, liberating gaseous oxygen. Considerable care is needed in the storage and use of hydrogen peroxide solutions to ensure that they are not contaminated or accidentally heated. No more than that immediately required should be withdrawn from store.

Hydrogen peroxide is an irritant to the eyes, nose and throat, and it irritates and burns the skin. Any parts of the body, particularly the eyes, which have received splashes of hydrogen peroxide should be well irrigated. Blisters caused by hydrogen peroxide should receive medical attention.

Hydrogen peroxide is also very poisonous if ingested, causing severe burns in the mouth, throat and stomach and should always be handled in specially labelled containers.

2.4.4.2 Alkalis

The main alkalis used in construction are sodium hydroxide, lime and aqueous ammonia, the latter having been considered in Section 2.4.2.

Sodium hydroxide or caustic soda is used in solution in water as a limestone cleaner, usually at concentrations of 5 to 20% (weight). It is also used in the formulation of polyphenol resin grouts.

Sodium hydroxide is also sometimes encountered in solid form as flakes or sticks. This is tipped into the pans of water closets where it dissolves in the water and is used to clear blockages. Whether in solid form or in solution, it is more corrosive to the skin than most acids (except hydrofluoric) because it reacts with proteins and fats to form deep and painful gelatinous tissue. The greatest hazard lies in eye contact. The same precautions should be taken in handling caustic soda solutions as apply in the case of hydrochloric acid.

Special care must be taken if solid caustic has to be dissolved in water before use. This liberates a great deal of heat.

Although caustic solutions do not react with carbonate containing building materials in the same way as acids, they react very vigorously with aluminium and zinc and their alloys, dissolving the metal and forming hydrogen gas and a mist of caustic solution.

Calcium hydroxide or slaked lime is not so much a weak alkali as a rather insoluble one (lime water) and is less used in solution than in the form of mixed pastes (mortar) and suspensions in water (lime wash). It has similar, although less pronounced, properties as caustic soda, and must be treated with care to avoid contact with the eyes and skin. The dust is also very fine, like cement and becomes easily airborne, where (like cement dust) it becomes a respiratory hazard.

Occasionally quicklime or calcium oxide will be available and have to be slaked with water on building sites to produce slaked lime. This is a more dangerous material which reacts violently with water and causes deep burns if allowed to come into contact with the bare skin. Its use needs careful supervision and the use of protective clothing and equipment.

2.4.4.3 Salts

The salts (so-called) considered here are sodium hypochlorite, sodium chlorate, sodium silicate, calcium chloride, zinc and magnesium fluorosilicates and ammoniacal copper carbonate.

Sodium hypochlorite is generally obtained as a solution and used as a general cleaner and disinfectant. It gives off the vapour of hypochlorous acid which is a strong irritant to the eyes, nose and throat. Liquid droplets in the eyes can cause serious burn injuries and eyeshields should be used when handling all but very dilute solutions.

Sodium hypochlorite reacts vigorously with acids to produce hypochlorous acid, oxygen and chlorine which can readily overcome anyone in the vicinity who is unable to escape (see chlorine, under Section 2.4.2). It must therefore be kept strictly separate in its own labelled container and care taken to prevent it coming into contact with acids, e.g. in disposing of surplus material left in containers.

Sodium chlorate[15] is used as a weedkiller and is dangerous both because of its reactivity and its toxicity. It is bought as crystals or powder which are dissolved in water before use. Its fire and explosion hazards, which are at least as serious as its toxic hazards are discussed in Chapter 3.2.

Sodium chlorate solution burns and irritates the skin and eyes, and its dust is very hazardous to inhale. Vegetation treated with the solution is toxic and precautions must be taken to keep domestic animals away. Wood, paper, old rags and other organic material which come into contact with it as packing material or when dissolving it, etc. should be burnt immediately.

Sodium chlorate reacts with mineral acids such as hydrochloric and nitric to give chlorine and chlorine dioxide, a brownish-yellow toxic gas which is also explosive.

It is essential to keep this material in its own labelled container well separated from other chemicals and never allow it to mix with them—particularly when disposing of surplus material.

When the material is handled rubber or PVC gloves, eye shields and completely impervious outer clothing and rubber boots should be worn. Contaminated footwear and clothing should be washed thoroughly before it is allowed to dry.

The solution may be sprinkled from a watering can or rose but should never be sprayed. Care must be taken to prevent organic material (e.g. sawdust, old rags, etc.) from coming into contact with the solution, apart from the vegetation which it is being used to destroy.

Sodium silicate covers a range of compounds of silica and sodium oxide, some of which are available as thick treacly syrups, whilst others are crystals.

Sodium silicate solution is used in construction for treating brick and concrete walls and floors and as an ingredient of special acidproof cements. It is very alkaline and droplets of sodium silicate solutions irritate and burn the eyes and skin. It should be treated similarly to caustic soda.

Calcium chloride is a neutral salt, very soluble in water, which is used as a cement accelerator. Strong solutions of calcium chloride irritate the skin and eyes, but the material is practically non-toxic.

Zinc and magnesium fluorosilicates, available as crystals, are used in solution as toxic washes on concrete and brickwork to control mould growth. The solutions irritate the skin and eyes and the materials are also toxic if ingested or allowed to enter the body via a cut or abrasion. If contact with the skin or eyes occurs, the affected part should be promptly and copiously washed with water.

Ammoniacal copper carbonate solution is used for similar purposes to the fluorosilicates. It is also toxic, and should be treated similarly, but has the added hazard of giving off ammonia gas, the hazards of which were discussed in Section 2.4.2.

2.4.5 Metal fumes and pigment dusts from welding, brazing, cutting and paint spraying[3,4,5,6,16]

Metal fumes evolved during welding, brazing and metal cutting are a constant hazard. Electric arc welders quite commonly suffer an attack of metal fume fever, which has flu-like symptoms with raised temperature, aches and pains and difficulty in breathing. A study of welders in a U.K. boiler construction company published in 1974 found that 31% of welders between the ages of 20 and 59 had at least one attack[16].

Metals and their compounds have different biological effects depending on their physical state (solid, liquid or gas), the valency (or state of combination) of the metal in the compound and how they enter the body. Many metals which are nearly, if not completely, harmless in themselves have highly toxic compounds. Organo-metallic compounds (such as tetraethyl lead) are generally more toxic than inorganic ones (such as lead carbonate).

The long-term OELs for metal fumes and airborne dusts (which apply generally to the oxides) as shown in *Table 2.4.1* range from 0.002 mg m^{-3} for beryllium (sometimes used as an alloying element in copper), to 10 mg m^{-3} for aluminium. Beryllium compounds affect many organs but are especially damaging to the lungs which may be severely inflamed and sometimes permanently damaged.

The composition and quantity of welding fumes much depends on the alloy being welded and the process and electrodes used. Shielded arc welding of aluminium, titanium and steels in an argon atmosphere forms relatively little fume, but creates intense radiation which can produce ozone. The arc welding of steels in an unprotected atmosphere generates considerable fume and sometimes carbon monoxide. The fumes generally contain oxides of iron, manganese, silicon and other constituents such as chromium and nickel depending on the alloys used.

When electrodes with fluoride fluxes are used, the fumes may be predominantly fluorides rather than oxides, which increases their toxicity. Thus, although the published OELs for welding fume (5 mg m^{-3} for 8 hour exposure and 10 mg m^{-3} for 10 minute exposure) are the same as for iron oxide, lower figures will apply if alloys containing significant amounts of manganese, nickel, chromium, vanadium or cobalt are used, or if fluorides are present.

The limits of 5 mg m^{-3} are not as a rule exceeded inside the welder's helmet, but the limits for the other metals mentioned may well be exceeded when welding alloys.

Iron oxide fume causes a mottling in the lung x-rays of welders similar to silicosis, but this does not appear to cause any disability.

Manganese oxide, and in particular the dioxide, produces severe effects similar to Parkinson's syndrome, including weakness, instability, difficulty in walking, spasmodic laughter and other grotesque symptoms. This is known as manganism.

Nickel and its compounds are allergenic and those exposed to nickel fumes may develop itching and dermatitis. In severe causes disorders of the central nervous system and allergic bronchial asthma may be encountered.

Chromium oxide causes bronchial asthma and may result in lung cancer after a long period of exposure. The main toxic hazard of chromium is through chromic acid and chromates. These are not normally present in welding fumes but chromates are used in some yellow and orange paint pigments. These may represent a hazard to painters when using spray guns, or when using dry sanders to remove old paintwork.

The main exposure of construction workers to chromium compounds, however, is probably through cement dust due to the presence of small traces of chromium in the rock or clay from which the cement is made (see Section 2.1.2.1).

Cadmium[17] In brazing, rods containing up to 20% cadmium are sometimes used. Coppersmiths using them for several years have shown evidence of kidney and liver damage. Cadmium also causes severe lung damage.

Beryllium[18] An alloying element in copper, used in electrical switchgear and non-sparking tools, has similar toxic effects to cadmium but at far lower concentrations. The brazing and grinding of articles made from beryllium alloys should only be done by specialist firms who are familiar with the hazards and precautions needed.

Vanadium is another alloying element in many steels. The oxide is more likely however to be encountered by cleaning, demolition and maintenance workers on oil fired boilers and chimneys, as many fuel oils contain vanadium which forms a fine oxide dust when the oil is burnt. It is mainly a skin and respiratory irritant but inhaled dust containing combined vanadium affects the central nervous system in a rather similar way to manganese.

Lead[19,20] When lead compounds, such as those present in lead paints, are subject to the high temperature of an oxyacetylene flame, they are reduced to metallic lead which vapourizes. The vapour then condenses in the air and oxidizes to form a very fine lead oxide fume. Demolition workers removing old cast iron arches etc. by flame cutting even in the open air have been found suffering from lead poisoning and have had to be admitted to hospital.

The first symptoms of lead poisoning are inability to sleep, fatigue and constipation. If severe exposure is continued, anaemia, colic ('painters' colic') and neuritis develop. The colic is painful and sometimes incorrectly diagnosed as appendicitis. Lead poisoning leads to many disorders, inflamed nerves, loss of teeth, sore gums and anaemia caused by destruction of the blood-forming tissues in the bone marrow.

Although lead paint is only used for special purposes today there is still a lot of old lead paintwork around which has to be stripped or demolished. Care should be taken to choose methods which do not create fume or dust. If this cannot be done, appropriate respirators will be required, together with the use of special working clothing and washing facilities. Work with lead is subject in the U.K. to 'The Control of Lead at Work Regulations, 1980' and the Approved Code of Practice for the said Regulations.

Mercury[21] is present in some electrical switchgear and mercury vapour lamps, and since it is highly toxic the handling and disposal of broken and worn out lamps and switchgear containing mercury needs care and a planned procedure.

Mercury has a sufficiently high vapour pressure, particularly in heated rooms, for the vapour arising from a spill to quickly create a health hazard. Mercury in particular attacks the central nervous system causing erratic behaviour (hence the saying, 'mad as a hatter') and it also attacks the liver and kidneys.

Zinc fume arises from the gas cutting or welding of galvanized steel and from brazing brass. Zinc compounds are also commonly used in paint pigments.

Zinc oxide fume causes similar 'flu-like symptoms to general welding fumes.

It is clear that the hazards of inhaling metal fumes from welding, cutting and brazing and the dusts of metal based pigments from paint spraying and sanding old paintwork need to be carefully reviewed by construction firms, particularly in the light of the health records of exposed workers. Identification of toxic compounds used both now and in the past (in the case of demolition) as well as monitoring the air in the breathing zone of exposed workers are necessary steps in solving these problems.

2.4.6 Other non-solvent organic compounds[3,4,5,6] (excluding those used as resin components)

Of the organic compounds used in construction, only solvents have so far been considered. Organic compounds, however, feature in a wide range of other uses in construction such as cleaning and preserving wood and masonry, killing weeds and insects, cement additives (e.g. as retardants and stabilizers for expanded cement), aerosol propellants, dispersing agents for emulsion paints, and as components of plastics and rubbers prepared in situ. It is obviously impossible to deal with all of these here individually, so a number of the best known ones have been selected for discussion. Most of the compounds are toxic and their vapours are sufficiently volatile to constitute a respiratory hazard. Several are readily absorbed through the skin into the bloodstream and many cause skin dermatitis, some being merely skin irritants whilst others are allergens.

Phenols and sodium phenate are used in solution in water as a general disinfectant and mould growth inhibitor on stone, concrete, brickwork and tiles. Phenol is readily absorbed through the intact skin and enough phenol to cause serious poisoning can be absorbed readily from clothing saturated with even a dilute solution. The vapour of phenol is also toxic and very irritating and can be particularly hazardous in hot weather and confined spaces. Good ventilation is essential where it is used.

The first symptoms of phenol poisoning (whether it enters the body through the skin or by respiration) may be dizziness and headache, followed by delirium, cold sweat and jerky breathing. Extensive absorption of phenol affects both the central nervous system and the respiratory system. This leads to muscular weakness, loss of consciousness and death from respiratory failure.

As with other very hazardous chemicals only sufficient phenol for the task in hand should be dispensed at any one time. Those handling phenol or its solutions should wear at least eye, if not full face, protection, rubber boots and rubber or PVC gloves, and PVC or other impervious outer clothing. To prevent burns and absorption, any phenol in contact with the skin should be immediately and completely removed by washing under a shower with soap and water. Contaminated clothing should be removed at once. If phenol solution enters the eye, the eye should be irrigated with running water for at least ten minutes, and medical attention sought promptly even though the injury does not appear serious at the time.

Chlorinated phenols, especially pentachlorophenol[22] are used as either dilute solutions in oil, aqueous solutions, or in the form of their sodium compounds. They are used in the same way as phenol as mould growth washes on stone and brickwork and also as timber preservatives. In the latter case pentachlorophenol may be sold mixed with tributyl tin oxide and insecticides such as dieldrin, a very toxic combination (see Section 2.4.1.5 for the last two compounds).

Pentachlorophenol itself is comparable to phenol in its acute toxic effects, being readily absorbed through the skin, although being practically non-volatile its vapour is less of a problem. It also affects the skin itself, causing dermatitis. The same precautions should be taken as for phenol.

Ethyl silicate is sometimes used as a water-proofing surface treatment agent for brick and concrete. It is a liquid of medium volatility (B.P. 165 °C) which irritates or burns the skin. Prolonged exposure to the vapour may lead to liver and/or kidney damage. The eyes and skin should be thoroughly washed if splashed or contaminated.

Ethanolamine is used in combination as a soap as an emulsifier in emulsion paints. It smells like ammonia. There are conflicting reports about its toxicity in the literature, but its published OEL puts it in the class of acute poisons, probably because it is converted to oxalic acid in the body. Isopropanolamine, which can be used in the same way as an emulsifier in emulsion paints, appears to be far less toxic and it would seem good policy to choose emulsion paints using isopropanolamine rather than ethanolamine.

Pyridine is a highly toxic and foul-smelling compound sometimes used in addition to methanol as a denaturant for ethanol to render it even more unpalatable. There appears, however, to be no good reason why ethanol used as a lacquer solvent in construction should have to contain pyridine when 'meths' containing less toxic denaturants is available.

Organo tin compounds, such as tri-butyl tin oxide, are used mainly in mixtures with other compounds in proprietary timber preservatives. Their volatility is medium to low.

They are skin irritants and allergens so that minute traces may cause rashes and dermatitis in sensitive persons. Their vapour may cause lung inflammation and lead to liver and kidney damage on severe exposure. Protective gloves and eye shields should be used by anyone handling these materials and breathing apparatus with an independent air supply should be used if they are sprayed. If eyes or skin have been affected they should be thoroughly irrigated with water. If a rash develops medical attention should be sought.

Chlorinated biphenyls and chlorinated naphthalenes are used mainly as electrical insulation waxes, but they are also sometimes found in paints, lacquers and adhesives. They all have low volatility but give off toxic vapours when heated, causing headache, dizziness, nausea and liver and/or kidney damage on prolonged exposure. In addition, they are also skin sensitizers. The same precautions apply as for organo tin compounds.

1,4-Dichlorobenzene is a solid of medium volatility used alone or as a component of paints etc. as an insect repellant and insecticide. Its hazards and the precautions to be taken are very similar to chlorinated biphenyls.

Dieldrin and DDT are other chlorinated hydrocarbons used as insecticides either as sprays (dissolved in a hydrocarbon) or as components of paints, etc. Prolonged exposure to the vapour or fine spray leads to liver and/or kidney damage and affects the central nervous system. They should be treated with extreme care and wherever possible less toxic materials (e.g. pyrethrums) should be substituted.

The precautions required are the same as for organo tin compounds. Chlorinated hydrocarbon insecticides which enter the body accumulate in fatty tissue and tend to remain there as they are chemically inert and insoluble in water and there is no mechanism for their elimination. One reported hazard of slimming is that when fat is burnt up in the body releasing energy, the concentration of chlorinated hydrocarbons (absorbed long ago) in the remaining fat may increase to a dangerous level, causing severe illness and sometimes death.

There are a number of other toxic insecticides which might be used in construction, particularly for the treatment of timber. The composition of such toxic washes should be checked, as well as that of the chemicals used to treat timber delivered to site, and expert advice sought where there is reason to suspect a toxic risk. Compounds of lead, arsenic, copper and chromium are among those used.

Nitroglycerine is an ingredient of dynamite, sometimes used in blasting. Besides its toxic hazards it is absorbed through the skin and is sufficiently volatile for its vapour to affect those in close proximity. It causes dilation of the blood vessels around the brain leading to headaches, and prolonged exposure can lead to digestive troubles, tremors, neuralgia, arteriosclerosis and strokes. Other nitrocompounds used in explosives, such as ethylene glycol dinitrate have the same effect.

All persons handling such explosives should avoid contact with bare hands as far as possible, and should wash their hands thoroughly after any unavoidable contact.

Oil Mists (mineral) are frequently present in the compressed air used for pneumatic tools and are, of course, exhausted to the atmosphere. Some oils may cause dermatitis and inflammation of the lungs, and even cancers after long exposure. The hazards can be much reduced by choosing lubricating oils from which the cancer-producing compounds have been removed. Workers' exposure to oil mists in the atmosphere needs, in any case, to be carefully controlled.

Coal tar pitch and, to a lesser extent, asphalt, when heated give off vapours which can cause sensitivity of the skin and respiratory organs, and cancer on long exposure. Skin contact and exposure to vapours from hot pitch and asphalt must be controlled, and personal hygiene, difficult though it is through their sticky nature, must be cultivated in all those handling and using these materials.

Paraquat is a non-volatile solid employed as such or in solution for use as a weedkiller. It is an extremely toxic compound and a strong irritant to the skin and mucous membranes. Traces of paraquat entering the body by any route produce delayed symptoms (up to three days) of nausea and vomiting and cause injury to the liver, kidney and central nervous system. The material must always be kept segregated in its own container under lock and key, and protective gloves should be worn as a minimum precaution. The material should only be applied on a still day, following the maker's instructions carefully. Solutions may be sprinkled but should not be sprayed.

Stearic acid and other fatty acids are used dissolved in heavy petroleum distillates as release agents in moulds and formwork for concrete. These are somewhat irritant to the skin and eyes. If spraying is used, full protective clothing and air line breathing apparatus should be used.

Calcium formate, used dissolved in water as a concrete accelerator, is the salt of the poisonous formic acid. It is an irritant to the skin and eyes and care must be taken to prevent powder becoming airborne.

Cresols, the main constituent of creosote used as a wood preservative, are rapidly absorbed through the skin which they burn and sensitize. Cresols, like phenol, if absorbed into the body primarily affect the respiratory and central nervous systems.

2.4.7 Components of resins produced on site[3,4,5,6]

Many plastics and synthetic rubbers are used in construction. Most are preformed prior to delivery to site and require only cutting, bending, welding or drilling. These preformed materials of high molecular weight are mostly non-volatile, insoluble in water and non-toxic. Even so, they are not entirely free from hazard:

- some fine plastic dusts, such as PVC, can be respiratory hazards[23];
- some contain unreacted toxic chemical raw materials, such as isocyanates;
- many decompose on heating strongly, to give toxic vapours. PVC and polyurethanes are particular risks here.

Many preformed resins are also applied in solution in a solvent as paints and varnishes. These are used not only on permanent structures, but also for the treatment of formwork and moulds for concrete to provide a smooth surface for the mould, or produce some desired effect on the surface of the concrete. Protective gloves and eye shields are needed when these are applied by brush. When they are sprayed[24] this should only be done on a still day by operatives using air line breathing apparatus and well away from other workers.

Besides these preformed plastics, several setting resins are formed 'in situ' on building sites. They are used in adhesives, surface coatings, structural and insulating foams, and for building up fibre-reinforced plastic objects. These are generally formed by mixing two components—one a preformed semi-plastic material, a paste or high-viscosity liquid, and the other a monomer which reacts with and 'cross links' the molecules of the semi-plastic material to form a three dimensional structure (which may be rigid or elastic). Various catalysts and other additives—colourants, stabilizers, fillers—are usually incorporated into one or other of the components, which are supplied as 'two pack kits' ready for immediate use.

Many of these components are volatile and all are reactive and toxic to a greater or lesser degree. In addition they can cause allergenic skin reactions (contact dermatitis) in people who have become sensitized to them by repeated exposure. The most volatile also pose respiratory hazards. It is quite impossible to cover all materials in this class here, but four of the most commonly used are discussed below:

- epoxy resins
- polyesters
- acrylic polymers
- polyurethanes

Epoxy resins are widely used in construction as water and chemical resistant surface coatings and as adhesives. Many are supplied as 'two pack kits'. Some are 'cold setting' whereas others require baking or stoving to set to the required strength and hardness. Only the cold setting epoxy resins are considered here.

The preformed semi-resin in epoxides is known as a diglycidyl ether, itself produced by reacting epichlorhydrin with a polyhydroxy compound, usually bisphenol A. This is usually done in the factory. The other reactant known as the curing agent is usually a poly amine—diethylamine triamine or triethylamine tetramine—although several others are sometimes used.

Most of the substances now supplied in two pack epoxy resin kits are of low toxicity. Some, however, are skin irritants and sensitizers causing rashes and dermatitis in sensitized persons on the slightest exposure.

Some of the main compounds used are:

Epichlorhydrin is a moderately volatile liquid, intensely irritating to the respiratory tract and skin, as well as being a skin sensitizer. The vapour inhaled over a period can damage the kidneys.

Bisphenol A (2,2-bis(p-hydroxy phenyl) propane) is a nonvolatile solid, generally considered a nuisance dust causing eye and nose irritation and possible dermatitis. Workers must be protected from high dust concentrations by ventilation or appropriate dust masks.

Diglycidyl ether, the condensation product of epichlorhydrin and bisphenol A or other similar compounds is a nonvolatile syrupy liquid which gives off irritating vapours when heated.

Diethylene triamine is a liquid of low volatility. It and similar amines are skin irritants and sensitizers, and any liquid that comes into contact with the hands should be washed off with soap and water.

Polyesters (and Styrene-Polyester Resins) are discussed next because in the usual two pack kits used in construction polyesters, polyurethanes and acrylic polymers all rely on the same or similar preformed polymers.

The term 'polyester' covers a wide range of condensation products between polyhydroxy alcohols (glycol, glycerol, etc.) and polybasic acids, which may be aromatic (phthalic, terephthalic), saturated (adipic, sebacic) or unsaturated (maleic). Polyesters from saturated dibasic acids include Terelene, a thermoplastic which is in its final chemical form before being spun into fibres. They also include 'Alkyds' which are thermosetting resins used for compression moulding of light switches and other electrical components and for stoving enamels and lacquers used on car bodies, refrigerators, etc. These are partly condensed chemically before being used, but undergo further reaction and setting when heated in a mould or oven. Polyesters made from unsaturated fatty acids such as maleic are condensed as far as possible in the factory to form viscous liquids or low melting solids. These react with styrene which forms cross linkages between the molecular chains so that the material sets to a tough hard solid. This may be done in the presence of woven or matted glass fibres to form glass reinforced plastics used to construct rigid structures such as car bodies, garden pools and boat hulls. It is also done to provide smooth and hard floor finishes. To accelerate the cross linking with styrene (or methyl styrene and similar compounds) various accelerators (catalysts), mostly organic peroxides such as benzoyl peroxide or cumene peroxide, are used, sometimes dissolved in the styrene.

The main toxic constituents of finished polyester resins as fabricated 'in situ' with glass or other reinforcement are styrene or methyl styrene and the peroxide catalyst. The uncured unsaturated polyesters without styrene or a catalyst are generally non-toxic. The cured resins may be skin irritants due to the presence of styrene and peroxide resins in them. Notes on these materials follow.

Semi-cured unsaturated polyesters may vary from viscous liquids to solids. They have negligible volatility and are not respiratory hazards, though care should be taken to avoid skin contact as they may contain sensitizing materials.

Styrene[25] *and Methyl styrene* are liquid aromatic hydrocarbons of medium volatility whose vapours are highly irritating to the eyes, nose and throat and mild narcotics. The vapours on prolonged exposure produce similar but more acute symptoms to those of aromatic solvents, i.e. headache, nausea and damage to the liver and nervous system. Good ventilation is essential when these materials are used.

Benzoyl and Cumene peroxides and other peroxides used as catalysts or accelerators for the curing of polyester-styrene resins are mostly solids of low volatility which decompose violently or explode if heated. They are skin irritants and potent sensitizers, small traces causing rashes and dermatitis to sensitive persons.

While discussing glass reinforced polyester resins, possible skin damage from the glass fibres must not be overlooked. These readily puncture the skin and facilitate the entry of toxic chemicals leading to septic wounds and sometimes worse.

Acrylic polyers. Again there are a wide range of polymers in this class ranging from Perspex or polymethyl methacrylate to fibres and soft rubbery materials applied as finishes to fabrics. These preformed plastics used in construction are generally non-toxic, although some of the monomers from which they are manufactured are highly toxic. Some hazards may arise from solvent vapours (e.g. chloroform) in joining together transparent Perspex parts.

One monomer at least in this series, methyl methacrylate, is used in the same way as styrene and usually mixed with it to produce cross linked polymers with unsaturated polyesters. By suitably choosing the proportions of styrene and methyl methacrylate, transparent cross linked polyester polymers can be made 'in situ' with the same refractive index as glass fibres, so that transparent glass fibre reinforced plastics can be made in which the glass fibres are quite invisible. These reinforced plastics have improved weather resistance compared with polyesters cross linked with styrenes alone.

Methyl methacrylate is an unsaturated ester and a liquid of medium to high volatility. Like styrene, its vapour is irritating to the eyes, nose and throat, and is reported to be a skin sensitizer.

The more toxic acrylic monomers are unlikely to be found on construction sites.

Polyurethanes constitute another large class of plastic materials from which fibres, tough structural polymers and foams (used for heat insulation and cushioning) are made. The types of polyurethane that may be produced on construction sites are two-pack polyurethane paints and varnishes and two-pack insulating foam kits. Two-pack floor finishes of the polyurethane type in addition to similar floor finishes based on epoxides and polyesters are often also encountered.

Urethanes are produced by a reaction between compounds containing the $-N=C=O$ (isocyanate) group and polyols. The main isocyanates[26,27] (which are highly toxic) are:

toluene di-isocyanate (TDI)
diphenyl methane di-isocyanate (MDI)
hexamethylene di-isocyanate (HDI)
naphthalene di-isocyanate (NDI)

Polyurethanes foams are made by mixing 'prepolymers' (which have active hydroxy groups), often polyesters similar to those used for styrene-polyester resins, with a di-isocyanate, a catalyst and a 'blowing agent', which releases 'freon' or carbon dioxide gas when the components are mixed. Soft foams for insulation and cushioning are made from toluene di-isocyanate and rigid foams, used in structural components to provide rigidity as well as insulation are made from diphenyl methane di-isocyanate (MDI).

Toluene di-isocyanate is a liquid of moderate to high volatility, whose vapour is very irritating to the eyes, skin and respiratory tract. It also acts as a lung sensitizer and may produce asthma-like symptoms on repeated exposure. Once sensitized, a person cannot stand repeated exposure. Careful precautions, including keeping, dispensing and mixing the material in closed containers, ensuring good ventilation, and the use of suitable respirators where necessary must be taken when handling toluene di-isocyanate. Contaminated skin and clothing should be washed thoroughly and at once.

Diphenyl methane di-isocyanate is a solid of low volatility which poses similar hazards to TDI. Care is needed to avoid creating dust or allowing dust to escape into the air.

Besides these two-pack resin kits, several resin applications involve the use or liberation of particular chemicals. These are discussed next.

Barium Peroxide is used as a curing agent for polysulphide type sealants, generally in the form of a paste. It is a strong oxidizing agent and irritant to the skin and eyes. Protective gloves should be worn and good ventilation is required if the formulation contains a solvent.

Acetic acid is evolved on curing some silicone sealants, which are usually supplied as a cream in tubes. It is an eye and skin irritant. Protective gloves should be worn and good ventilation is needed.

Urea-formaldehyde[9] foamed resins are applied to wall cavities for thermal insulation of houses by specially trained contractors. The hazards of formaldehyde residues to other building workers and future occupants were discussed in Section 2.4.2 under 'Formaldehyde'.

Resorcinol, a non-volatile solid with similar toxic properties to phenol, is used in solution with formalin in the preparation of certain grouts. Workers engaged on this operation face the combined hazards of formaldehyde and resorcinol. Good ventilation is required and protective gloves should be worn, and care taken to avoid skin contact.

Rubbers, pitch (coal tar) and bitumen are used separately or in combination in solution in special solvents (toluene, chlorinated hydrocarbons) and as emulsions in water in preparing or laying damp-proof membranes. Operatives face the combined hazards of the solvent and the pitch or rubber. Protective gloves and eye shields are generally required, together with good ventilation. No smoking should be allowed if organic solvents are used.

2.4.8 Other airborne particles[3,4,5,6]

A number of substances giving rise to airborne dusts and other particles have already been discussed, under previous sections, particularly Section 2.4.5 and their OELs given in *Table 2.4.1*. There remain a number of other relatively innocuous dusts, often of vague and undefined composition, and two types of harmful airborne particles to which construction workers are frequently exposed—asbestos and silica.

2.4.8.1 *Airborne and respirable dusts*

It was shown in Chapter 2.3 that the health hazards of any particular dust depend to a great extent on its particle size. Relatively coarse dusts, even if they reach the nose or mouth, are largely trapped, filtered and washed out in the nose and upper respiratory tract before ever reaching the lungs.

To take account of this the British Medical Research Council has defined respirable dust[28] as a fraction of the total airborne dust which can be sampled using instruments (elutriators and cyclones) which monitor the size range which most affects the respiratory organs. These instruments are available from Casella London Ltd and Rotheroe and Mitchell Ltd.

TABLE 2.4.5. Occupational exposure limits for other airborne dusts and fibres

Type	Substance	OEL's Dust mg/m³ Total	OEL's Dust mg/m³ Respirable	Fibres/ ml	TWA over period Hours	Notes on occurrence in construction	Mean health hazards
Nuisance dusts	Non-silicaceous mineral dusts	10	5		8	Lime and limestone clays most soils; marble	Respiratory congestion without permanent damage
	Wood dust (non-allergenic)	10	5		8	Circular saws and other wood working machines	
Semi-nuisance dusts	PVC dust	10	5		8	Sawing and machining PVC	
	Barium sulphate	—	2		8	Sanding paintwork with BaSO₄ pigment	
	Talc	10	1		8		Respiratory congestion
	Mica	10	1		8		
	Diatomaceous earth	—	1.5		8	Rare	
	Coal dust, less than 5% quartz	—	2		8	Tunnels in carbonaceous strata	
Mineral dusts containing silica	Mineral dusts containing quartz or crystalline silica	30 % quartz + 3	10 % respirable quartz + 2		8	Cutting and grinding granite; sandstone; furnace bricks; and terazzo floor tiles and flint	Silicosis
	Mineral dusts containing cristobalite or tridymite	15 % silica + 3	5 % respirable silica + 2		8	Rare	Silicosis
Asbestos	Dust consisting of or containing crocidolite			0.2	4		Pneumoconiosis; Mesothelioma; Lung and bronchial cancer
	Dust consisting of or containing amosite but not crocidolite			0.5	4	Removing old insulation	
	Dust consisting of or containing other types of asbestos but not crocidolite or amosite			1.0	4		
	Tripoli	0.3		0.1		Polishing granite and other hard stone surfaces	

For many dusts, two OELs are given, one for the concentration of total airborne dusts, and the other which is usually 50% of this figure, for respirable dust only. Dusts falling under this category for which OELs are published are given in *Table 2.4.5*.

2.4.8.2 'Nuisance dusts'

Nuisance dusts are generally those for which there is a long history of human exposure and which appear to have little adverse effect on the lungs. They do not appear to produce significant disease or toxic effect when exposures are kept under reasonable control. It cannot be said that they are totally inert and produce no response in the lungs. Any reaction on the lungs, however, should have the following characteristics:

- the 'architecture' of the lung spaces remains intact;
- collagen (scar tissue) is not formed to a significant extent;
- the tissue reaction is potentially reversible.

The generally accepted OELs for nuisance dusts are 10 mg m^{-3} for total dust and 5 mg m^{-3} for respirable dust.

The following materials commonly encountered in construction have been classed as nuisance dusts[3]:

Alundum	Magnesite
Calcium carbonate	Marble
Calcium silicate	Mineral wool fibre
Cellulose (paper fibre)	Plaster of Paris
Cement, Portland	Rouge
Corundum	Silicon
Emery	Silicon carbide (carborundum)
Glass (fibrous or dust)	Starch
Graphite (synthetic)	Tin oxide
Gypsum	Titanium dioxide
Kaolin	Zinc stearate
Limestone	Zinc oxide dust

The OELs of some of these substances might be queried. Thus Portland cement dust, depending probably on its chromium content, may be a powerful sensitizer, producing asthma-like symptoms.

Referring to *Table 2.4.5*, wood dust (non-allergenic) is treated as any other nuisance dust, yet it is difficult to find authoritative data on which wood dusts are allergenic and which are not. The following woods are prime suspects in the medical literature[29]:

Boxwood
Cottonwood
Western red cedar
Cork

The clear link discussed in Chapter 2.1 between furniture workers and cancer of the nasal cavity now appears to be caused by the dust of oak and beech wood[29]. This is more likely to be the very fine dust produced by sanding machines rather than coarser sawdust.

2.4.8.3 'Semi-Nuisance Dusts'

The four dusts shown in *Table 2.4.5* as semi-nuisance dusts, with OELs for the respirable portion of between 1 and 2 mg m^{-3} are probably of only passing interest in construction, being only rarely encountered there. Apart from the first one, a special 'rider' attached to the others that if they contain more than 5% quartz they are to be treated as mineral dusts containing silica.

2.4.8.4 Mineral dusts containing silica

It appears, fortunately, that not all minerals comprised of or containing a high proportion of silica are a potent cause of silicosis. The key to this seems to be that they should be composed of quartz or crystalline silica, and that the crystals should have sharp edges. Sand itself is nothing if not quartz, yet fine natural sand, the grains of which have been well rounded by wind and sea, is not generally considered as a severe silicosis hazard. Or is this just a matter of particle size? One might suppose that if desert sand were a severe silicosis hazard the population of much of North Africa who are exposed to frequent sandstorms would have become extinct long ago. Yet sand blasting and the use of sandstone grinding wheels are well recognized silicosis hazards. This may be because fresh crystal faces and edges are constantly being formed in these operations.

To return to practical issues, sufficient evidence has been accumulated that construction workers engaged in cutting and grinding granite and other rocks with a high quartz content and furnace brick workers who use bricks of high silica or chromia content are particularly at risk from silicosis. Generally such workers are provided with little or no protection in the form of ventilation or respirators.

2.4.8.5 Asbestos[30,31]

Airborne asbestos fibres are now recognized as the cause of a great deal of respiratory disease. Work with asbestos is controlled in the U.K. by the Asbestos Regulations, 1969. However, steps have recently been announced by the U.K. Health and Safety Commission to reduce still further the OELs for the different types of asbestos. Indeed, it seems very probable that the use of the two most harmful forms of asbestos, crocidolite and amosite will soon be banned altogether in the U.K.

The greatest risk lies in the removal of asbestos insulation from buildings and in the demolition of old buildings. This is now done almost entirely by specialist contractors who are well aware of the risks involved. Under Regulations which are expected to be made shortly, it is likely that the removal of old asbestos insulation from buildings will onlybe carried out by licensed contractors who are properly equipped with the required protective clothing, respirators and monitoring equipment.

It is not, however, always appreciated that even the cutting and drilling of asbestos cement board, pipes and other articles (where the asbestos fibres are securely trapped by the cement) results in the liberation of asbestos fibres into the atmosphere, and may create respiratory hazards. Those responsible for such work should familiarize themselves with the

Regulations and Approved Code of Practice. Proper arrangements should be made for local exhaust ventilation where such work is done, as well as for trapping and safe disposal of the dust collected.

2.4.9 Precautions[1,4]

The storage and use of 'chemicals' and the various other substances discussed in this chapter call for a number of precautions to be implemented. These may be of a general nature or of a specific nature for the particular chemicals and substances used. The precautions required overall must take into account not merely the direct health hazards of the materials, but also hazards arising from their flammability, reactivity and possible explosiveness. The precautions given in CIRIA's guide[1] cover all these points. Here we deal mainly with precautions called for on account of direct health hazards, since fire and explosion hazards are considered in Part 3.

CIRIA list the precautions required for safe working with chemicals under five headings:

    controlled storage facilities
    appropriate working conditions
    protective measures
    personal hygiene
    careful disposal of waste

To these we would add two more, though this is not to imply that CIRIA have neglected them:

- training and informing workers about the properties, hazards and safe use of the materials they are handling; and
- monitoring the working environment for concentrations of hazardous materials.

The second of these additional requirements is as yet less appreciated in the construction industry than in certain manufacturing industries, and is dealt with in Chapter 2.7.

An important point made more than once by CIRIA is:

"Use a harmful chemical only if no safer alternative is available and ensure that all necessary safety precautions can be taken."

At the same time the safety of the alternative needs to be carefully assessed before discarding one known to be hazardous but whose hazards are at least known and appreciated—'Better the Devil you know'.

Points relevant to the health hazards of 'chemicals' and other substances are discussed below under each of CIRIA's five headings.

2.4.9.1 Storage

All chemicals and other substances described in this chapter need to be stored safely and securely. The store should be constructed of non-combustible materials and be dry, well ventilated and secured with adequate locks when not attended. The store should be supervised by a trained person, with good eyesight and sense of smell.

The contents of the store should be regularly checked to ensure that containers are securely closed and that both they and their labels are in good condition.

Volatile materials, particularly toxic ones, must be stored in well ventilated areas in case of leaks. Chemicals which can react with one another must be kept well apart, e.g. acids and alkalis. Containers of corrosive chemicals should be placed in polythene or PVC drip trays and stored in a well ventilated place away from direct sunlight and where they are not liable to be knocked over.

A check should be made to see whether there are any chemicals which are liable to freeze and rupture their containers in cold weather. If so, they should be kept in a store which is permanently heated.

A washroom with running water should be near the store and should be provided with a first aid box including an eye wash bottle. *Tables 2.4.1* and *2.4.3* give any special additional storage requirements for particular chemicals and solvents.

2.4.9.2 Working conditions

The following safety measures should always be taken:

- train all persons handling chemicals and other potentially harmful substances in their safe usage;
- withdraw from store only the minimum amount of chemical or harmful substance needed for immediate use;
- provide good ventilation, with particular attention to volatile materials with low OELs;
- provide all necessary protective clothing and equipment for the work in hand, including eye shields, gloves, rubber boots, PVC overalls and appropriate respiratory protection;
- take appropriate steps to restrict the presence of airborne dusts and mists, particularly toxic ones of low OEL, in the working atmosphere. This may call for the use of wet methods, or the use of equipment (e.g. sanders) with dust extractors attached;
- wash down any spills of corrosive water soluble liquids;
- prohibit eating or drinking while handling chemicals or other substances described here, and especially prohibit smoking when handling toxic or flammable substances;
- all operators should wash exposed skin well with soap and running water at the end of the day, before all meals, and immediately after skin contact with chemicals harmful to the skin or readily absorbed into the bloodstream through it.

2.4.9.3 Disposal of surplus and unused chemicals

A plan for the disposal of unused chemicals should be made in advance of their use, with a list of all chemicals and similar substances showing how each should be disposed of. This may require consultation with local health authorities. Many toxic and hazardous chemicals may not be deposited with builders' rubble, but require special arrangements to be made for disposal at an authorized waste disposal site.

Burning of unwanted timber treated with toxic preservatives as well as of PVC and polyurethane plastic materials, is potentially hazardous. If it is done on site the bonfires should be located well away from and downwind of areas where the public are present or people are working, and should not be lit on calm (or very windy) days.

Relatively small quantities of water soluble inorganic acids, alkalis and salts and water soluble organic materials of low toxicity (lower alcohols, ketones and organic acids) may be disposed of through surface and municipal drains provided they are well diluted and washed down with large amounts of water. Any significant quantities of acids should be neutralized with lime or soda ash before disposal in this way. Highly toxic compounds, such as cyanides, and compounds of arsenic, lead, mercury, should never be disposed of in this way. Hydrocarbon, chlorinated hydrocarbon and nitroparaffin solvents which are insoluble in water should also on no account be disposed of either in surface or municipal drains. Hydrocarbons and nitroparaffins for the most part may be very cautiously burnt, but chlorinated hydrocarbons (many of which are non-combustible) should be removed to an authorized waste disposal site. Herbicides, insecticides and toxic compounds used in the preservation of wood and masonry should also only be disposed of in this way.

2.4.9.4 *Personal protective clothing and equipment, general*

Personal protective clothing and equipment is used for a wide range of hazards, of which chemical health hazards form only a part. The subject is therefore discussed as a whole in Chapter 2.8. Monitoring of chemical and other hazards is discussed in Chapter 2.7. Since, however, this chapter contains tabulations of most of the chemicals and other substances which constitute health hazards (*Tables 2.4.1, 2.4.3, 2.4.4* and *2.4.5*) it is natural to try to include some guidance here as to the personal protection called for when using particular chemicals and toxic substances. This is far from easy, since the need or otherwise for personal protection depends on many other factors. These include the concentrations (in water or other media) of the hazardous substances employed, the manner in which they are employed (e.g. by spatula or trowel, by brushing, sprinkling or spraying, or by mechanical device not requiring human intervention), the quantities involved and the conditions in which they are used (indoors in a still environment or outdoors in possibly windy conditions). The need for personal protection thus depends ultimately in most cases on personal judgement. CIRIA recommend various protective measures in their tabulations of chemicals used in construction[1]. The present *Tables 2.4.1* and *2.4.3* also contain suggestions for the use of eye protectors, gloves, rubber boots, PVC overalls and respirators.

The first point to be borne in mind is that most operatives are reluctant to wear or employ most forms of personal protection, particularly respiratory and eye protection. They feel, usually rightly, that such equipment interferes with their sensory perception and bodily coordination, restricts limb movement, lowers work output and actually renders them more accident prone.

Even concentrated acids and other highly reactive and toxic chemicals were in my school days, and still to a large extent are, handled by schoolchildren during practical chemistry classes without personal protection of any kind. There probably were cases of health damage, but the only victim I recall was our school chemistry teacher with a penchant for research. He spent much of his spare time distilling mercury in glass apparatus of his own construction in order to concentrate a particular isotope. His nervous twitches and eccentric mannerisms appear in retrospect to have been typical of mercurial poisoning.

2.4.9.5 *Respiratory protection*

Let us start by considering the need for respiratory protection, since the OELs given in *Tables 2.4.1*, *2.4.3* and *2.4.5* apply primarily to respiratory hazards. It seems reasonable to assume that respiratory protection will only be required where it proves quite impracticable to keep the airborne concentrations of the various vapours, fumes, dusts and mists below the officially published OEL figures.

So little hard information is available on this point in the construction industry that the main need at present appears to be more monitoring and hence more information on concentrations of hazardous airborne materials, before respiratory protection of various sorts is prescribed. In the absence of sampling and analysis, some estimates can be made of the likely concentrations of, say, the vapours of paint solvents inside buildings based on the room size, the rate of paint application and estimates of the number of air changes per hour. Such estimates can and should be made, but they are no substitute for monitoring under real conditions, the results of which are bound to contain surprises.

In most cases it would appear 'a priori' that acceptable concentrations of hazardous airborne materials can be met by normal ventilation, supplemented in some cases by the provision of special local ventilation and extraction systems. There are four areas, however, where because either the high concentrations of hazardous materials or their extreme toxicity, special personal respiratory equipment is required. These are:

1. *Paint and lacquer spraying*. Here there are the dual hazards of solvent vapour, which mixes in all proportions with air, and finely divided dusts and droplets of pigment, resins and drying oils. The type of equipment recommended here in nearly every case is the *air line respirator* which provides a supply of clean air to the operator's nose and mouth via a tube.

2. *Shot or sand blasting, cutting, grinding or polishing of rock, masonry, brick or stone with high contents of silica or other toxic minerals (e.g. chromia)*. Here crude dust filters are often employed, although there is a general lack of information about their effectiveness and about airborne dust concentrations generally. It may well be that airline respirators should be used more on these operations.

3. *Welding and brazing special alloys containing highly toxic elements, beryllium, cadmium, chromium, cobalt and vanadium, etc.* It is said[16] that the concentrations of welding fume behind a welder's face shield are normally below the 8 hour OEL figure of 5 mg m^{-3}. This, however, is unlikely to be the case when the alloys contain significant percentages of the above elements, and here it seems the welder requires extra respiratory protection. A supply of clean air to the welder's face shield via a flexible tube seems to be the preferred solution. The incoming air may also serve to cool the welder's head and face and give some protection against radiant heat.

4. *Removal of asbestos insulation from old buildings during maintenance or demolition.* The preferred solution here seems to be the use of a high efficiency dust respirator equipped with a lightweight portable electric motor, battery and fan which is carried by the operator. It is hoped that only specialist contractors licensed and approved by the HSE will in future be allowed to do this work.

2.4.9.6 Hand protection

Next we can consider hand protection, which generally means wearing gloves, generally of rubber or PVC, although barrier creams have also limited application.

Of all parts of the body, the hands are most likely to come into contact with substances used at work. This applies particularly to solids, pastes and putties, as well as to liquids applied by brush. Although much can be done by training to eliminate hand contamination, this can rarely be done completely. We are thus left with the irreducible minimum requirement that protective impervious gloves should be worn when handling all materials likely to damage the skin or be absorbed through it into the bloodstream and affect other body organs.

2.4.9.7 Eye protection

The eyes are such important and sensitive organs that they must not be unnecessarily exposed to risk. Those wearing spectacles and contact lenses of course have some protection, whilst a variety of plain safety spectacles, goggles and eye shields are available. One still, however, has to reckon with the fact that for a person with good eyesight who has no occasion to wear glasses for sight correction, the wearing of any sort of eye protection is often strange, uncomfortable and alien.

Legally there is an obligation on employers to protect the eyes of their workers against injury from particles or fragments thrown off in the course of the process under section 65 of the Factories Act, 1961. This obligation is clarified and extended by 'The Protection of Eyes Regulations, 1974', which contains a schedule of processes. The majority of these processes relate to mechanical injury from flying particles, and only the following have a possible relation to hazardous chemicals employed in construction:

> "11. The operation, maintenance, dismantling or demolition of plant or any part of plant, being plant or part of plant which contains or has contained acids, alkalis, dangerous corrosive substances, whether liquid

or solid, or other substances which are similarly injurious to the eyes and which has not been so prepared (by isolation, reduction of pressure, emptying or otherwise), treated or designed and constructed so as to prevent any reasonably foreseeable risk of injury to the eyes of any person engaged in any such work from any of the said contents."

"12. The handling in open vessels or manipulation of acids, alkalis, dangerous corrosive materials, whether liquid or solid and other substances which are similarly injurious to the eyes, where in any of the foregoing cases there is a reasonably foreseeable risk of injury to the eyes of any person engaged in any such work from drops splashed or particles thrown off."

"14. Injection by pressure of liquids or solutions into buildings or structures or parts thereof where in the course of such work there is a reasonable foreseeable risk of injury to the eyes of any person engaged in the work from any such liquids or solutions."

Whilst the eyes may be injured by a number of gases and vapours in concentrations well above the OELs, work in such atmospheres would require respiratory protection as well. Such respiratory protection should include eye protection. Eye protection has, therefore, not been shown as a requirement for handling any of the organic solvents listed in *Table 2.4.3*, although it is shown in *Table 2.4.1* as a requirement for handling all acids, alkalis and other chemicals where there is a risk of damage from particles or droplets entering the eye.

2.4.9.8 Use of rubber boots and PVC overalls

Rubber boots are seen as a requirement mainly for the use of hazardous chemicals sprayed on the ground, or likely to be used on floors, or where there is a risk of splashes, burning through ordinary footwear.

Rather similar considerations apply to PVC overalls. Chemicals which may penetrate ordinary clothing and cause a significant hazard on reaching the skin are included as those calling for the use of impervious PVC overalls.

Some chemicals and solvents which may damage or penetrate the skin and shown as calling for rubber or PVC gloves are not, however, shown as requiring PVC overalls. This is either because they are unlikely to penetrate ordinary clothing, or because if only a small splash reaches the clothing it is likely to evaporate rapidly before reaching the skin.

Finally, it must be emphasized that the personal protection shown in *Tables 2.4.1* and *2.4.3* are only suggestions to stimulate thought and not intended to be final absolute requirements.

References

1. CIRIA, *A guide to the safe use of chemicals in construction*, Construction Industry Research and Information Association, London (1981).
2. HEALTH AND SAFETY EXECUTIVE, *Guidance Note EH 40/83 Occupational Exposure Limits, 1983*, HMSO, London (1983).

3. TYRER, F.H. and LEE, K., *A Synopsis of Occupational Medicine*, John Wright, Bristol (1979).
4. NATIONAL SAFETY COUNCIL, *Accident Prevention Manual for Industrial Operations*, NSC, Chicago (Frequently updated).
5. STELLMAN, J.M. and DAUM, S.M., *Work is Dangerous to your Health*, Random House, Transatlantic Book Service (1980).
6. KINNERSLY, P., *The Hazards of Work*, Pluto Press (1980)
7. HEALTH AND SAFETY EXECUTIVE, *Guidance Note GS 5, Entry into Confined Spaces*, HSE, London (1980).
8. HEALTH AND SAFETY EXECUTIVE, *Guidance Note EH 20, Phosphine—Health and Safety Precautions*, HMSO, London.
9. HEALTH AND SAFETY EXECUTIVE, *Toxicology Review 2, Formaldehyde*, HMSO, London (1981).
10. E.C. Directive 78/610/EEC, *Vinyl Chloride Monomer*.
11. HEALTH AND SAFETY EXECUTIVE, *Toxicology Review 4, Benzene*, HMSO, London (1982).
12. HEALTH AND SAFETY EXECUTIVE, *Guidance Note EH 7, Petroleum-based Adhesives in Building Operations*, HMSO, London.
13. HEALTH AND SAFETY EXECUTIVE, *Toxicology Review 6, Trichloroethylene*, HMSO, London (1982).
14. HEALTH AND SAFETY EXECUTIVE, *Guidance Note EH 5, Trichloroethylene*, HMSO, London.
15. HEALTH AND SAFETY EXECUTIVE, *Guidance Note CS 3, Storage and Use of Sodium Chlorate*, HMSO, London.
16. BLACKADDER, E., Toxic Releases and their Biological Effects, Symposium paper from *Symposium Series 47*, Institution of Chemical Engineers, Rugby (1976).
17. HEALTH AND SAFETY EXECUTIVE, *Toxicology Review 7, Cadmium*, HMSO, London (1981).
18. HEALTH AND SAFETY EXECUTIVE, *Guidance Note EH 13, Beryllium—Health and Safety Precautions*, HMSO, London.
19. HSC Approved Code of Practice, *Control of Lead at Work*, HMSO, London (1980).
20. HEALTH AND SAFETY EXECUTIVE, *Guidance Note EH 20, Control of Lead: Outside Workers*, HMSO, London.
21. HEALTH AND SAFETY EXECUTIVE, *Guidance Note EH 17, Mercury—Health and Safety Precautions*, HMSO, London.
22. HEALTH AND SAFETY EXECUTIVE, *Toxicology Review 5, Pentachlorophenol*, HMSO, London (1982).
23. HEALTH AND SAFETY EXECUTIVE, *Guidance Note EH 32, Control of Exposure to Polyvinyl Chloride Dust*, HMSO, London.
24. HEALTH AND SAFETY EXECUTIVE, *Guidance Note EH 9. Spraying of Highly Flammable Liquids*, HMSO, London.
25. HEALTH AND SAFETY EXECUTIVE, *Toxicology Review 1, Styrene*, HMSO, London (1981).
26. HEALTH AND SAFETY EXECUTIVE, *Guidance Note EG 16, Isocyanates: Toxic Hazards and Precautions*, HMSO, London.
27. HEALTH AND SAFETY EXECUTIVE, *Guidance Note MS 18, Isocyanates—Medical Surveillance*, HMSO, London (1977).
28. HAMILTON, R.J. and WATTON, W.H., *Inhaled Particles and Vapours*, ed. C.N. Davies, Pergamon, Oxford (1961).
29. HAMILTON, A. and HARDY, H.L., *Industrial Toxicology*, 3rd ed., Publishing Sciences, New York (1974).
30. HSC Approved Code of Practice and Guidance Note, *Work with Asbestos Insulation and Sprayed Coatings*, HMSO, London (1983).
31. HEALTH AND SAFETY EXECUTIVE, *Guidance Note EH 10, Asbestos: Control Limits and Measurement of Airborne Dust Concentrations*, HMSO, London.

2.5

Man made physical health hazards

The physical health hazards covered in this chapter comprise noise, vibration, several hazardous types of electromagnetic radiation and electricity. Other physical health hazards arising out of the environment, such as heat and cold, as well as work in cramped and unnatural body positions, are considered in Chapter 2.6.

2.5.1 Noise[1]

Noise, which is usually defined as unwanted sound, is both an environmental, as well as an occupational, health hazard. Construction employs a whole orchestra, or rather arsenal, of noisy machinery, including pile drivers, pneumatic concrete breakers, circular saws, riveters, cartridge guns, compressors right down to the humble hammer. On the environmental level, local authorities in the U.K. have wide powers under Sections 60 and 61 of the Control of Pollution Act, 1974 to control noise emanating from construction and demolition sites, in order to protect persons in the locality from its effects. Construction workers who suffer far more from such noise are, however, far less well protected by present legislation.

The most serious health hazard caused by noise is occupational deafness, which may result from continuous exposure to moderately heavy noise levels, or relatively short exposure to high levels of impulsive noise such as the firing of guns or the bursting of shells. Sudden noises also affect the circulation of the blood, the 'startle reaction', causing release of adrenalin into the bloodstream, tightening of the blood vessels and fatigue and headache. Noise also probably interferes with the digestive system, and it seriously interferes with sleep, even when the person is not actually woken. This can affect construction workers on offshore platforms where they may be working and resting on alternate 12 hour shifts for two weeks or more at a stretch in a very confined and noisy environment[2]. The tiredness of workers deprived of proper sleep is a contributory cause of accidents.

Noise is also a more direct contributory cause of accidents in interfering with communications. When the Loddon motorway viaduct near Reading collapsed during construction on 24 October, 1972, killing three men and seriously injuring seven, eye witnesses who scrambled clear said that their warning shouts were drowned by the noise of concrete pumps and vibrators[3]. Partial deafness caused by previous noise exposure is also a serious hindrance to communications.

A good deal of research has been done on the effect of noise on working efficiency. Here the results are more complicated. Noise of any sort, even light noise, generally tends to have a deleterious effect on work requiring a high degree of mental concentration and abstract thought, such as solving a difficult mathematical problem[4]. Light noise, on the other hand, particularly background music, can actually improve the performance of workers engaged in boring or repetitive work.

These effects are readily appreciated, without going into the theory and physics of noise, since all except the totally deaf recognize differences in the quality and loudness of noise. Expressions such as 'a deafening thunder clap' are both common and accurate. But the sensation of loudness, being subjective, is open to different assessment by different observers and is difficult to reduce to numbers. Real progress in understanding and controlling noise only started when the science of sound and hearing (acoustics) had been placed on a quantitative basis, with appropriate units, and when instruments had been developed which give consistent and reliable noise readings which are not dependent on the senses of the observer.

To understand and deal with noise problems in a modern setting we have to delve a little into acoustics[5].

2.5.1.1 Sound and its velocity

Sound is a form of wave motion consisting of pressure pulsations which radiate in all directions in the medium through which they pass. Sound waves travel in straight lines but can bend round small and medium sized obstacles (up to, say, 5 m in size) with some loss of intensity. Like all forms of wave motion, sound has to start from somewhere and requires energy (usually mechanical or electrical) to initiate and sustain it. The sound from a noisy machine represents some loss in energy from the machine.

When sound waves travelling through a medium such as air reach the boundary of another medium (such as a brick wall) they are partly reflected by the boundary, partly absorbed and converted to heat at the boundary, and partly continue as sound waves through the new medium.

Sound waves have characteristic velocities in any medium which depend on its elasticity and density, both of which depend on temperature. In air, the velocity increases with temperature, from 332 m/s at 0 °C to 350 m/s at 30 °C. The velocity is not affected by changes in atmospheric pressure near the earth's surface but sound cannot pass through a perfect vacuum as there is nothing to transmit the pressure waves.

The velocity of sound in water and most solids is much higher than in air, as *Table 2.5.1* shows.

Because of the different velocities of sound waves in different media, and in the same media (e.g. air) at different temperatures, sound waves are refracted and, to some extent change direction when passing from one

TABLE 2.5.1. Sound velocities in various media

Media	Sound velocity (m/s)
Air at 20°C	344
Cork	500
Water at 20°C	1461
Clay rock	3480
Steel	5000
Aluminium	5104

media to another, or when passing through a single media which has a temperature gradient. Thus sound, as we say, travels further in the early hours of a cold still morning when the air temperature is lowest close to the ground, than it does on a sunny afternoon, when the ground is hotter than the air above it. This is because in the first case part of the sound waves which move forward and upwards at a shallow angle to the horizontal travel faster than those moving horizontally near the ground. Hence they move through an arc and return to ground level some distance away to reinforce the waves moving horizontally closer to the ground. This can be of some importance when considering the impact of construction noise on the environment.

2.5.1.2 Sound frequency

Sound waves are characterized by their frequency and intensity. The frequency is the number of pulsations (cycles) per second, known as Hertz (Hz).

Audible sound frequencies lie in the thousand-fold range of 20 to 20 000 Hz. A two-fold frequency range corresponds to one octave. Thus the whole audible range corresponds to 10 octaves ($2^{10} = 1024$). Taking the musical note C, its frequencies on the scientific scale over the audible range are 16, 32, 64, 128, 256, 512, 1024, 2048, 4096, 8192 and 17384 Hz.

The following list gives some idea of the frequency ranges of some common sounds:

thunder	20–40 Hz
voice of a bass singer	70–320 Hz
voice of a soprano singer	260–1040 Hz
normal speech	120–350 Hz
a piano	32–3300Hz

Pressure waves with lower frequencies than 20 Hz are known as infrasonic, those with frequencies higher than 20 000 Hz are ultrasonic. Bats' cries (a form of animal 'radar') are in the ultrasonic region with a frequency of about 48 000 Hz. There are many industrial as well as military applications of ultrasonics, including non-destructive thickness measurement of metals and sonar depth sounding.

The frequency is closely related to the period (τ) and wavelength (λ) of a sound wave.

The period (seconds) is the time taken for one complete pressure cycle:

$$1/\tau = f(\text{frequency}) \tag{1}$$

The wavelength λ is the distance between consecutive parts of the wave motion which are in phase:

$$f \times \lambda = c \qquad (2)$$

(frequency × wavelength = velocity).

2.5.1.3 Sound power, intensity, pressure and level[5]

Sound is a form of energy and thus the rate at which sound is produced is measured in watts. The *sound power* of a source is denoted by P which is usually a very small fraction of a watt. It cannot be measured directly but is calculated from the *acoustic intensity* or the *sound pressure* measured at a distance from the source.

The *acoustic intensity*, designated by I, is the rate at which sound energy passes through unit area of a surface at right angles to the direction of the sound. I is measured in watts per m².

If a compact sound source of power P watts is able to radiate sound uniformly in all directions through a uniform medium without loss of power, the acoustic intensity at a distance R metres from the source is given by:

$$I = P/\pi R^2 \text{ watts/m}^2 \qquad (3)$$

where πR^2 is the surface area of a sphere of radius R.

This equation shows that the acoustic intensity varies inversely with the square of the distance from the source.

Since sound waves are pressure fluctuations, the intensity of sound can also be measured as the mean of the pressure fluctuations at any point. The root mean square is used instead of the average, since the pressure is positive in one half of the cycle and negative in the other half, and the average would be zero. This is known as the *sound pressure*, designated by p and expressed as Newtons per metre² (N/m²). The *acoustic intensity* and the *sound pressure* are simply related and merely different ways of measuring the sound intensity.

All three quantities, the sound power of a source, the acoustic intensity and the sound pressure can vary over an enormous range. The smallest intensity which can generally be detected by a healthy human ear is 10^{-12} W/m². This corresponds to a sound pressure in air under usual room conditions of 2×10^{-5} N/m². Because of the wide range a logarithmic scale is used to express the sound power level of sources as well as the sound pressure level or acoustic intensity at any point.

The unit almost exclusively used is the decibel, which is one tenth of the more fundamental unit the 'bel', which is more directly related to power or pressure. The decibel (denoted dB) is used both for the power level of a sound source and for the acoustic level or sound pressure level observed at a distance from it. To avoid confusion it is important to qualify figures quoted in decibels so that it is clear which is meant.

The *sound power level* of a sound source in decibels is designated by L_w which is given by:

$$L_w = 10 \log_{10} P/P_0 \qquad (4)$$

where P_0 is the reference power level taken as 10^{-12} W (unless otherwise stated). In bel, the sound power level would simply be $\log_{10}(P/P_0)$.

The sound pressure level (L_p) referred to simply as the sound level, which is a measure of the sound pressure or acoustic intensity, is given in decibels by the equation:

$$L_p = 20 \log_{10}(p/p_0) \tag{5}$$

where p_0 is the reference sound pressure to which L_p is related and taken as 2×10^{-5} N/m^2, unless otherwise stated.

The sound power level of a source is calculated from the sound level and the distance from the source at which it is measured. When the sound is radiating uniformly from a point source on a flat reflective surface, L_w and L_p are related by the equation:

$$L_w = L_p + 20 \log_{10} R + 8 \tag{6}$$

This applies to most sound sources in construction.

If the sound source was high in the air and radiating uniformly in all directions, a figure of 11 would apply in equation (6) instead of 8.

From equation (6) it follows that the sound levels L_{p1} and L_{p2} at distances R_1 and R_2 metres from the source are related by the equation:

$$L_{p1} - L_{p2} = 20 \log_{10} R_1/R_2; \tag{7}$$

where R_2 is twice R_1

$$L_{p1} - L_{p2} = 6 \text{ (approximately)}$$

i.e. doubling the distance reduces the sound pressure level by 6 decibels.
Example
A pneumatic breaker is being used to break up an old road on a new and otherwise bare construction site. The nearest house to it is 160 metres away. If the noise level caused by the breaker alone is 91 dB(A) at a point 10 metres from it, what would be the noise level caused by the breaker at the facade of the nearest house?
Answer
160 metres = 10×2^4 metres
For each doubling of the distance the noise level is reduced by 6 dB(A). Hence at the facade of the nearest house the noise level equals:

$$91 - 4 \times 6 = 67 \text{ dB(A)}$$

The sound levels in decibels, together with the corresponding acoustic intensities and sound pressures for various environmental noises are given in *Table 2.5.2*.

Sound begins to be felt as well as heard at about 120 dB and at 140 dB it begins to cause pain.

Continuous sound levels of 150 dB and higher have a very high energy content and are rarely found, due to the large energy input needed to create them. They are sufficient to break most unreinforced glass windows. Human exposure at these levels is very dangerous and painful, causing permanent loss of hearing in a matter of seconds. Still higher sound levels are possible discontinuously as impulse noise, such as an explosion or thunder at close quarters. *Figure 2.5.1* illustrates the amplitude (pressure)

TABLE 2.5.2. Sound levels, acoustic intensity and sound pressures for typical environmental noises

Sound level (dB)	Acoustic intensity (W m^{-2})	Sound pressure (N m^{-2})	Environmental noise
0	10^{-12}	2×10^{-5}	Most sensitive hearing threshold
10	10^{-11}		Leaves rustling
20	10^{-10}	2×10^{-4}	Quiet country lane
30	10^{-9}		Library
40	10^{-8}	2×10^{-3}	Quiet conversation
50	10^{-7}		Quiet street
60	10^{-6}	2×10^{-2}	Normal conversation
70	10^{-5}		Average traffic on street corner
80	10^{-4}	2×10^{-1}	Inside bus
90	10^{-3}		Inside underground train
100	10^{-2}	2×10	Circular saw
110	10^{-1}		Powered earth moving scraper, woodworking shop
120	1	2×10^1	Pneumatic road breaker / Threshold of pain
130	10		Jet engine at 30 m limit without ear muffs
140	100	2×10^2	Jet with after burner at 15 m
150	1000		Sound can burn skin / Unsafe without acoustic helmet
160	10×10^3	2×10^3	Cartridge stud gun
170	100×10^3		
180	10^6	2×10^4	Anti-tank rocket close to limit of sound waves / Shock waves (faster than sound)

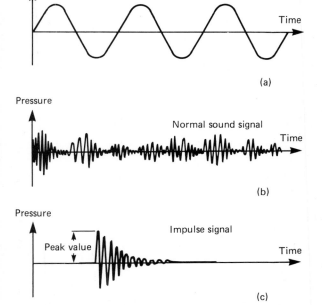

Figure 2.5.1 Amplitude and frequency of soundwaves (a) continuous tone, single frequency; (b) continuous tone, several frequency components; (c) discontinuous impulsive sound.

and frequency of continuous sound waves of single frequencies, of different frequency components and also of discontinuous impulsive sound[5]. There is an upper limit to the sound level of normal sound waves even of an impulsive type, since the pressure amplitude becomes a significant fraction of atmospheric pressure. Thus a sound level of 180 dB would have peak over and under pressures of $2 \times 10^4 \, \text{Nm}^{-2}$ (about 3 p.s.i.), which is enough to cause severe structural damage to most buildings. This is near the limit of intensity for sound waves. Waves of higher energy are no longer sound waves but shock waves, which are supersonic and travel faster than the speed of sound. They result mainly from large explosions and their intensities are no longer measured in decibels but on other scales.

It is important to remember that the decibel scale is a logarithmic one. Because of that the sound from two or more adjacent sources expressed in decibels cannot simply be added. Rather we have to take the sound level of each source in decibels, divide each by 10 to give it in bel, take the antilogarithm of each of them, add the antilogarithms and then take the logarithm of their sum and multiply it by 10 to get the combined sound pressure level in decibels.

Example:

A band performing in the open air consists of a singer, a drummer, a guitarist and a clarinet, without amplification. From a point in mid-audience their average individual sound levels when playing are as given in the first column of *Table 2.5.3*. Thus the sound level when all perform together would be 83 dB as shown in the Table.

TABLE 2.5.3. Calculation of combined sound level of a small band from a point in mid audience

	Sound level (dB)	*Anti-log.* (bel)	*Combined sound level* (bel)	(dB)
Singer	75	$10^7 \times 3.16$		
Drummer	78	$10^7 \times 6.31$		
Clarinet	69	$10^7 \times 0.79$		
Guitarist	72	$10^7 \times 1.58$		
Total		$10^8 \times 1.184$	8.265	83

2.5.1.4 Hearing and the dB(A) scale

The human ear

Figure 2.5.2 shows the structure of the human ear. It consists of three main parts, the outer ear or auricula, the middle ear or tympanum and the inner ear or cochlea. (The outer ear is merely a channel which acts as a resonator at about 3000 Hz and is separated from the middle ear by the ear drum.) The ear drum (tympanic membrane) is a stiff conical diaphragm which vibrates in step with the sound waves received. The middle ear is filled with air and contains three small bones, known as the hammer (malleus), anvil (incus) and stirrup (stapes) which transform the air-induced vibrations of the eardrum into more powerful mechanical vibrations. The inner ear is a

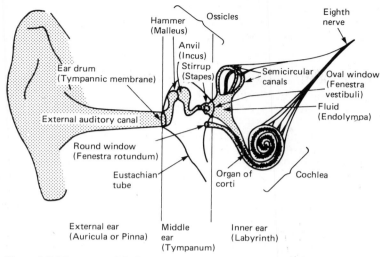

Figure 2.5.2 Structure of the human ear. Wave motions in the air set up sympathetic vibrations which are transmitted by the eardrum and the three bones in the middle ear to the fluid-filled chamber of the inner ear. In the process the relatively large but feeble air-induced vibrations of the eardrum are converted to much smaller but more powerful mechanical vibrations by the three osicles, and finally into even stronger fluid vibrations. The wave motion in the fluid is sensed by the nerves in the choclea, which transmit neural messages to the brain.

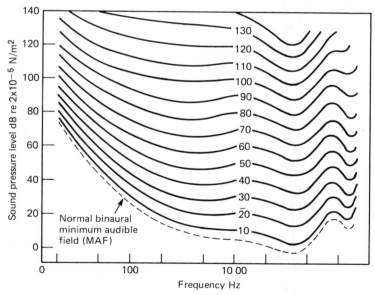

Figure 2.5.3 Normal equal loudness contours for pure tones. They can be applied when (1) the source of sound is directly ahead of the listener (2) the sound reaches the listener in the form of a free progressive sound wave (3) the sound pressure level is measured in the absence of the listener (4) the listening is binaural (5) the listeners are otologically normal persons in the age group 18–25 years inclusive. Courtesy A.J. Pretlove, University of Reading, reproduced from reference 5.

hollow coil of bone filled with a fluid and divided along its length by a membrane. Attached to this membrane are about 24 000 hair-like cells connected to delicate nerve filaments which pass via the auditory nerve to the brain.

The dB(A) scale

Whilst the decibel scale is an absolute one based only on the sound energy, irrespective of frequency, it does not correspond to human hearing perception at all frequencies, since human hearing is most sensitive to sound at frequencies of 3000 to 4000 Hz, and less sensitive at higher and lower levels. This is illustrated in *Figure 2.5.3* which shows a series of 'constant loudness contours' as perceived by young people (age 18–25) of normal hearing, for different 'absolute' sound levels in decibels and frequencies between 20 and 20 000 Hz.

A weighted decibel scale, the A scale, written dB(A) has therefore been devised which gives the same reading for sounds of the same apparent loudness at any frequency. Three other weighted scales, the B, C and D scales, have also been devised for special purposes (e.g. the D scale for aircraft noise), but these need not concern us here.

Most sound level meters measure sound levels in absolute (dB) terms. Weighting network circuits reduce the response of the meter to low and high frequency sounds and increase the response to medium frequency sounds and give readings on the dB(A) scale.

Equations (5) and (6) apply equally to the dB(A) scale as to the absolute scale when the concept of an A weighted sound power level is used.

Thus

$$L_{w(A)} = L_{p(A)} + 20 \log_{10} R + 8 \qquad (6)(A)$$

for a point sound source on a flat reflective surface, and

$$L_{w(A)} = L_{p(A)} + 20 \log_{10} R + 11 \qquad (6')(A)$$

for a point sound source at a considerable height from the ground (much greater than R).

Also

$$L_{p1(A)} - L_{p2(A)} = 20 \log_{10} R_1/R_2 \qquad (7)(A)$$

where $L_{w(A)}$ = the A weighted sound power level in dB(A), and $L_{p(A)}$ = the A weighted sound level in dB(A).

In making a complete noise survey it is also necessary to know the distribution of the sound over various frequency ranges. Sound filters which only allow sound within a limited frequency channel (e.g. 1 octave or ⅓ octave) are attached to the meter and sound level measurements are then carried out over the whole frequency range in narrow bands, giving in effect an audiogram of the sound. This greatly assists in identifying the sources of the sound and the types of control measures needed. This, however, can be a difficult task involving considerable expertise as well as the use of additional equipment including at least a microphone, a calibrator for the sound level meter and a wind shield. An oscilloscope, a tape recorder and an accelerometer are also very desirable accessories. A good discussion is given by F.H. Middleton[6].

Before getting too deeply involved in such investigations it should not be forgotten that the human ear and brain, particularly those of a skilled mechanic familiar with the machine, form a first class system for machinery noise analysis. A good mechanic with unimpaired hearing can often quickly identify various noise sources in a machine after listening to it whilst operating its various controls. He is less capable, however, of assessing or remembering absolute noise levels and for this the noise level meter is essential.

Various rotary equipment emit characteristic noise tones, often superimposed on 'broad band noise'.

The tonal noise of gears and fans is related to their shaft speed and the number of gear teeth or fan blades by the equation:

$$f = nx/60 \qquad (8)$$

where f is the sound frequency Hz, n is the rotational speed of the shaft in rpm, x is the number of gear teeth or fan blades.

In addition there may be a tone characteristic of the shaft speed itself (where $x = 1$). Most casings, frames, chambers and pipes have natural frequencies at which they resonate. These natural frequencies may be altered when their points of attachment to other parts are altered, but unlike the tones of gears and fans which vary with the shaft speed, these natural frequencies of resonance remain the same when the speed of the machine is changed.

A comparison of three noise band spectra of the same sound, octave band, ⅓ octave band and narrow band, is given in *Figure 2.5.4*. The last of these shows four major peaks each likely to be associated with a particular noise source, as well as several minor peaks, whereas even the ⅓ octave waveband cannot identify more than one or at most two of these peaks.

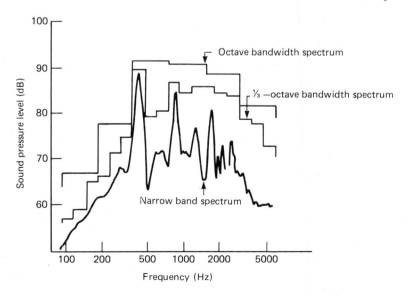

Figure 2.5.4 Comparison of octave, ⅓ octave and narrow band spectra of the same sound.

2.5.1.5 Hearing loss and audiograms

An instrument, known as an 'audiometer' is used to test the quality of human hearing (*Figure 2.5.5*). This enables a person's hearing to be measured at various frequencies over a wide range (about 5 octaves), and compared with the average hearing of a young person of normal and unimpaired hearing. The difference in hearing perception at any frequency can be expressed in decibels and is known as the hearing loss or hearing level, and a plot of hearing loss against frequency is an audiogram.

Figure 2.5.5 Audiometers such as the Bruel and Kjaer 1800 are used increasingly in industries where high noise levels prevail. Routine tests are fully automatic and take about 8 minutes. Frequency is represented along the x-axis and the y-axis represents the hearing threshold of the individual under test who automatically registers his own hearing threshold by operating the handswitch. This system provides an immediate documentary record with which subsequent records can be compared.

Such an audiogram is shown in *Figure 2.5.6*. Noise induced hearing loss usually starts with and is characteristic of a frequency band around 4100 Hz[7]—the 'C_5 dip'—as opposed to hearing loss resulting from age and middle ear disease which are more or less uniform over the whole frequency range.

Both sound level meters and audiometers are discussed later in Chapter 2.7 under 'monitoring health hazards'.

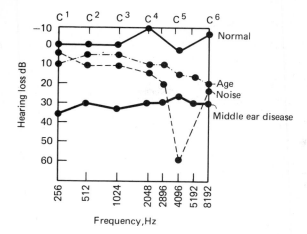

Figure 2.5.6 Audiograms in hearing loss due to age, noise and middle-ear disease (from *The Diseases of Occupations* by Donald Hunter, EUP, London).

Much, but by no means all, hearing loss is caused by exposure to high noise levels, usually over a long period. There are, however, other reasons for hearing loss, of which five of the most common are now discussed.

- age induced loss, or *presbycusis*. This is difficult to distinguish completely from noise induced defects, from which most people in industrialized countries probably suffer to some extent. Thus a survey on one primitive Amerindian tribe showed that their old people suffered far less hearing loss than the average town dwelling American.
- ringing in the ears, or *tinnitus*. This has many causes and is often taken as a warning symptom of noise induced hearing loss.
- conductive deafness, due to defects in those parts of the ear which conduct the sound waves to the inner ear. These include thickening of the eardrum, stiffening of the joints of the small bones of the middle ear, or wax in the external ear. Conductive deafness affects all frequencies evenly, but since sound is still conducted through the head to the inner ear, the loss is limited to about 50 dB.
- cortical deafness is due to defects in the brain centres and mainly affects the elderly.
- nerve deafness may be due either to a defect in the auditory nerve or to loss of sensitivity in the sensory cells in the inner ear. There is at present no medical cure, and the loss is different at different frequencie. Noise induced hearing loss is a form of nerve deafness and is most marked at about 4000 Hz.

Noise induced hearing loss depends on:

- the noise intensity
- frequency content of the noise
- the length of exposure
- the individual response

Three types of loss may be distinguished:

- acoustic trauma, or instantaneous damage from a short exposure to high intensity noise, such as nearby explosions, which could result in a ruptured eardrum, damaged hair cells in the inner ear, or dislocation of the bones of the middle ear.
- temporary threshold shift (TTS), or reduction in hearing sensitivity may occur after a relatively short exposure to excessive noise. This may recover completely on removal to and rest in a quiet environment.
- permanent threshold shift (PTS). As the exposure to excessive noise increases, the recovery time from TTS gradually becomes longer until the threshold shift becomes permanent and the hearing mechanism becomes irreversibly damaged. PTS generally begins in the 4000 Hz octave band and gradually spreads through other bands until total deafness results.

Unfortunately, workers exposed to high noise levels are seldom aware of their deafness until it is severe. The situation was well described by J. Sataloff[8]:

"A hearing impairment may cause no handicap to a chipper or a riveter while he is at work. His deafness may even seem to be to his advantage since the noise of his work is not so loud to him as it is to his fellow workers with normal hearing. Because there is little or no verbal communication in most jobs that produce intense noise, a hearing loss will not be made apparent by inability to understand complicated verbal directions. However, when such a workman returns to his family at night or goes on his vacation, the situation assumes a completely different perspective. He has trouble understanding what his wife is saying . . . This kind of situation frequently leads at first to a mild dispute and later to serious family tension."

If an audiometer is not available a rough and ready check on a person's hearing can be made by taking him into a very quiet room. Standing as far away as possible and speaking in a whisper, ask him to reply when he hears and understands you, whilst you approach him slowly. A person with good hearing should hear you at 8 m, while someone in the early stages of noise induced deafness will not hear you until you are much nearer, perhaps as close as 1 m. Other signs of hearing loss are:

- inability to hear and understand speech in a noisy environment;
- sensations of head noises (buzzing, droning, ringing or whizzing in the ears) after working in a noisy environment (tinnitus);
- temporary deafness after exposure to high noise levels for short periods;
- being told by one's family that one is going deaf.

2.5.1.6 *Recommended limits of noise exposure at work*

In the U.K., following the Robens report[9], the *Code of Practice for reducing exposure of employed persons to noise*[10] was published, in which the following maximum acceptable noise levels were set out:

Noise 249

1. If exposure is continued for 8 hours in any one day and is to a reasonably steady sound, the sound level should not exceed 90 dB(A).
2. If exposure is for a period other than 8 hours or if the sound level is fluctuating, an 'equivalent continuous sound level' (Leq$_{(A)}$) may be calculated and this value should not exceed 90 dB(A).

   Leq$_{(A)}$ is, in fact, an average sound level experienced by the worker, usually over his daily working period. However, since the scale is a logarithmic one, the arithmetic average cannot be used. The period has to be divided into a number of short steps of equal length and the sound level determined for each of these. Each sound level is converted to bel and the antilogarithm of each is found. The antilogarithms are added and averaged. The average of the antilogarithms is then converted into decibels ((A) scale) to give Leq$_{(A)}$.

   Expressed mathematically, Leq$_{(A)}$ is given by the equation:

   $$\text{Leq}_{(A)} = 10 \log \frac{1}{T} \int_0^T \frac{L_{(A)}}{10} dT \tag{9}$$

   where T is the time of exposure.
3. If the non-continuous exposure cannot be adequately measured and controlled, any exposure at a sound level of 90 dB(A) or more should be regarded as exceeding the accepted limit and requiring the use of ear protection. Places where this level is likely to be exceeded should be clearly identified.
4. Overriding limits: the unprotected ear should not be exposed to sound pressure levels exceeding 135 dB, or instantaneous sound pressure levels exceeding 150 dB in the case of impulsive noise. Other parts of the body should not be exposed to sound pressure levels in excess of 150 dB. The method of calculating the equivalent sound level for noncontinuous exposure is based on research carried out by Burns and Robinson of the National Physical Laboratory and the Medical Research Council, published in 1970[11].

For every increase of 3 dB(A) in sound level above 90 dB(A) the maximum recommended exposure is halved, as shown in *Table 2.5.4*.

TABLE 2.5.4. Maximum recommended exposure to noise at various levels per working day

Noise level dB(A)	90	93	96	99	102	105	108	111	114	117	120	126	129
Max. exposure hours, minutes', seconds"	8	4	2	1	30'	15'	7'30"	3'45"	1'52"	56"	28"	7"	3"

If a person is exposed to varying noise levels during a working day some above and some below 90 dB(A), the following simple method may be used to determine whether the exposure is excessive:

1. Estimate the daily exposure at each sound level.
2. Express each duration as the fraction of the duration permitted by *Table 2.5.4*.
3. Add together all the fractions; if the total is more than 1, the exposure is excessive.

TABLE 2.5.5. Sound power levels and sound levels experienced by workers for various construction activities and plant

Operation and category of work	Activity or detail of work	Plant observed		Number of observations	Sound power level-dB(A)		Average sound level-dB(A)	
		Type and detail	Size		Highest	Lowest	at 2m	at 10m
Demolition	Foundation breaking	Excavator with breaker attachment	200 kg m per blow	3	124	119	107	93
	Dropping ball demolition	Tracked crane	123 kW	1		121 107	93	
Earthworks breaking out	Breaking out concrete for drainage	Pneumatic breaker	35 kg	2	121	118	104	91
	Breaking road surface	Pneumatic breaker	35 kg	7	123	114	103	90
ground clearance	Levelling and clearing ground	Loader, tracked	37 kW	5	118	110	99	85
trench excavation	Trenching	Excavator, tracked	45-83 kW	5	116	104	96	82
mass excavation	Ground excavator	Bulldozer (silenced)	201-290 kW	3	120	114	103	89
	Ground excavator	Excavator, tracked	34-92 kW	8	115	106	97	83
	Ground excavator	Loader, wheeled	37-410 kW	5	123	104	100	86
	Earth removal	Loader, tracked	31-72 kW	7	116	104	97	83
haul roads	Earth removal	Scrapers	220-475 kW	3	123	122	108	94
filling, levelling and compacting	Levelling ground and spreading fill	Bulldozer (silenced)	46-306 kW	11	118	104	99	85
rock drilling	Ground consolidation drilling	Pneumatic rock drill	120 mm piston	3	132	122	113	99
Dredging		Ship, chain bucket		1		124	110	96
Piling		Diesel hammer	3100-6219 kg m	5	136	125	117	103
	Steel, sheet, tubular and H-section							
	Sheet steel, precast concrete and driven casing	Drop hammer	2.5-6 tonnes 0.5-2 m drop	11	129	111	104	92
	Impact bored, cast in situ	Tripod winch	12-25 kW	5	112	103	93	79

Category	Description	Equipment	Specification	n					
	Bored, cast in situ	Crane mounted auger	Crane eng. 58-113 kW Donkey eng. 37-150 kW	10	124		111	101	87
	Sheet steel	Jacked	220000 kg force to each pile	3	106		94	85	71
	Tubular steel casing cast in situ	Electrically driven vibratory extraction	24 Hz	1		125		109	97
	Pinning reinforcing	Modified air hammer	15 kg	1		118		104	90
Concreting: fixing reinforcement steel concrete pumping	Various	Lorry mounted concrete pump	97-130 kW	6	118		106	95	81
placing concrete finishing/ scabbling	Concrete mixing	Truck mixer	5-6 m^3 capacity	8	116		96	96	82
	Grinding foundation slab	Concrete grinder	225 mm blade	1		115		101	87
	Chipping concrete	Pneumatic chipper	4-11 kg	6	119		103	95	81
Roadworks stationary	Removal of old road surface	Tractor mounted compressor & pneumatic breaker or hammer	39-41 kW	2	122		114	104	90
mobile	Road surfacing	Asphalt spreader	53-90 kW	2	114		101	92	78
mobile	Removal of old road surface	Road raiser and lorry	97 kW	1		115		101	87

Example:
Exposed at 90 dB(A) for 4 hours, followed by 99 dB(A) for ¾ hour.
Exposure at 90 dB(A) = ½ that permitted by *Table 2.5.4*.
Exposure at 99 dB(A) = ¾ that permitted by *Table 2.5.4*.
Total ½ + ¾ = 1¼. Hence exposure is excessive.

Today these calculations can to some extent be obviated or simplified by the use of a sound meter with an integrating attachment which gives a direct reading or printout of $Leq_{(A)}$ over the required period of exposure. It is often impracticable, however, to leave a sound meter in an exposed position on a busy building site for eight hours or longer. Thus the meter with $Leq_{(A)}$ output may be used to measure the mean sound level over a number of shorter sampling periods (e.g. 5 to 10 minutes) during the working day, and the overall $Leq_{(A)}$ is then calculated.

At the time of writing the Code is not yet an 'approved Code', i.e. it is a voluntary Code and not legally binding, although it is expected that it will soon become obligatory. The maximum of 90 dB(A) set by the British Code is probably not adequate to protect the hearing of all workers. If a proposed EEC Directive[12] comes into force the maximum legal limit in all member countries will be 90 dB(A) initially, reducing to 85 dB(A), over a period of 5 years from the date of ratification. Shrill noises with frequencies over 2000 HZ are considered to be more harmful than low pitched noise.

2.5.1.7 Noise levels on construction sites

An extensive survey of noise levels on construction and demolition sites was made by Beaman and Jones of CIRIA (1977)[13]. The object of the survey was to provide a guide to the construction industry in meeting the requirements of the Control of Pollution Act, 1974. Part III of this Act empowers Local Authorities to limit the noise from construction and demolition sites, without, however, providing much guidance on how much noise should be allowed at the site boundary. The survey showed that on the quietest site the measured boundary levels ranged from 52 to 62 dB(A) Leq at different points on the boundary, but on the noisiest site the range was 68 to 92 dB(A) Leq.

A typical noise limit might be 73 dB(A) Leq over a 12 hour period at a point 1 m from the adjacent facade of the nearest residential building, but this is entirely a matter for the discretion of the Local Authority. The noise levels to which construction workers are exposed has received less attention, no doubt because there is as yet no legal requirement, despite the official recommendation of 90 dB(A) Leq over an eight hour working day. The data given by Beaman and Jones does, however, provide a useful insight on the noise levels to which many construction workers are exposed. Similar data for sound power levels for a variety of site plant and equipment is also given in BS 5228:1975.[14]

Table 2.5.5 gives the highest and lowest observed sound power levels, taken from Beaman and Jones' report, for various noisy activities in construction and demolition, in which a range of different types of construction plant were used. The Table also gives averaged sound levels which the plant operator would experience at a distance of 2 m and other workers would experience at a distance of 10 m from the noise source, as calculated from equation 5(A).

The figures for the plant operator are indicative only since in the case of hand held power tools, such as pneumatic breakers, rock drills and poker vibrators, the operator will be closer than 2 m to the noise source, whereas in others such as piling, the operator may be further away.

With the exception of the jacked piling operation (which is included in the Table for comparison with conventional methods of piling), the average sound levels of all operations at 2 m from the sound source exceed 90 dB(A), with most sound levels at 2 m over 100 dB(A), and two are above 110 dB(A). Thus virtually all operatives of the types of plant shown in *Table 2.5.5* need hearing protection, as do a fair proportion of other site workers who are obliged to work within 10 m of the source. Whilst some relief is given by the fact that some of the operations shown in the Table are not carried out over the whole length of the working day, this is balanced by the fact that on many construction sites, workers are exposed to the noise of several excessively noisy activities at the same time, all taking place within a limited area.

2.5.1.8 *Noise reduction on construction sites*

The reduction of noise on construction sites to levels where operators are not subjected to an $Leq_{(A)}$ (based on an 8 hour day) in excess of 90 dB(A), is proving a slow business in the U.K. Progress appears to be faster in some other countries, particularly Sweden, which has a lower limit of 85 dB(A), and tighter regulatory controls. At present noise induced hearing loss is not classed as an occupational disease so far as the U.K. construction industry is concerned.

Whilst the HSE have discretionary powers under the Health and Safety at Work, etc. Act, 1974 to compel employers to prevent their workers being exposed to potentially dangerous noise levels at work (i.e. $Leq_{(A)}$ of 90 dB(A) or more over an 8 hour day), this is only spelt out clearly in the case of woodworkers, under Regulation 44 of the Woodworking Machines Regulations, 1974. This is quoted below:

> "*44. Noise.* Where any *factory* or any part thereof is mainly used for work carried out on *woodworking machines*, the following provisions shall apply to that *factory* or part, as the case may be:
> (a) where on any day any person employed is likely to be exposed continuously for 8 hours to a sound level of 90 dB(A) or is likely to be subject to an equivalent or greater exposure to sound
> > (i) such measures as are reasonably practicable shall be taken to reduce noise to the greatest extent which is reasonably practicable, and
> > (ii) suitable ear protectors shall be provided and made readily available for the use of every such person;
>
> (b) all ear protectors provided in pursuance of the foregoing paragraph shall be maintained and shall be used by the person for whom they are provided in any of the circumstances in paragraph (a) of this Regulation;

(c) for the purpose of paragraph (a) of this Regulation the level of exposure which is equivalent to or greater than continuous exposure for 8 hours to a sound level of 90 dB(A) shall be determined by an *approved* method."

British Standard BS 5228:1975[14] and Digest 184[15] of the U.K. Building Research Establishment provide guidance on the control of noise in construction and demolition. A useful review[16] on the subject is provided in a paper by P.J. Hallman of the Building Research Establishment and presented at a seminar at the Public Works Exhibition 'Quieter Construction and Demolition' in November 1976. This considers four categories of construction plant:

site vehicles
piling operations
power tools
stationary plant

Site Vehicles

To substantially reduce noise levels from site vehicles requires attention to silencing, enclosure of engine, panel damping, as well, sometimes, as radical redesign of the engine. Manufacturers of site vehicles have been working actively on this problem for several years and a new excavator, dumper, shovel loader or bulldozer of today is substantially quieter than its predecessor of ten years ago, as well as subjecting its operator to less vibration. These new vehicles do not, however, solve the noise problem of older machines which still have several years of useful life. The first problem is to identify the dominant noise source. There are four particular areas where a substantial reduction in noise level may be achieved by expert modification:

- the engine cooling system, particularly the air fan, fan guard and radiator grill, as well as the water pump;
- silencers on air inlets and exhausts;
- the engine canopy, which on many vehicles does not cover the whole engine. This may have to be replaced by a complete canopy of damped steel, together with lined apertures and ducts for cooling air and a belly plate under the engine to reduce reflected noise;
- sound insulation of the operator cab. This may include sealing all gaps between the engine compartment and the cab, fitting rubber gaiters round all control levers, and lining the cab with acoustic absorbent material. Resilient engine mounts can also reduce noise and vibration levels.

Table 2.5.6 shows the noise levels of various site vehicles before and after acoustically engineered modification in all these areas. The overall reduction represents the sum of a number of smaller reductions resulting from particular modifications.

Some manufacturers of site vehicles, particularly of a pedestrian operated vibratory roller which is not discussed here, offer noise reduction kits which incorporate the types of modification discussed here. The effect of the age, general condition and state of maintenance of any vehicle on its

TABLE 2.5.6. Noise levels (dB(A)) inside driver's cab of standard (1974) and modified site vehicles

Vehicle 1	Articulated dumper	Articulated wheeled shovel loader	Hydraulic crawler excavator	Cabbed fork lift truck
Standard machine	93–94	113	103.5	103
Fully modified machine	84–85	84	88	85
Total reduction	9	29	15.5	18

noise emission also needs to be appreciated. Significantly higher noise levels than those shown in *Table 2.5.6* for 'standard machines' are often found as a result of defective silencers, uneven firing of engines, worn gears and bearings.

Piling
Most conventional pile driving by impact hammer exposes operators and persons obliged to be in the vicinity to noise levels well in excess of 90 dB(A), as *Table 2.5.5* shows.

Two broadly defined methods of quieter piling are available to contractors:

- the use of hydraulic rams, which apply sufficient force to push the pile into the ground (or extract it) without impact or ground vibration;
- the complete enclosure of the piling hammer and piles being driven.

The first method, exemplified by the Taywood Pilemaster[16,17] operated by Terresearch Ltd., consists of a heavy rig containing a set of hydraulic rams arranged side by side in a crosshead, each ram bearing on and secured to the top of a pile. The rig is lowered on top of the piles by a crane and pressure applied by the rams to one or two piles at a time. The force is derived partly from the weight of the rig and partly from the pull against the other piles to which the rig is attached. The main remaining noise sources are the electric generator (if mains power is not available) and the motor of the power pack in the rig. The total noise level of the Pilemaster, together with its attendant crane and generator, has been measured at 62 dB(A) at 15 m. The method is most suited to clays and other cohesive soils, but is less effective in granular soils.

Several organizations have developed quieter piling rigs on more conventional lines in which the hammer is enclosed and the piles either enclosed or damped by a clamping device. Examples are the 'Hush piling rig' developed by Sheet Piling Contractors Ltd and a modified and shrouded drop hammer developed by W.A. Dawson Ltd[16]. These can generally reduce the sound levels of individual hammer blows at 7 m from the pile to 90 dB(A) or less. It is, however, difficult to achieve reductions much in excess of 10 dB(A) merely by shrouding conventional piling hammers if the piles are left free to resonate and transmit airborne noise.

For a fuller discussion of quieter piling methods References 16 and 17 should be consulted.

Power Tools

A variety of power tools are used in construction in which the motive power may be provided by compressed air, an electric motor, a hydraulic system, a petrol engine or an explosive cartridge. Tools may be reciprocating and/or percussive or rotary. Compressed air tools tend to be noisier than other types due to the air exhaust. They are also generally preferred for certain duties, especially concrete breakers and rock drills, which are two of the worst noise offenders.

Pneumatic concrete breakers and rock drills have three main noise sources:

- exhaust air leaving the side ports of the body of the tool;
- impact noise of the piston radiating from the tool casing;
- ringing noise resulting from impact of the tip of the steel tool on a rigid surface;

Noise from the first two sources may be reduced by the use of fabric and rubber silencers strapped or bolted to the tool. In the absence of an effective exhaust silencer a suitable length of exhaust hose will both reduce the noise and distance it from the zone of the operator. Early designs of air exhaust silencers suffered from icing problems, but these have now been overcome.

Damped steel tools are now widely available which reduce the third noise source.

The fitting of an exhaust silencer together with the use of a muffled steel tool was found to reduce the sound level at 7 m from unsilenced pneumatic concrete breakers by about 10 dB(A) from the range 90 to 102 dB(A)[14]. Swedish pneumatic power tool manufacturers have pioneered the development of pneumatic power tools with reduced noise and vibration.

TABLE 2.5.7. Maximum sound power levels of hand held power picks and breakers

Weight class of pick or breaker kg	Maximum sound power level dB(A)	
	Stage 1. For immediate implementation	*Stage 2. For implementation 3½ years after Stage 1*
Less than 20	110	108
20 to 35	113	111
Over 35	116	114

Based on a draft EEC Directive[18] (Code 82/112) at present held up because of problems of third world countries, the following maximum sound power levels, given in *Table 2.5.7*, are likely to be enforceable in member countries of the EEC for hand held power picks and breakers of all types, whether pneumatic, petrol driven, etc. These limits apply to the noise of the breaker or pick only and exclude the noise of the tool on the concrete or other material being broken. Special test methods have been devised to make these measurements. In the U.K. the HSE is endeavouring to ensure that equipment made in the U.K. meets these limits.

The use of nylon gears and air valves in rotary and percussive power tools in place of metal ones has reduced their noise considerably. The extent to which engineering research may need to be taken to reduce noise is indicated by the provision of 'rifling' in the bore of cartridge operated fixing tools. This allows the gas formed when the cartridge explodes to be dissipated more gradually, thus reducing the noise from this type of equipment.

Stationary Plant[16]

Most compressors, generators, pumps, welding sets and concrete mixers and pumps used on site are driven by internal combustion engines, which themselves can be major sources of noise. The sources of engine noise and the modifications required to reduce it follow similar lines to those described for site vehicles. Both sound-reduced compressors and generators are available, as well as proprietary acoustic enclosures for reducing noise radiation from existing ones.

The noise level from a small standard 2.4 m^3/min rotary vane air compressor with a 3 cylinder diesel engine drive was reduced from 95 dB(A) at a distance of 1 m to 81 dB(A) by a number of modifications which apparently added about 8% to the selling price of the compressor. These consisted of:

1. Fitting a better proprietary silencer to the air intake;
2. Fitting a better exhaust system;
3. Replacing a metal cooling fan by a plastic one with air foil-shaped blades;
4. Removing projections which caused turbulence from the fan path;
5. Fitting flexible mountings between the engine/compressor assembly and the chassis;
6. Designing a well-sealed compartment for the compressor/engine assembly.

The high-pitched whine from carpenters' circular saws was studied by the Building Research Establishment. They found that it could be reduced from 110 dB(A) at the operator's position to 90 dB(A) by placing spring-loaded felt-tipped damping pegs in contact with the disc at certain critical positions on the disc.

2.5.1.9 Noise level monitoring and personal hearing protectors (see Chapter 2.7 for detailed discussion)

Today there are available a range of compact, robust and reliable sound level meters of varying degrees of sophistication (*Figure 2.5.7*). The majority have built-in filters for the 'A' and other scales. Ancillary equipment, such as recorders, integrators and computing circuits, are available for giving direct readings of time-averaged sound levels. Analysers which give the contributions of narrow frequency bands of one octave or one third of an octave to the overall sound level can be obtained. Many of these can be handled, set up and used by skilled and professional workers after a short course of training. Construction companies need to have people on their staff who can use such equipment, which are generally stationary and tripod-mounted, in order to check that the sound levels

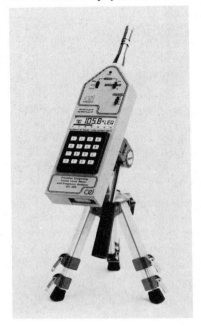

Figure 2.5.7 A precision computing sound level meter and frequency analyser (CEL Instruments Ltd).

resulting from their operations meet the requirements of Local Authorities under the Control of Pollution Act (1974).

For measuring the average sound level to which a worker is exposed over the period of his working day, personal noise dose meters are available. These fit directly into the worker's pocket and give a readout as the percentage of the permitted dose ($Leq_{(A)}$ of 90 dB(A) over an 8 hour day). The U.K. Code of Practice[10] requires the microphone of such dose meters to be kept within 1 m of the employee's head position whilst at work. Most dose meter microphones can be clipped onto the lapel of an overall or the brim of a safety helmet.

Many construction workers are exposed to 8 hour $Leq_{(A)}$ levels in excess of 90 dB(A) in the course of their work. This situation cannot be remedied without expensive modification of existing construction plant and equipment. Thus some form of ear protectors must clearly be provided. This is not a satisfactory long-term solution, since ear protectors are rarely comfortable when worn for long periods. Perspiration and the feeling of isolation encourage their removal.

Ear protectors are of two main types, ear plugs and ear muffs. Ear plugs are inserted into the auditory canal, whereas ear muffs are designed to cover the external ear completely. The attenuation which may be achieved by typical ear plugs and muffs is shown in *Figure 2.5.8*. Ear plugs can generally be used by workers exposed to noise levels in the range 90–105 dB(A). Ear muffs should be worn by those exposed to noise levels in excess of 105 dB(A). Special attention should be given to the protection of workers exposed to noise levels in excess of 115 dB(A), to ensure that muffs can be provided which will give adequate attenuation of the sound.

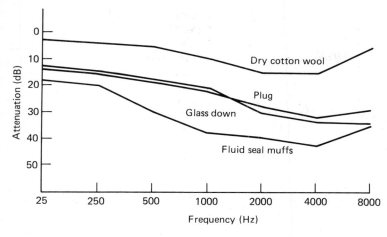

Figure 2.5.8 Typical performance of ear protectors

Parallel to the provision of hearing protectors, there is an obvious need to carry out regular audiograms on workers who may be exposed to high noise levels and to keep records. This once used to be a highly skilled job requiring the services of an expert. Now, with the equipment currently available, workers' hearing can be tested with reliability and confidence by most nurses and others after a short course of training.

2.5.2 Mechanical vibration

Mechanical vibration (a term used to distinguish it from electromagnetic and other forms of vibration), referred to henceforth simply as vibration, is a universl phenomenon, and usually associated with health and vitality. Vibro massages, horse riding and even the more pedestrian jogging, are supposed to 'tone up' the skin or muscles or shake up the liver, or be beneficial in some way. The heart and lungs vibrate and without this vibration we should be dead.

It is only recently that exposure of the human body to too much vibration has become recognized as a health hazard, and that vibration of certain frequencies and in certain directions is more hazardous than others. Thus the first recognition of 'white finger disease' among pneumatic tool operators came in 1911[19].

Many construction workers are exposed to vibration, sometimes of dangerous levels. This can arise either from the vehicles they drive or operate, usually seated in a cab (whole body vibration), or from the power tools they hold in their hands (hand and arm vibration). In some cases these vibrations arise from the use of engines and machines where it is simply an unwanted and undesirable side effect, and plays no useful role whatever (e.g. most site vehicles). In other cases vibration plays an essential role in the performance of the tool or machine (concrete vibrators, vibratory road rollers, concrete breakers and rock drills).

Vibration transmitted to the human body is more difficult to measure than noise and its effects are also more difficult to predict. Individuals suffer most from particular frequencies of vibration, but a frequency which most affects Mr Smith will not be the same as that which affects Mr Brown and vice versa.

As yet there are no universally accepted limits to the vibration which workers may be exposed to, nor is there much official guidance.

Before discussing the health problems caused by vibrating vehicles, power tools, etc. to construction workers, some of the main characteristics of vibration are discussed.

2.5.2.1 Physical characteristics of vibration

Most readers will be familiar with some characteristics of vibration from courses of mechanics. Thus the following points are intended more as reminders than as basic instruction.

Vibration has been defined as a 'to and fro' movement, and two distinct types have to be recognized, 'free vibration' and 'forced vibration'.

Free vibration is the natural vibration of something set in motion by an initial impulse, whereas forced vibration is the movement of something (e.g. a cocktail shaker) under a superimposed external oscillatory motion—often created by a rotating shaft with an eccentric cam.

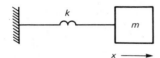

Figure 2.5.9 Simple mass/spring system.

All objects possessing mass and elasticity or springiness have the capacity to vibrate freely, the simplest form of which is known as 'simple harmonic motion'. An example of this is the spring. When oscillating the returning force at any point is proportional to the displacement from its natural position. This proportionality can be expressed by a constant k known as the stiffness of the spring (*Figure 2.5.9*).

$$\text{Restoring force} = -kx \tag{1}$$

The restoring force accelerates the mass m so that

$$m \frac{d^2x}{dt^2} = -kx \tag{2}$$

and

$$\frac{d^2x}{dt^2} = -\left(\frac{k}{m}\right)x \tag{3}$$

where t is time and d^2x/dt^2 is acceleration (e.g. in metres per second^2).

The natural frequency of simple harmonic motion, expressed as f_0 (measured in Hertz or cycles per second) is given by the equation:

$$f_0 = \frac{1}{2\pi}\sqrt{k/m} \tag{4}$$

The vibrating system (spring attached at one end to a fixed object and to the vibrating object at the other) has a certain vibrational energy which is proportional to the maximum displacement (amplitude) or the peak acceleration. If no energy were lost the vibration would continue indefinitely with the same amplitude. In practice there is always a gradual loss of energy through friction or its conversion to sound waves. This loss of energy, known as damping, results in a vibration of gradually decreasing amplitude, of frequency very slightly less than the natural frequency f_0 for an undamped system. The change in the displacement of the vibrating object with time for the damped system is shown in *Figure 2.5.10*.

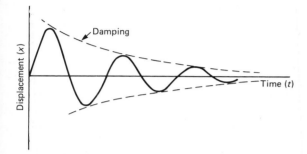

Figure 2.5.10 Exponential decay due to damping.

Since most solid objects found in construction (e.g. beams, window panes, pipes, metal sheets), have both mass and springiness, they can vibrate freely at characteristic frequencies. Rectangular sheets of metal or glass have two, or sometimes more natural frequencies. Their natural vibrations are thus more complex. The natural frequency of vibration of such objects also depends on where they are fixed (to some relatively immovable object) and what part is free to vibrate.

The human body is not usually thought of as one that can vibrate freely, with characteristic frequencies. Since, however, all parts have mass and some elasticity, natural vibration is possible, although it is heavily damped. Approximate natural frequencies of various limbs and organs of the human body are given in *Table 2.5.8*[20]. Wide variations are, however, possible depending on the body weight, dimensions and other characteristics.

Forced vibration in the simplest case (like that of a cocktail shaker) happens where the object simply follows the vibratory path and frequency which is superimposed on it by mechanical means. Where the object is one which can vibrate freely on its own at a characteristic frequency the

TABLE 2.5.8. Natural frequencies of vibration of various parts of human body for vertical vibrations (subject standing or sitting)

Organ or vibrating system	Natural frequency of vibration (Hz)
Eye (in eyeball)	60
Head/neck	20
Shoulder/arm	4
Thorax/abdomen	3–6
Spine	5
Pelvis	5
Legs	5

situation becomes more complicated. The movement then depends both on the natural frequency of vibration of the system and on the vibration of the applied force. When the frequency of the applied force matches the natural frequency of the system, the coupling becomes very efficient and large displacements are possible. This is known as resonance. Resonance can damage the vibrating system, as all will appreciate when a rotating shaft passes through its critical speed (i.e. its natural frequency of vibration). Resonance can also damage organs of the human body.

To summarize, the main characteristics of vibration are the frequency, displacement and direction; since, however, the displacement at any frequency in the case of harmonic vibration is proportional to the peak acceleration, we often find vibration characterized by its frequency and acceleration instead of displacement.

2.5.2.2 Whole body vibration

This is found at frequencies of 0.1 Hz to 30 Hz. It occurs when the subject sits, stands or lies on a vibrating seat, floor, structure, bunk, etc. Depending on the frequency, acceleration, time of exposure, and other circumstances, the result may be loss of coordination, giddiness, seasickness, spinal disorders, varicose veins, piles, headaches, constipation, bone damage and nervous disorders.

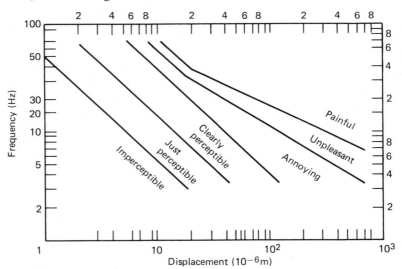

Figure 2.5.11 Reiher–Meister scale for vertical vibration on standing subjects

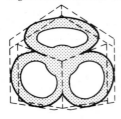

Figure 2.5.12 The organ of balance. Three circular canals attached to the inner ear (from *The Biology of Work*, by E.G. Edholm, Weidenfeld and Nicholson, London).

A subjective response of standing subjects to vertical vibration is shown in *Figure 2.5.11* which is known as the Reiher–Meister Scale.

Low frequency vibration caused by movement of a ship at sea, sometimes of cars and aeroplanes and of tall slender structures, in high winds, affects the human balancing organ. This consists of three semi-circular canals set at right angles to each other, close to the inner ears (*Figure 2.5.12*). The canals are filled with fluid and are attached to nerve endings which are sensitive to changes in pressure. Movements such as those described cause corresponding movements of the canal fluid which stimulate the nerve endings, and cause giddiness, etc.

Reciprocating engines, particularly diesel, give rise to vibrations at several frequencies which affect drivers and others exposed to them in different ways. Vibrations of the natural frequency of the spine (about 5 Hz)[20] which is transmitted through the driver's seat is a major cause of back trouble in drivers of construction vehicles. At this frequency the movement of the head is amplified by resonance of the spine, but a person standing on a platform vibrating at 5 Hz is less affected as his legs absorb much of the vibration. Excessive vibration at frequencies of 10 to 20 Hz can damage the lungs, cause rectal bleeding, blood in the urine, constipation and heart failure in extreme cases. Vibration at 20 Hz can cause resonance of internal head organs causing headache and psychoneurotic symptoms[21]. Vibration in the range 30–40 Hz may cause the eyes to resonate and affect vision. An American survey[22] by the National Institute of Occupational Safety and Health found that drivers of heavy construction equipment were specially prone to heart disease, obesity, muscle and bone problems.

Success in preventing driver vibration in construction vehicles is primarily achieved in the design stage and depends to a large extent on good suspension of the cab or cabin. In cases where an independently suspended cab is impracticable, a suspended seat helps, although this only reduces vertical vibrations, while the driver still has to copy with vibrating instruments and controls. Wear and tear and lack of maintenance in any vehicle increases vibration.

2.5.2.3 Hand and arm vibration [19,20,21]

This falls in the frequency range of 4 to 2000 Hz. Hand and arm vibration is suffered by the users of reciprocating and some rotary hand-held tools, particularly pneumatic drills used for breaking up road surfaces, concrete and masonry. It can lead to several types of occupational disease:

1. 'White fingers' or 'dead fingers', known as Raynauds phenomenon, is the commonest in the U.K. It results in loss of circulation which usually starts in cold weather, and leads to loss of feeling and partial loss of the use of fingers, and in extreme cases to gangrene and amputation. It is not always caused by vibration, and women are more prone to it than men.
2. Osteoarthritis of the arm joints, especially the elbows. This seems to be the commonest in Germany where it is compensatable. It may be followed by muscle and nerve diseases.

3. Hardening of the soft tissues, particularly the palms of the hands.
4. Small areas of decalcification of the bones.

The intensity of hand and arm vibration depends on a number of factors. Health measures include:

- pre-employment medical screening and employment only of fit persons;
- limitation of spells of work;
- use of specially-designed tools to minimize vibration and recoil.

Regarding the last point, models equipped with a device known as the Pons cylinder[19], as well as the manipulators and robot tools systems developed in Czechoslovakia and Sweden (Atlas Copco) deserve brief mention.

Shock absorbers and the isolation of handles of tools, and the supporting of hand-held tools reduce the level of exposure, so long as these measures do not interfere with the control of the tool. Gloves are of doubtful value in reducing hand vibration, since the operator has to grasp the handle more tightly through the glove. They are, however, useful in keeping the hands warm and promoting blood circulation. This itself serves to reduce susceptibility to Vibration-induced White Finger (VWF).

2.5.2.4 Vibration standards, measurements and legal control

The international Standards Organisation (ISO) has issued several vibration standards including one[24] for whole body vibration; The British Standards Institute has a similar draft standard[25] (DD 32:1974), as well as one on hand-arm vibration (DD 43:1975)[26].

Measurement of vibration is more difficult than of noise, although similar principles are involved. Some methods are discussed in Chapter 2.7.

The three criteria which are considered most important to health are:

1. The acceleration of the surface, measured in metres/second^2 (m/s^2).
2. The frequency of the vibration, measured in cycles/second or Hz.
3. The duration of exposure.

The direction of vibration, whether vertical or horizontal, is also very important.

The BSI draft for whole body vibration suggests three different bases for different groups of people. These bases are:

1. *The reduced comfort boundary*, to be used for example in design of passenger accommodation in buses.
2. *The fatigue decreased efficiency boundary* (FDE), to be used in critical and hazardous tasks such as driving vehicles and delicate work, where fatigue lowers performance.
3. *The health and safety exposure limit*, set at half the limit of voluntary tolerance for healthy males fastened to vibrating seats.

These limits for whole body vibration are shown graphically in *Figure 2.5.13*. This shows that people are most susceptible to vertical vibration in

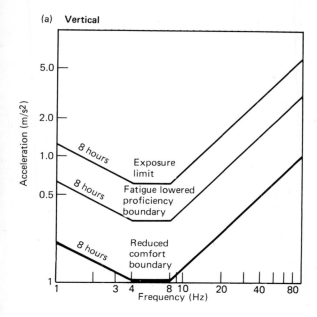
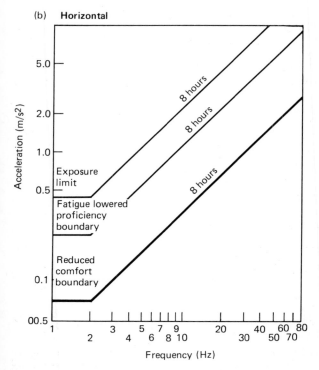

Figure 2.5.13 Standard for exposure to whole body vibration.

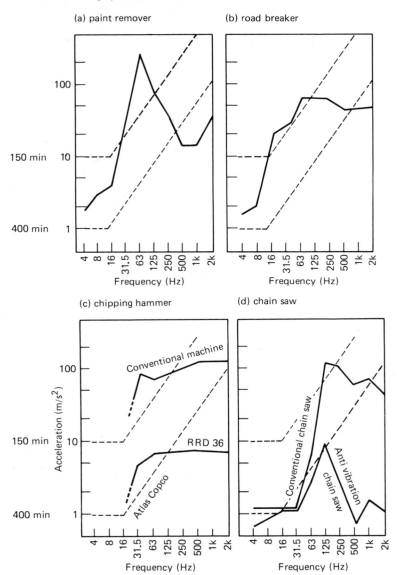

Figure 2.5.14 Standard for exposure to hand-arm vibration, with various appliances superimposed.

the frequency range 4 to 8 Hz, and to horizontal vibration in the frequency range 1 to 2 Hz.

The draft standard for hand-arm vibration sets two limits, one for a cumulative exposure of 400 minutes, which is the maximum exposure which should be allowed per working day, and one for a cumulative exposure of 150 minutes. The maximum susceptibility is at frequencies from 4 to 16 Hz. The limits are shown in *Figure 2.5.14*, which also shows performances of a paint remover, a conventional road breaker and comparisons between conventional and specially designed chipping hammers and chain saws.

Occupational diseases caused by vibration are not as yet included in the U.K. prescribed disease list, which qualify for statutory benefit, although writers cramp, and beat hand from which many miners suffer, which follows the 'frequent jolting of the hand in the use of pick and shovel', are on the prescribed list[19]. There have, however, been a number of cases where compensation was won as a result of court actions brought under Common Law, and in consequence a number of recent claims have been settled out of court.

In addition to the health hazards already discussed, vibration is also a hazard to safety and mechanical integrity. It can lead to fatigue and failure in metals, loosen nuts and bolts on machines and structures, and dislodge tools and solid objects placed on structures with risk of injury to those below.

Vibration also affects people's ability to read instruments and written instructions, as every car or coach passenger knows. This can lead to hazards in the operation of plant and machinery. Vibrational frequencies of 3 Hz are said to be the most dangerous for instrument reading[27]. Most vehicle drivers are familiar with the occasional lurching which occurs at low speeds, with the weight of the foot and body alternately depressing and releasing the accelerator pedal.

2.5.3 Electromagnetic radiation[28,29]

The vast spectrum of electromagnetic waves known to man is shown diagrammatically in *Figure 2.5.15*, and ranges from wavelengths of 10^8m (comparable to the earth's diameter) down to 10^{-16}m or 10^{-7} nanometres (nm) (less than the size of a proton). Within this range visible radiation occupies a narrow band of 400 to 700 nm (1 nm = 10^{-9}m).

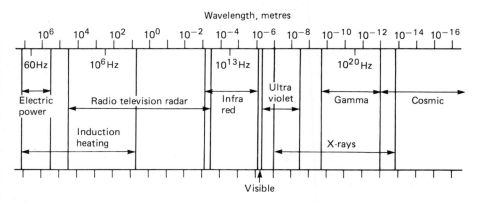

Figure 2.5.15 The electromagnetic spectrum.

Whilst most bands of electromagnetic radiation have characteristic hazards, only those commonly encountered in construction are discussed here. They are:

- ultra violet (UV) radiation, particularly arising from welding, with wavelengths ranging from 10 to 400 nm (although mostly between 100 and 400 nm);
- visible light with wavelengths from 400 to 700 nm;
- infra-red (IR) radiation, also arising from welding and other hot operations, with wavelengths from 700 to 10^6 nm, but mainly between 700 and 10 000 nm;
- laser light beams, a special form of intense light of a single frequency ('monochromatic'), generally emitted in pulses in the form of narrow pencil light beams, and used in construction and surveying for alignment purposes and for accurate measurement. This is mostly in the visible and near infra red range;
- ionizing radiation from sealed radioactive sources, mainly used as an aid to weld inspection. All radiation with wavelengths shorter than about 1 nm is capable of removing electrons from atoms to produce ions, but the radiation from the radioactive sources, discussed here, has wavelengths between 10^{-4} and 0.2 nm.

Brief comments on protection are given in this secion, but a more detailed discussion on personal protection is given in Chapter 2.8.

2.5.3.1 UV Radiation

Ultraviolet radiation has a more violent action on the body than visible or infra red radiation, and being invisible it can produce severe burns without warning. The main source of ultraviolet radiation exposure for construction workers is arc welding, but carbon arcs and mercury vapour lamps also produce sufficient UV radiation to be a hazard to those near them. The commonest exposure to ultraviolet radiation is from direct sunshine, especially at high altitudes and low latitudes, where there is less air in the path of the sun's rays to absorb it.

The principal hazards of ultraviolet radiation are to the eyes and the skin. The short-term effect on the eyes is conjunctivitis, an often acute and painful inflammation but usually without lasting effects. Prolonged exposure may, however, lead to cataracts.

Eye protection for welders is mandatory under 'The Protection of Eyes Regulations, 1974'. Processes involving the use of an exposed electric arc or an exposed stream of arc plasma require the use of 'approved shields or approved fixed shields'. However, for welding of metals by means of apparatus to which oxygen and flammable gas or vapour is supplied under pressure, 'approved eye protectors' may be used as an alternative to approved shields or approved fixed shields. Eye protection is discussed in more detail in Chapter 2.8.

Ultraviolet light can cause tumours which occasionally turn malignant on exposed areas of the skin. Welders and outdoor workers in strong sunshine are at risk unless protected. Those with fair skins are usually most affected by UV light.

Uv radiation also increases the skin effects of some chemicals used in construction, such as cresols (in creosote). This was discussed in Chapter 2.3 under 'photosensitization' and in Chapter 2.4 under 'chemical health hazards'.

The skin may be protected from UV light either by the use of skin creams containing UV absorbents, or when this proves insufficient, by suitable clothing. Wool absorbs UV radiation more effectively than cotton and woollen clothing is therefore recommended for electric arc welders.

2.5.3.2 Visible light (except lasers)

Light or lighting is usually thought of more as a safety than a health hazard, although both high and low levels of lighting can cause eye problems. High lighting levels can cause degeneration of the retina and impaired vision, whereas work in poor lighting or semi-darkness is the cause of miners' nystagmus, an involuntary rolling of the eyes, associated with depletion of the eye sensitive rods in the retina. Other symptoms resulting from poor lighting are vertigo, headache and sensitivity to bright light.

Artificial lighting, generally in the form of tungsten halogen floodlights, is used to extend the duration of the working day, especially in winter, on many construction sites. Multiple lighting from several directions is needed to avoid shadows cast by obstructions. Recommended design and minimum illumination levels on construction sites are given in *Table 2.5.9*[28].

TABLE 2.5.9. Recommended design and minimum lighting levels on construction sites[28]

Type of work	Recommended level, lux	
	Design	Practical minimum
Movement of people & vehicles, material handling & transport by crane & dumper truck	20	5
Rough work & site clearance	50	20
Bricklaying, scaffold & shuttering erection, reinforcing, concreting	100	50
Fine craft work inside buildings: joinery, plastering, painting, electrical & plumbing installation	300	200

Stroboscopic effects can be a serious lighting hazard for users of rotating or oscillating machinery since they can give the impression that fast-moving machinery is stopped or moving slowly. These are most serious with high pressure mercury, halide and sodium lamps with transparent envelopes. Lamps with fluorescent coatings (phosphors) give less flicker, whilst tungsten filament lamps give least cyclic variation. The use of special or rectified electric circuits can greatly reduce stroboscopic effects. To achieve effective and trouble-free lighting is only possible when those who design and instal it have had training in illumination.

2.5.3.3 Infra-red radiation

Infra-red radiation comprises a large part of the heat radiated from hot objects, including fires, furnaces and hot metal. It mainly affects the skin and eyes.

The burning sensation, reddening and discolouration of the skin generally provide adequate warning of skin injury. Less warning, however, is given of possible eye damage. The action of the lens of the eye is to focus IR and other radiation on to the back of the eye or retina, causing a heat cataract. Eye fatigue and headache are early symptoms. This has been common among hand blowers of glass and was one of the earliest (1908) compensatable industrial diseases in the U.K. Fortunately, this appears to be a rather rare condition among construction workers, although it may be found in those constantly using blowlamps, infra-red ovens and gas fired heaters. Persistent eye exposure to intense sunlight without eye protection can also cause heat cataracts.

Even workers who are not engaged in a scheduled process for which eye protection is legally required (see Chapter 2.8) should, if they feel they are exposed to excessive radiation, be provided with safety spectacles with appropriate filter lenses, and should have their eyes tested regularly.

2.5.3.4 Lasers[30,31,32]

The laser (Light Amplification by Stimulated Emission of Radiation) is a recent invention and the first successful experimental units were not operating until 1960. Since then, development, especially in the military field, has been very rapid, while in the civil field a number of different types of laser of a wide range of power and hazard potential are now used.

Such applications include:

measurement, especially of distances
communications
welding
cutting
eye surgery
display and advertisement
scientific research

The main uses of lasers in construction are for surveying and alignment (of bridges, walls, tunnels, etc.).

Laser waves are of a single colour frequency and are almost parallel to one another, all travelling in the same direction. Atoms which can emit light are pumped full of energy, e.g. by the application of a high voltage or a powerful impulse of radiation, and stimulated to fall to a lower energy level when they give off light waves which are directed to produce the coherent laser beam. The subject is highly technical and can only be understood by those who have studied physics to an advanced level.

Most laser beams are highly collimated, i.e. they are extremely narrow and only diverge very slightly. These rays lose little of their original energy intensity, even after travelling for considerable distances.

The main dangers of lasers are to the eyes of the viewer, although the more powerful ones may also cause skin damage, produce hazardous concentrations of ozone in the surrounding air (Chapter 2.4), heat objects in their path with consequent fire hazards (Chapter 3.2), whilst their high voltage electrical circuits are potential hazards (Section 2.5.4).

The vulnerability of the eye to lasers results principally from the focussing of a laser beam entering the eye by the cornea and lens on to a very small spot on the retina where its intensity may be increased by a factor of 100 000. Looking directly along a laser beam can therefore burn small areas of the retina, destroying the optic rods and cones and causing partial loss of vision (blind spots). Whilst it is usually a straightforward matter to ensure that persons are not exposed visually to direct laser beams, it is less easy to prevent their exposure to reflected beams. All flat objects reflect laser beams to a greater or lesser extent, and the reflected beam from a glass or polished metal surface can be almost as dangerous as a direct beam.

To safeguard personnel against laser hazards the British Standards Institution have published standards for classifying and labelling lasers according to their degree of hazard. These classes, together with their definitions and principal labelling requirements, are given in *Table 2.5.10*. The very wide range of power and hazard potential of industrial lasers is clear from this table. Lasers of classes 1 and 2 present little hazard potential, whilst class 3A lasers primarily give rise to eye hazards arising from the use of optical aids, e.g. telescopic theodolites, which can concentrate the beam before it enters the eye. Lasers of higher class than 3A find application in cutting and welding, but generally in carefully controlled factory conditions and not on construction sites.

The lasers now most used in civil engineering are helium–neon (He–Ne) continuous wave lasers with radiant outputs between 0.5 and 10 mW, and emitting radiation in the visible red part of the spectrum at a wavelength of 655 nanometres (nm). Most of these are of classes 2 and 3A. These are used in civil engineering both for alignment and as a reference datum for measurement.

Another type, the gallium arsenide (GaAs) pulsed laser, with output power of about 10 mW, emits radiation in the near infra-red region at wavelengths between 850 and 950 nm, and is used solely by surveyors as a distance measuring (ranging) instrument.

Visible He–Ne laser beams are used both in the static mode and in rotating beam systems. In the static mode they may be used vertically as an optical plumbline, or horizontally or in other directions as reference datum lines for measurement.

In rotating beam systems, the beam is usually made to rotate in a horizontal or vertical plane. Used in a horizontal plane the laser is a useful measuring and surveying instrument for up to a radius of about 300 m. In the vertical plane, laser beams are used for alignment during the erection of walls, towers, structural steelwork, etc.

Recommendations by the HSE for the use of lasers in construction may be summarized as follows:

- Lasers used in construction should be confined to classes 1, 2 and 3A.
- When using class 2 products, the beam should, where reasonably practicable, be terminated at the end of its useful path. The laser should not be aimed at vehicles or personnel, particularly at head height.

TABLE 2.5.10. Definitions and labelling requirements for laser products (abstracted from BS 4803:1983, Parts 1, 2 and 3)

Laser class	Definition	Max. output power	Notes on eye protection and other hazards (as given in BS 4803:1983)	Main labelling requirements, labels fixed to product & legible from safe place at 2 m from product			
				Hazard warning Label Yellow supplementary label	Prohibition Label Red supplementary label	Label Blue	Mandatory supplementary label
1	Inherently safe because of low power or engineering design		No protection needed	No	No	No	
2	Low power devices emitting visible radiation (in range 400 nm to 700 nm) and no more than accessible emission level (AEL) of class 1 for other spectral regions	1 mW average over 0.25 sec for continuous wave (CW) lasers	Not inherently safe but eye protection normally afforded by aversion responses, including the blink reflex.	Yes 'CLASS 1 LASER PRODUCT'	Yes 'Do not stare into beam'	No	
3A	Devices emitting visible radiation and no more than AEL of class 1 for other spectral regions. Irradiance at any accessible part of beam shall not exceed 25 Wm^{-2}	5 mW for CW lasers	Protection for unaided eye is normally afforded by aversion responses, including the blink reflex, but direct intrabeam viewing with optical aids may be hazardous	Yes 'LASER RADIATION CLASS 2 LASER PRODUCT'	Yes 'Do not stare into beam'	Yes	'Obtain safety officer's approval for use of optical instruments'
3B	Devices emitting visible and/or invisible radiation	0.5 W for CW lasers. 5 joules per m^2 for pulsed lasers	Direct intrabeam viewing may be hazardous but under following conditions they may be safely viewed via a diffuse reflector	Yes 'LASER RADIATION CLASS 3A LASER PRODUCT' 'LASER RADIATION CLASS 3B LASER PRODUCT'	Yes 'Avoid exposure to beam'	No	

4	High output devices capable of producing hazardous diffuse reflections	They may cause skin injuries and could also constitute a fire hazard. Their use requires extreme caution	Yes	'LASER RADIATION CLASS 4 LASER PRODUCT'	Yes 'Avoid eye or skin exposure to direct or scattered radiation' No

a) a minimum viewing distance of 50 mm between screen and cornea; and
b) a maximum viewing time of 10 secs; and
c) a minimum diffuse image diameter of 5.5 mm

- Class 3A laser products should only be used when, due to high ambient light levels, more power is required than is available from Class 2 products. The following additional precautions should then be taken:
  1. a laser safety officer should be appointed
  2. only trained personnel should instal, adjust and operate the equipment
  3. areas where Class 3A lasers are used should be controlled, posted with standard laser warning signs, and access restricted to persons who have been advised on the precautions to be taken.
  4. class 3A lasers should, where possible, be aligned mechanically or electronically rather than manually.
  5. precautions should be taken to ensure that persons do not look directly into the beam.
  6. special light filters should be used when the beam has to be viewed with optical instruments unless an exception has been made by the laser safety officer after detailed assessment of the hazard.
  7. if the beam irradiance at the boundary of the controlled area exceeds the maximum permissible exposure level under any viewing conditions, including the use of 80 mm viewing optics, the laser beam should be terminated within the controlled area.
  8. the position of any laser beam path should, where reasonably practicable, be either well above or well below eye level.
  9. precautions should be taken to ensure that the laser beam is not unintentionally directed at reflecting surfaces such as mirrors, lenses, etc. and also that such reflecting surfaces are not accidentally introduced into the beam path.
  10. when not in use the laser should be stored securely to prevent access to it by unauthorized persons.

In the rare instances where lasers of classes 3B and 4 have to be used in construction, this should only be done by adequately trained personnel after careful evaluation by a specialist laser safety officer and after all necessary precautions have been taken.

Construction firms should only purchase lasers manufactured and labelled in accordance with BS 4803, 1983, or an equivalent national or international standard. Lasers already in use and not classified or so labelled should be classified according to this standard with the help, where necessary, of the British Standards Institute, and the appropriate labels affixed.

When eye protection is required, special goggles, glasses with side pieces, or anti-laser eye shields should be used. These must be certified by their makers to provide the attenuation required (10 fold, 100 fold or 1000 fold) at the wavelengths of the laser being used. Eye protection designed for a particular laser of one wavelength is generally worse than useless when a different type of laser of a different wavelength is employed. It gives a false sense of security and may encourage direct viewing of the beam with the very real risk of serious eye damage.

2.5.3.5 *Ionizing radiation*[33]

Electromagnetic radiation with short wavelengths (10^{-13} to 2×10^{-10} m) is produced both by x-ray machines and by radio-active isotopes. The resulting rays are termed 'x-rays' or 'gamma rays' depending on which source they come from, but there is little or no difference between them when they have the same wavelength and energy.

The rays are termed 'ionizing' because they have the power of separating electrons from their atomic orbits, thereby producing charged nuclei or ions and free electrons. These ions may react with molecules in ways which are not normal in life. They can kill living cells, cause malignant tumours and change genes. In consequence they are a very serious health hazard, although the harm they may do is seldom apparent until some time after exposure.

Potentially harmful exposure to ionizing radiation is measured in 'rems' which measure the cumulative dose received by an exposed person from zero time.

We are all exposed to very low levels of ionizing radiation from radio-active materials in the ground and rocks, and in the bricks and stones of our houses, without apparently suffering from ill effects. There is still controversy as to how much additional ionizing radiation man may be exposed to in work-related processes, but the general answer is 'as little as possible'. X-rays and radio-active sources have many practical uses, but because of their hazards they are only used where there is no practicable alternative, and even then to the least extent possible and under closely controlled conditions.

The subject is a vast and complex one, but fortunately little of it has so far crept into the life of construction workers. Masons working with granites and other stone with higher-than-average radioactive content are exposed to rather more than the normal background radiation level. Demolition workers have occasionally been exposed to radioactive materials arising from the previous occupancy of the building (e.g. a factory previously making or using luminous paints). Lightning conductor engineers are sometimes involved with radioactive sources which were applied to the tops of lightning conductors to provide a conducting path for the electricity through the air. Such operations need expert advice, monitoring and control.

The most widely recognized use of radioactive materials in construction, however, is the 'gamma radiation' used in the inspection of welds and in some metal thickness gauges. The radioactive materials employed in this way are sealed inside special capsules and known as 'sealed sources'. They may only be handled by professional, trained inspectors who, in this case, are known as 'classified workers', and their use is controlled by the Ionizing Radiations (Sealed Sources) Regulations, 1967.

These state the maximum doses which may be received by exposed workers in various parts of the body during any year. Protection from any radiation source depends on limiting the time of exposure, maintaining the maximum distance from the source, and the use of shielding between the source and the person exposed which absorbs the radiation.

The Regulations give detailed requirements for the competent supervision, recording and monitoring of radioactive exposure and the health of classified workers who alone are allowed to work with sealed radioactive sources. Personal 'dose meters' are attached to the clothing of classified workers which give an indication of the cumulative radiation exposure over a predetermined period. These may take the form of 'film badges' or thermoluminescent dose meters.

The radiation source in gamma radiography consists of a small metal capsule inside which the radioactive material is sealed. The capsule is fitted into a holder which is kept in a larger protective container which absorbs a high proportion of the radiation. The holder is specially designed to ensure that the capsule cannot be dislodged and knocked out of it. Radioactive sources must never be picked up by the hands. Those used on construction sites are usually removed and used by the radiographer on a long handling rod, or by mechanical, electrical or pneumatic means.

As well as the risk to the trained radiographer, whose exposure and health is carefully monitored and who is specially trained to appreciate the hazards of radiation, there may also be a health risk to other workers in the vicinity when a radioactive source is removed from its shielding container and used. Warning lights and a permit to work system should be used for workers who may be incidentally exposed to gamma radiation. As far as possible, however, weld inspection and other operations involving the use of radioactive sources should be carried out at night-time or when other workers are not liable to be exposed.

The complex technicalities of ionizing radiation are not discussed here, mainly because of the small number of construction workers likely to be affected. Detailed expert guidance is readily available.

Occasionally x-ray machines may be used on construction sites for radiographic weld inspection in place of sealed radioactive sources. They are, of course, much bulkier than radioactive sources, and they only emit ionizing radiation when they are switched on.

2.5.4 Electricity[34]

Electricity in unskilled hands can injure and harm in many different ways, so it is not surprising that the subject keeps popping up in different chapters of this book, i.e.

Chapter 1.4 Hazards of Electrical Work
Chapter 1.5 Roofing workers—risks of overhead cables
Ground workers—risks of buried cables
Chapter 3.3 Fire and Explosion risks

The direct effect of electrical shock on the human body, if not immediately fatal, is usually one from which the casualty recovers quickly. Electrical shock is thus usually considered as an accidental injury (under the heading of safety) rather than an occupational disease (under the heading of health). Nevertheless, the effects of electricity on the human body need to be appreciated.

There are four main types of electrical injury:

shock
burns
falls caused by shock
arc eye or conjunctivitis, largely confined to electric arc welding, and considered in Section 2.5.3 under 'Electromagnetic radiation'

An electric current flowing through the human body causes shock and injury, the severity of which depends on several factors:

- the magnitude of the current (expressed in milliamps)
- whether the current is a d.c. or a.c., and its frequency in the case of a.c. current
- how long the current flows through the body
- the path of the current through the body, and particularly whether it passes near the heart or other nerve centres
- personal factors such as the health and mental state of the injured persons.

The magnitude of the current itself depends on:

the applied voltage
the resistance of the electrical path through the body

The resistance in turn depends on the area of the skin in contact with electrically live objects, the moisture content of the skin, and again the voltage. The resistance of the body, particularly the skin, falls as the voltage is increased.

The effects of three different types of electric current, d.c., 50 Hz a.c. and 10 000 Hz (high frequency) a.c. current on a man's body are shown in Table 2.5.11[35]. A woman's body is more vulnerable than a man's, and the same effects in a woman are found at about 75% of the currents shown in Table 2.5.11.

TABLE 2.5.11. Effects of electric current on a man's body

Current in milliamps			Effect
d.c.	a.c. 50 Hz	a.c. 10 000 Hz	
0-5	0-1	0-9	No sensation
6-55	1-8	10-55	Mild shock
60-80	9-15	60-80	Painful shock and hand unable to let go of live parts in case of a.c. current
80-100	16-20	80-100	Some loss of muscular control
110-350	20-45		Severe shock and loss of muscular control
400-800	50-100		Possible heart failure (ventricular fibrillation)
over 800	over 100		Usually fatal

The most striking feature of this table is the much greater effect of a main's frequency a.c. current on the human body than d.c. or high frequency a.c. current of the same magnitude.

Electric shock is a general term for the disturbance of the normal functioning of the nerves and the muscles which they control caused by the current. Mains frequency a.c. current of 50–60 Hz produces the maximum excitation of nerve endings. Neither d.c. nor high frequency a.c. current stimulate them to anything like the same extent, although high frequency current, which is pulsed at lower frequencies (10 to 500 Hz) may have the same effect as mains frequency.

The commonest symptoms of electrical shock are stabbing and numbing pains at the parts of the body in contact with the conductors and sometimes along the path of the current. Further effects are the involuntary contraction of muscles whose nerves are along the path of the current. This may cause a man to grip and be unable to release a live conductor or a tool in his hand, or it may cause back and leg muscles to contract violently so that he springs backwards and falls. It may also cause muscle injury.

Most fatal electrical accidents are caused by a disturbance known as 'ventricular fibrillation', of the normal heartbeat. This prevents the heart pumping blood effectively and interferes with the supply of oxygen to the brain. The heart may then stop altogether. Unless help is given quickly death will follow.

The most important step in saving the life of an electrocution victim is to get his or her heart, as well as the respiratory system, functioning again properly. This requires some form of heart massage, as well as artificial respiration. The older methods of artificial respiration (such as the Schafer), which also involve massage of the heart, have proved particularly effective here. However, a considerable proportion of the casualties took periods in excess of 10 minutes and up to one hour to revive. So it is important to persevere with this treatment for at least an hour if needed.

The length of time for which a current passes through the body depends mainly on whether the victim has grasped a live conductor and is able to let go or not, and how long it takes to switch off the power, either manually or by an automatic cut-out. The magnitude of the (a.c.) current causing heart failure (ventricular fibrillation) appears from research on animals to vary inversely as the square root of the time for which it is applied. Thus a mains a.c. current of 200 milliamps passing for one second through a man's body will have about the same effect as one of 20 milliamps passing for 100 seconds—both with a high probability of causing heart failure.

Fordham Cooper, former HM Electrical Inspector of Factories, states[34] that shock is most dangerous when the current passes from one hand to the opposite foot, although hand to hand shocks are also dangerous, while shocks from foot to foot are much less likely to be fatal. Shocks involving the head but avoiding the heart may stun without causing heart failure and are used in slaughterhouses, as well as in electroconvulsive therapy. Such shocks appear to be rare as industrial accidents.

The physical state of health of the victim and the condition of his heart are obviously important factors in determining the effects of electric shock. But it seems that his mental state is no less important. Even mild electrical doses can produce hysteria and exaggerated reactions in some people, particularly if the shock was totally unexpected or if the victim had a phobia about electricity. Other cases have been reported of people unable to let go of a live conductor yet who released themselves on seeing a switch opened, when it was not the switch controlling the circuit to which they were exposed.

2.5.4.1 The importance of voltage in AC supply

For a given electrical resistance, the current passing should vary directly with the applied voltage. The resistance of the human body is not, however, constant like most conductors, but decreases as the voltage is increased. The current therefore increases at a rate higher than in direct proportionality to the voltage. This is shown in the tentative figures given by Fordham Cooper for hand to hand contact in *Table 2.5.12*.

TABLE 2.5.12. Calculated electrical characteristics of human body at 50 Hz in dry conditions (from W Fordham Cooper)

Volts	12.5	31.3	62.5	125	250	500	1000	2000
Resistance R, ohms	16500	11000	6240	3530	2000	1130	640	362
Current milliamps I = V/R	0.8	2.84	10	35.2	125	443	1560	5540

TABLE 2.5.13. Approximate threshold shock voltages at 50 Hz ac

Case	Minimum threshold of:	Volts r.m.s.
A	Feeling	10–12
B	Pain	15
C	Severe pain	20
D	'Hold on'	20–25
E	Death	40–50
F	Range for fibrillation	50/60–2000

Other authorities give the internal electrical resistance of the human body (hand to foot) as about 400 to 600 ohms, whilst stressing that the principal resistance lies in the thin layer of the outer skin. This depends critically on whether the skin is dry or wet, being in the order of 10 000 ohms or even considerably higher for dry skin, but 1000 ohms or less for wet skin.

The danger of electric shock from mains frequency a.c. current is thus critically dependent both on the voltage and on whether the person exposed to the current is working in wet or dry conditions.

An important factor for hand to foot contact is the footwear worn. Rubber and moulded plastic boots and shoes are generally much better insulators than leather ones.

Shocks from 220 volt a.c. supplies are severe and are often fatal if the skin is wet, whereas those from 50 volt a.c. supplies in similar conditions are generally mild and very rarely fatal. The effect of 12.5 volts applied across the human body is usually so slight that it can only be felt by sensitive persons. *Table 2.5.13* (also from Fordham Cooper) gives the a.c. voltages required for various thresholds of shock.

2.5.4.2 Treatment for electric shock

The importance of artificial respiration and heart massage (especially in the older methods of artificial respiration which involve some degree of heart massage as well), has already been stressed. When breathing has ceased this must be started immediately, using whatever method is known to the person at hand, as soon as the victim has been released from the live parts, and without waiting to move him to a more appropriate place. It must also be appreciated that a person revived in this way may remain in a critical condition for some time and his heart and breathing may fail again and need to be revived before he fully recovers. Thus he should not be removed to hospital immediately after revival unless accompanied by someone in the ambulance or other form of transport who can watch him closely and apply artificial respiration and massage again en route if necessary. Even in cases of serious electric shock, after effects such as headaches, nervous pains or mental disturbance are fortunately rare, provided there has been no serious physical injury. However, in cases where the shock causes muscle injury, fragments of myoglobin from the injured muscle may be transported by the blood to the kidneys, which may thereby be damaged.

2.5.4.3 Burns

Burns resulting from electrical accidents often produce more serious injuries than electric shock. This occurs particularly with direct currents, both of low voltage (below 80) and also with very high voltages. Four types of burns, listed in *Table 2.5.14*, are described by Fordham Cooper.

TABLE 2.5.14. Types of electrical burns

Type	Features
Contact burns	Resulting from touching live conductor. Local and sometimes very deep, reaching the bone. Position on body important in reconstructing an accident. Deep burns which destroy tissue below the skin without much visible surface injury can be very serious, leading to other complications such as blood clots which lodge in the lungs or valves of the heart. The patient should be kept under medical supervision until the danger of such side effects has passed.
Arc burns	Often extensive and covering large areas of the body, particualrly from high-voltage flash-over. Injury usually sterile. Require normal burn treatment.
Radiation burns from short circuit	Similar to UV and IR radiation effects discussed in Section 2.5.3.
Vapourised metal	Fusing of small conductors or open fuses vapourises metal (copper, tin, silver) which may impregnate face or hands. Main risk is to eyes, but these are easily protected by spectacles, etc. Injuries require medical treatment and supervision, especially as the metals which are toxic may enter the bloodstream and attack other body organs.

2.5.4.4 Accidental electrical injuries in construction

An analysis of reported electrical accidents for one year on construction sites in the U.K. is given in *Table 2.5.15*.

It appears that not all the accidents reported here actually caused injury. Part 1 of Schedule 1 of the Notification of Accidents and Dangerous Occurrences Regulations, 1980 makes it obligatory to report a number of dangerous occurrences wherever they occur. These include (under paragraph 3), any 'electrical short circuit or overload attended by fire or explosion which resulted in the stoppage of the plant involved for more than 24 hours and which, taking into account the circumstances of the occurrence, might have been liable to cause *major injury* to any person'.

By far the most frequent cause of accidents is through contact with buried cables, although none of the 75 reported accidents were fatal. The main cause of fatal accidents, on the other hand, was contact with overhead cables, whether by crane, materials, tools or direct contact with the body. Surprisingly few accidents were caused through the use of

TABLE 2.5.15. Electrical accidents on U.K. construction sites over one year

Equipment involved	Building		Engineering	
	Fatal	Total	Fatal	Total
Portable tools (class 1)	–	1	–	1
Lamps	–	1	–	–
Testing sets, inc. lamps, instruments & test leads	–	–	1	1
Plugs, sockets, couplings & adaptors	–	–	–	1
Cables & flex for portables (other than test sets)	–	2	1	3
Electric hand welding (ex. eye flash)	–	–	–	1
All other portable apparatus	–	3	–	2
Transformers and reactors	–	–	–	1
Other switch, fuse & control gear above 650V	–	–	–	1
Circuit breakers not exceeding 650V	–	1	–	3
Switches & links not exceeding 650V	–	–	–	5
Fuse gear not exceeding 650V	–	–	1	9
Fixed lamps	1	2	–	1
Cables and accessories (ex. cables for portable apparatus & buried cables)	–	11	–	16
Buried cables	–	39	–	36
Contact by cranes & similar machines	4	5	2	5
Direct contact by persons, materials, tools	3	4	–	–
Batteries	–	–	–	1
Radio, TV, electronic instruments & power packs	–	–	–	1
Apparatus not classified	–	–	1	1

portable electric tools and cables, or through portable electric welding sets. This good safety record is due partly to the now nearly universal use in construction of low voltage circuits (110 volts) with centre pole earthed (cpe) for portable electric tools, and partly to the high degree of training of electricians. Special plugs, sockets and other fittings have been designed exclusively for use with these low voltage circuits.

The provision of portable transformers, switches, etc. on building sites has caused problems, but suitable ones are available to BS 4363 and BS 4343. The whole matter is covered by CP 1017, 'Distribution of Electricity on Construction and Building Sites'.

2.5.4.5 Further points concerning shock protection

- Portable electric tools are protected by a fuse in the plug. It is essential not only that the correct fuses are fitted, but also that after a fuse has blown, the cause be established before another is fitted.
- Insulating rubber footwear to BS 2506:1964 gives good protection to the wearer against shock from a live to earth short from normal mains voltage, whilst allowing sufficient conductance to allow the rapid discharge of static electricity to earth.
- Several types of sensitive ground fault interrupters are available which rapidly break any circuit in the event of quite small leaks to earth. These will protect an individual in a live to earth short, but not in a live to live contact.

2.5.4.6 Electric shock protection for welders

Open circuit voltages on arc welding sets, whilst generally not above 110 V a.c. cannot be ignored as a shock hazard. The welder may easily become exposed to this voltage while changing electrodes, setting up work or changing working conditions, and the danger is particularly great in hot or humid weather when the welder is sweating.

Welders must be trained to keep their bodies insulated both from the work and the metal electrode and holder, and should never allow an electrode or any metal part of the electrode holder to touch their bare skin or any wet clothing.

- electrode holders and cables should be well insulated;
- dry gloves and clothing should be worn;
- welders should wear insulating footwear such as that to BS 2506:1964;
- electrodes should never be changed with bare hands or wet gloves, or when standing on wet floors or grounded surfaces;
- if a cable becomes worn, the exposed part must be suitably covered with a rubber or plastic sleeve or insulating tape;
- cables must be kept dry and free of oil and grease;
- welding cables must be kept well away from power supply cables and high tension wires;
- cables must be protected against accidental damage or entanglement.

References

1. PENN, C.N., *Noise Control*, Shaw and Sons, London, 1979.
2. BRITISH MEDICAL ASSOCIATION SCOTTISH COUNCIL, *The Medical Implications of Oil Related Industry*, BMA, Edinburgh, **4**, (1975).
3. KINNERSLY, P., *The Hazards of Work and How to Fight Them*, 6th imp., Pluto Press, London, 63 (1980).
4. POULTON, E.C., 'Skilled Performance and Stress' in *Psychology at Work*, edited Peter B. Warr, Penguin, London (1971).
5. PRETLOVE, A.J., 'Fundamentals of Noise' in *Noise and Vibration Control for Industrialists*, edited by S.A. Petrusewicz and D.K. Longmore, Pan Elek, London (1974).
6. MIDDLETON, A.W., *Machinery Noise*, Engineering Design Guide 22, Oxford University Press, published for the Design Council.
7. HUNTER, D., *The diseases of occupations*, 2nd ed., English Universities Press, London, 790–796 (1957).
8. SATALOFF, J., *Hearing Loss*, 2nd Ed., J.P. Lippincott, Philadelphia, Pa (1980).
9. LORD ROBENS COMMITTEE, *Health and Safety at Work*, HMSO, London (1972).
10. DEPARTMENT OF EMPLOYMENT, *Code of Practice for Reducing the Exposure of Employed Persons to Noise*, HMSO, London (1972).
11. BURNS, W., and ROBINSON, D.W., *Hearing and Noise in Industry*, HMSO, London (1970).
12. EEC Proposed Directive, COM(82) 645 final, (15.10.82) Brussels.
13. BEAMAN, A.L. and JONES, R.D., *Noise from Construction and Demolition Sites—measured levels and their prediction*, Building Research Establishment, Watford, (1976).
14. BS 5228:1975 *Code of Practice for Noise Control on Construction and Demolition Sites*, British Standards Institution, London.
15. BUILDING RESEARCH ESTABLISHMENT, *Demolition and Construction Noise*, B.R.E. Digest 68/75, Watford.
16. HALLMAN, P.J., *A Review of Noise-Reduced Construction Plant*, Building Research Establishment, Watford, (CP 68/76).
17. PAGE, E.W.M. and SEMPLE, W., 'Silent and vibration-free sheet pile driving', Proceedings of the Institution of Civil Engineers, **41** (November), 475–497 (1968). and **43** (June), 331–343 (1969).
18. E.E.C. Draft Directive 82/112, published in the Official E.E.C. Journal (1975).
19. DALTON, A.J.P., *et al.*, *A worker's guide to the health hazards of vibration and their prevention*, British Society for Social Responsibility in Science, London, 55 (1977).

20. MURRELL, K.F.H., *Ergonomics*, Chapman and Hall, London (1979).
21. KING, I.J., 'Noise and Vibration', Chapter in *Occupational Hygiene*, edited by Waldron and Harrington, Blackwell Scientific, London (1980).
22. NIOSH, *The Health of Heavy Equipment Operators*, Archives of Environmental Health, p. 141, May/June (1976).
23. ILO, *Occupational Safety and Health Series 42. Building Work, A Compendium of Occupational Safety and Health Practice*, International Labour Office, Geneva, 227 (1979).
24. ISO 2631:1978, *Guide for the Evaluation of Human Exposure to Whole Body Vibration*, International Standards Office, Geneva.
25. DD 32:1974, *Guide to the Evaluation of Human Exposure to Whole Body Vibration*, British Standards Institution, London.
26. DD 43:1975, *Guide to the Evaluation of Exposure of the Human Hand–Arm System to Vibration*, British Standards Institution, London.
27. EDHOLM, E.G., *The biology of work*, Weidenfeld and Nicolson, 121 (1967).
28. WALDEN, H.A. and HARRINGTON, J.M. *Occupational Hygiene*, Chapters 8, 10 and 11, Blackwell Scientific London (1980).
29. NATIONAL SAFETY COUNCIL, *Accident Prevention Manual for Industrial Operations*, Chicago (frequently revised).
30. BEESLEY, M.J., *Lasers and their Applications*, Taylor and Francis, London (1971).
31. COX, E.A., *Lasers in the Surveying and Construction Industry*, (available from the Health and Safety Executive, London).
32. BS 4803:1983 (3 parts), *Radiation Safety of Laser Products and Systems*, British Standards Institution, London.
33. DEPARTMENT OF EMPLOYMENT, *Ionising Radiations: precautions for industrial users*, HMSO, London (1969).
34. FORDHAM COOPER, W., *Electrical Safety Engineering*, Butterworths, London (1979).
35. DALZIEL, C.F., 'Effects of electric current on man', *Electrical engineering* (Feb., 1941).

2.6

Other health hazards

As well as the chemical and physical health hazards discussed in Chapters 2.4 and 2.5, construction workers face a number of others which are generally absent in manufacturing industry. Most of these can be grouped under the following headings:

- biological hazards, arising both from the animal kingdom and from microscopic and sub-microscopic organisms;
- weather and the 'microclimates' in which construction workers operate;
- work in cramped and otherwise unnatural body postures;
- living conditions and welfare facilities, or rather the lack of them, particularly for migrant workers;
- work related mental stresses arising largely from the conflicts between the need to earn a living and the risks of the job. These stresses often lead to irrational behaviour and alcoholism.

2.6.1 Biological hazards

Biological health hazards originate from living things or are living things themselves which are capable of harming, or causing disease in humans. They include animals (both wild and captive), insects and body parasites, such as intestinal worms, fleas, lice and scabies, various kinds of micro-organisms such as bacteria, viruses and fungi, and poisonous plants.

Most infectious diseases which are common to the general public such as measles, chickenpox, German measles and mumps are caused by micro-organisms. Only those which have a special significance for construction workers are discussed here. They usually only affect particular groups of workers in particular circumstances and locations.

2.6.1.1 Animal hazards (excluding insects and visible parasites)

Under this heading are included wild and captive animals, birds, reptiles, snakes and fishes. (Some are also hosts of parasites and micro-organisms which may attack humans.)

In most industrialized countries today, where man's dominance over the animal kingdom is taken for granted, stories of construction workers being attacked by animals generally cause only a sceptical smile. According, however, to HSE's report 'Construction 81-82'[1], not less than 25 construction workers were injured by animals in the U.K. in 1982. Some were attacked by security dogs, others by angry wasps, and in one case roof workers were dive bombed by angry pigeons whose nests they were disturbing. Clearly construction workers cannot ignore these hazards, although no general remedies can be given. Solutions must be found to fit individual cases. In much of the developing world, however, where man's dominance over nature is less secure, animal hazards to construction workers cannot be taken lightly. This is not surprising when we consider that protection of the family from predatory animals has been one of man's main reasons for constructing homes and shelters. It applies even more to the nests, burrows and lairs created by the animals themselves. Construction workers in Africa have long faced the hazards of wild lions, elephants, and monkeys. A graphic description of the danger from lions in the early days of railway construction in Kenya is given in Patterson's classic *The Maneaters of the Tsavo*[2], once a favourite schoolboy adventure story.

Even in France, the dangers of snakebites to construction workers prompted a special survey and report[3], with an illustrated catalogue of the snakes liable to be encountered, measures to stop the spread of venom, how and when to inject serum and the treatment of wounds. Safety measures include warnings not to expose ones feet but to wear high-sided boots, to beat or shake bushes, make a noise, and have snake bite kits with ampoules of the appropriate serum at hand. Stacked timber and other stacked materials should be moved with a bar or similar appliance, and not by hand, in regions infested with venomous snakes. Workers should have instructions in the use of snake bite kits, and if bitten should use the kit as directed and remain still.

Many other wild animal hazards to construction workers exist, including bears, crocodiles, sharks, pumas, condor eagles, hornets and scorpions. These also have to be protected from man and the whole subject tends to lie rather outside the scope of this book.

2.6.1.2 *Parasitic hazards[4], insects and spiders*

Fleas and lice of various kinds, scabies and parasitic worms are health hazards to entire communities and subject to control by local health authorities and inspectors.

Hookworm

Hookworm disease or ankylostomiasis was one time prevalent among brick and ground tunnel workers in the U.K. and arises when the worker's bare skin comes into contact with contaminated earth containing the larvae of the worm. The main early signs and symptoms are skin eruptions and intense itching. If unrecognized and untreated the patient later suffers anaemia, shortage of breath and bowel upsets. Although the disease is not common today in the U.K., safety specialists and supervisors of men in the occupations referred to should be alert to the signs and symptoms of the disease and refer those suffering from them for medical treatment. The disease is readily cured by modern drugs.

Dangerous Insects
Workers in areas infested with dangerous insects should keep as much of the body covered as possible, wearing tight-fitting clothing, gauntlets and leggings. They should also use appropriate insect repellants.

Fleas
Cases have been reported of demolition workers being attacked by cat fleas from colonies of wild cats inhabiting derelict buildings. Cat fleas are a different species to those which normally bother humans, but if the cats are destroyed they may transfer to the human host. Proprietary sprays and powders containing pyrethrin and other insecticides are available for exterminating fleas and similar insects, although time and patience and knowledge of their life cycles are needed. Remember that even flea bites require medical dressing to prevent them turning septic, and some fleas, such as the Indian rat flea, transmit plague and other diseases.

Jiggers
The chigoe, or jigger, a species of flea originating in South America, is particularly dangerous, as the female lays eggs beneath the skin of the human host which cause serious and extensive sores, particularly of the feet, and fever, which may be fatal if untreated.

In regions infested with chigoes, workers should:

- avoid low vegetation as far as possible
- avoid sitting on the ground or on logs
- dust legs and arms with sulphur and take sulphur tablets
- use insect repellants such as dimethyl phthalate
- take a hot bath or shower every day
- get medical treatment at once if a bite becomes inflamed

Ticks
Ticks are parasitic bloodsucking insects which cause irritation and anaemia and may also spread parasitic protozoa which cause red water fever and other fevers. They lay eggs in long grass which hatch into larvae which attach themselves to the legs of passing animals and humans.

In tick-infested regions workers should:

- inspect the body and clothing at least once a day, removing ticks found on the body, if possible without allowing the skin to be punctured
- take care that ticks do not get into clothing or beds at night
- get medical treatment if they become feverish

Spiders
The bite of most spiders is poisonous, but only a few can pierce the human skin. Their bite is painful, akin to a bee sting, and occasionally fatal. Whilst rare in Britain, the dreaded 'black widow' and 'brown recluse' are found in other temperate regions, principally North America. Several poisonous spiders are found in the tropics, notably the burrowing trap-door spiders

and funnel web spiders of Australia and the wolf spider of Brazil. In regions infested with harmful spiders, workers should:

   wear gloves
   inspect all objects before handling them
   inspect outdoor toilets before using them

Site managers in such regions should check whether serums are available, and, where necessary, keep a stock on site.

Malarial mosquitos, tse-tse flies, sand flies
In regions infested with malarial mosquitos, sand flies or tse tse flies, workers should be issued with appropriate prophylactic pills and insect repellants, and educated about their need and use. Site and safety managers, health and medical authorities may need to implement a programme of pest control on and around construction sites. Window openings of site huts and particularly sleeping quarters should be protected with wire gauze.

2.6.1.3 Micro-organisms[4,5]

Most micro-organisms which cause disease or illness among workers fall into three classes:

   bacteria
   fungi, including yeasts and moulds
   viruses

Bacteria
Bacteria are microscopic single cell living organisms which multiply by dividing into two parts. Some feed on dead organic matter which they help to decompose. Others which feed on living organisms are termed parasites and are responsible for many diseases in man which include anthrax, tuberculosis, leprosy, tetanus, typhoid and cholera.

The following bacterial diseases are prescribed as 'occupational diseases' in the U.K. (see Chapter 2.2):

Anthrax—associated with handling animals and particularly their skins, hides and hair
Glanders—associated with handling horses
Leptospirosis or Weil's Disease—associated with those who work in areas where rats may be present
Tuberculosis—associated with workers who come into contact with people or animals suffering from the disease
Brucellosis—associated with handling animals and their products
Farmer's lung—associated with handling mouldy hay or other vegetable matter

These diseases are, however, prescribed only in relation to the worker's employment. The only one of these likely to be accepted as occupational in the case of construction workers is Leptospirosis which principally affects workers in coal mines, tunnels, abattoirs and sewers. Its initial symptoms are headache, muscular pain, nausea, vomiting, fever and jaundice.

Although not recognized as occupational in the U.K., tetanus or lockjaw is usually caused by soil contaminated with the organisms entering a deep cut or wound. Its symptoms vary from exaggerated muscle responses to the well known lockjaw. Ground workers and tunnel workers appear to be more at risk from it then most other construction workers.

Fungi

Fungi are plants which obtain their food from the living or dead tissues of other plants or animals. They vary in size from invisible yeasts to moulds which include edible mushrooms. The main fungal diseases in the U.K. are Tinea or Ringworm, a skin infection, and Thrush or Candidiasis, an infection of the mouth and intestines. Neither are recognized as occupational.

Viruses

Viruses are the smallest of the micro-organisms and for the most part are invisible under the most powerful light microscope. They can be viewed only indirectly by means of an electron microscope. Viruses grow and multiply only within the living cells and tissues of plants, animals and bacteria, and many cause diseases. These include mumps, smallpox, German measles, rabies, hepatitis, influenza, psittacosis and ornithosis.

Of these only viral hepatitis is recognized as an occupational disease in the U.K. This is a disease of the liver which often follows the ingestion of contaminated food or drink. The symptoms include fever and jaundice.

Ornithosis or 'pigeon fancier's lung' is caused by breathing the dust from the droppings of infected birds. Roof workers, painters, steeplejacks and demolition workers are among those occasionally exposed to this disease. Normally the illness is mild and similar to influenza, but in severe and untreated cases it resembles pneumonia and may result in death. It is also a disease to which some workers may become sensitized, and develop an allergic reaction even to traces of bird droppings. In such instances the person should be excluded from work involving contact with birds and their droppings. Workers heavily exposed to bird droppings are advised to wear dust masks and wash hands, face and contaminated clothing thoroughly after work.

Another disease transmitted by micro-organisms of which construction workers should be aware, is Humidifier Fever. This is transmitted by various organisms thought to include bacteria and fungi which can thrive in the humidifiers of air conditioning systems. The organisms are inhaled by those in air conditioned hotels, offices and factories. Symptoms are similar to influenza and include breathlessness, chest tightness, raised temperature and a general malaise. The illness is particularly noticeable when a worker returns to work after weekends or holidays.

The prevention of Humidifier Fever depends on the proper maintenance of humidifier systems of air conditioning units. Access by micro-organisms to humidifiers should be prevented by properly fitting and maintaining air filters. The humidifier water needs to be frequently changed and the whole humidifier system frequently cleaned. Air conditioning systems should also be monitored for micro-organisms.

Preventative measures against all types of diseases transmitted by micro-organisms include high standards of personal hygiene, including proper washing, toilet and changing facilities. Also sampling and monitoring of the whole working environment including the area immediately around individual workers should be undertaken. Samples should be microscopically examined to give an indication of the degree of contamination which may be present. This work can be undertaken by hygienists and laboratory technicians with a good training in bacteriology. Methods of monitoring and personal protection are discussed in Chapter 2.7.

2.6.1.4 Poisonous plants

Fortunately few poisonous plants are actually liable to attack us and in any case their hazards are as a rule easily avoided.

The most commonly encountered poisonous plants in the U.K. are stinging nettles and thistles, although in North America poison ivy, poison oak and poison sumac are common weeds. Poisonous plants should, as far as possible, be eradicated from construction sites during preliminary site clearance. As most of them are allergenic, persons known to be hypersensitive to them should not be employed on sites where they are present.

Apart from these plants found growing wild, many woods used in construction have caused health problems. African boxwood (Gonioma kamassi), which contains an alkaloid poison similar to curare, used by South American Indians, once caused a good deal of illness following the inhalation of dust created when the wood was machined. The symptoms of tiredness, mental fatigue, headache and watering eyes are similar to those of hayfever. Although still recognized as an industrial disease, it is now rare as this wood is little used today.

Splinters of several woods, including pitch pine, cause festering wounds, and it is hard to categorize any wood as completely harmless. Even the dust of English oak and beech is strongly suspected as the cause of cancer of the nose among wood working machinists in the furnuture industry. Care needs to be taken to eliminate wood dust from the atmosphere breathed by wood workers, and also to treat even minor wounds caused by wood splinters immediately. Even when the woods themselves are harmless and non-toxic, there is often the likelihood that the wood in question has been treated with a toxic preservative or paint.

2.6.2 Weather [6,7]

Weather cannot be described by any single parameter, but depends on temperature, wind velocity, humidity, precipitation (of rain, snow and hail) and sunshine. Human life itself can only exist within a limited range of weather conditions, whilst an even narrower range is needed for comfort. The second motive for building, after protection of the young from wild animals, is protection from the weather, and the creation of controlled microclimates in which men can live and work in comfort, no matter what the conditions are outside.

Man's need for a particular range of weather conditions arises from the fact that he is a warm blooded animal with a body core temperature of 37 ± 0.5 °C which is controlled by a built-in thermostat. This temperature has to be maintained if the heart, lungs, digestive organs and central nervous system are to function properly. The skin and body extremities however can vary from the core temperature by greater margins and are usually found to be in the range 30 to 35 °C. They have to be cooler than the body core temperature in order that the heat produced along with the energy generated for blood circulation, lung ventilation, body movement and physical work can escape to the surroundings.

The preferred environmental temperature for various tasks is given in Table 2.6.1.

TABLE 2.6.1. Preferred environmental temperatures for various tasks[8]

Type of work	Heat production calories per 24 hours		Preferred environmental temp. °C
	Men	Women	
Sedentary, mental	2200	1800	20–23
Sedentary, light	2700	2200	19
Standing, light	3000	2500	18
Standing, medium	3600	3000	17
Standing, heavy	4000		10–15

Even this, however, is a simplistic picture, since the rate of heat loss from the body to the environment depends on air humidity and velocity as well as temperature. The preferred environmental temperature for most construction work lies between 10 and 18 °C, but building workers often have to work in temperatures outside this range and only occasionally enjoy the comfortable microclimates they are creating for others. In the U.K. medical studies have confirmed ones obvious impression that cold and wet weather in winter adversely affect the health of construction workers. A report by Geoffrey Taylor[9] which included the study of 300 workers mainly between the ages of 20 and 50, suggested that older men left the construction industry because of diseases related to cold—rheumatism, bronchitis, and heart disease. Low body temperature was linked with higher accident and illness rates, lower bonus earnings, productivity and efficiency. Cold, of course, also increases the risk of some other occupational diseases (such as 'dead fingers' caused by holding vibrating tools).

In hot countries, such as the Arab-Iranian Gulf area and most of India, the opposite is true; summer is the worst time, when hard physical work becomes impossible at high ambient temperatures, and general listlessness and rapid heart rate are the first symptoms which soon lead to heat stroke.

The various factors which define the weather are first discussed separately.

2.6.2.1 Temperature

The environmental temperature can be measured in several ways, three of which are commonly used in connection with human working conditions:

dry bulb temperature
wet bulb temperature
black bulb globe temperature

The dry bulb temperature is simply the temperature measured by an ordinary mercury-in-glass thermometer held or suspended in the air. It is, however, neither sensitive to the humidity of the air, which affects the rate at which the body can lose heat by perspiration, nor to thermal radiation.

The wet bulb temperature is the temperature of a mercury-in-glass thermometer whose bulb is covered by a wick saturated with water. Its temperature is always lower than the dry bulb temperature except when the air humidity reaches 100%. The difference between the dry and the wet bulb temperature is a measure of the air humidity which can be readily established using hygrometry tables.

The black bulb globe temperature is measured by a thermometer whose bulb is at the centre of a copper globe which is painted matt black, and is most sensitive to thermal radiation and air movement.

The three thermometers mounted on a single stand are illustrated in *Figure 2.6.1*.

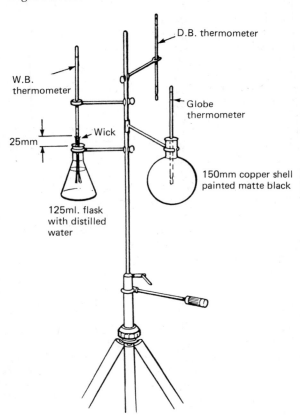

Figure 2.6.1 Wet bulb globe temperature apparatus (National Safety Council, U.S.A.)

For measurement of the environment temperature as it affects the worker's body, the wet bulb temperature index (WBGT)—a combination of three temperatures just considered—is used. This is given by the following equations:

$$\text{WBGT} = 0.7 \text{ globe temperature} + 0.2 \text{ wet bulb temperature} + 0.1 \text{ dry bulb temperature (for outdoor use)}$$

$$\text{WBGT} = 0.7 \text{ globe temperature} + 0.3 \text{ wet bulb temperature} \text{ (for indoor use)}$$

The main importance of this index is in assessing the thermal environment and limits of work in hot climates.

The environmental temperature zone as measured by the WBGT in which a person can perform the same task whilst wearing the same clothing in comfort is quite narrow, i.e. only 2–3 °C. As temperatures on most construction sites may change by far more than this—sometimes more than 15 °C—during the working day, there is a need which is not always appreciated, for construction workers to be able to shed or add one or more layers of clothing during an average working day. The upper WBGT at which hard physical work can be done effectively by a very lightly clad and acclimatized worker is about 32.5 °C.

TABLE 2.6.2. WBGT and recommended work/rest regimes[6] in hot weather

	Light	Work load Moderate	Heavy
Continuous	30.0	26.7	25.0
75% work, 25% rest per hour	30.6	28.0	25.9
50% work, 50% rest per hour	31.4	29.4	27.9
25% work, 75% rest per hour	32.2	31.1	30.0

Table 2.6.2 shows recommended work/rest regimes for various WBGT values as given by the American Conference of Governmental Industrial Hygienists in *Threshold Limit Values for Physical Agents*[6].

The effects of high ambient temperatures on the health of the worker can be very serious. The body temperature can only be controlled so long as clothing and air conditions allow sweat to evaporate rapidly enough for the heat produced by the body to be lost. If the control system is overstressed various symptoms such as swollen ankles, cramps and prickly heat begin to appear, leading in severe cases to a rise in body temperature to 42 °C, with heat stroke, coma and sometimes death[10,11,12].

For work in severe cold, the Wind Chill Index has been devised. This is a measure of the rate at which heat is lost per unit area of the human body, which usually ranges from 90 to 1400 kilocalories per hour per square metre. At values higher than 1400 the index is related to the time taken for exposed flesh to freeze, which ranges from 20 minutes at 1400 to 1 minute at 2400. The index represents a combination of temperature, wind speed and radiation loss. Constant values of the index are shown plotted against temperature and wind speed in *Figure 2.6.2*. This shows the critical importance of wind speed. At −40 °C in still air the index has a value of 1100. A man suitably clothed can work in these conditions with his face

Weather 293

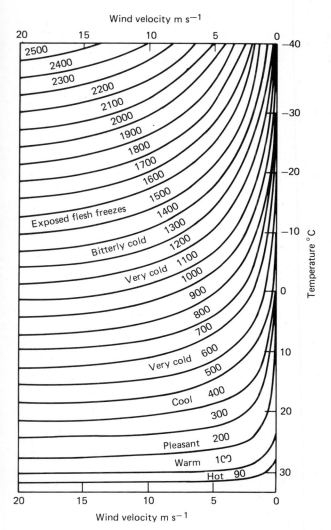

Figure 2.6.2 Wind chill index. Heat loss from the human body for various air temperatures and wind velocities (from F.S. Gill in *Occupational Hygiene*, edited by H.A. Waldron and J.M. Harrington, Blackwell Scientific, 1980).

exposed. In a slight breeze of 2 m/sec at the same temperature the index would rise to 1600, when the same worker would suffer severe frostbite. For values of the index above about 700 which corresponds to a wind speed of 2 m/sec at 0 °C or 8 m/sec at 10 °C, good head covering becomes essential for workers as well as good heat insulating clothing.

2.6.2.2 Windspeed

Even in the U.K. climate, construction workers unless suitably clad can suffer serious ill health through exposure to cold winds. This includes

294 Other health hazards

hypothermia (general lowering of body temperature), frostbite, dulling of mental faculties, slower reactions and greater proneness to accidents.

The ideal windspeed for most construction work appears to be about 0.5 m/sec, a very light breeze. As the windspeed increases, the hazards of construction work increase and first one and then another job becomes very hazardous, if not impossible. High winds are thus both a safety and a health hazard. Windspeed limits should, therefore, be set beyond which work on exposed roofs, ladders and scaffolds should be halted.

2.6.2.3 Rain, hail, snow and frost

Rain is a further and obvious health and safety hazard for construction workers. U.K. employers are required to provide rainproof outer clothing and headgear for those workers who have to work in foul weather, but this itself can cause problems in wearing the right amount of clothing underneath to achieve a good heat balance[13]. Work in any case can only proceed in fairly light rain, and nearly all work has to be suspended in a heavy downpour.

Hailstones can be a serious hazard to those working outside in some countries. Thus, in March, 1982, about 40 people caught in the open in Kanpur, Northern India, were killed by large hailstones in a sudden storm. Snow and frost interrupt most work, and generally bring a crop of accidents. One sound precaution is to turn over all scaffold boards immediately after snow and frost.

2.6.2.4 Weather forecasts and meteorological data

Construction supervisors and workers, like farmers, require reliable weather forecasts, and no site office should be without a suitable radio set kept permanently tuned to a local station which issues regular weather forecasts and storm warnings. It is also recommended that they should have a few simple basic meterological instruments such as the three thermometers necessary to determine the WBGT (*Figure 2.6.1*) and a wind direction and speed indicator (*Figure 2.6.3*). A small automatic weather

Figure 2.6.3 Anemometer and wind direction indicator—particularly important where exposed work on high structures is to be carried out (Munro Sestrel).

station, which gives audible and visual indication of changes in temperature, wind speed and direction, can be valuable on large construction sites in enabling workers to be given the opportunity of putting on or discarding a sweater, jacket, etc. and of providing danger warnings to roof workers, steel erectors, painters and others whose work and safety is affected by wind.

2.6.2.5 Night work, weather protection and clothing

Night Work
It is sometimes necessary to work at night (e.g. in hot climates), or in bad visibility. Adequate artificial lighting has then to be provided. Problems and standards of artificial lighting were discussed in Chapter 2.5.
Weather Protection
Much construction work today is carried out under temporary covering provided by polythene film, particularly during bad seasonal weather, although this, of course, is easily damaged or destroyed by high winds, along with its supports, so it too can cause accidents.
Working Clothes
Working clothes for construction workers are closely related to weather conditions and are discussed in Chapter 2.7. A good deal of research has been carried out in recent years on clothing for outdoor workers. Suggested reading on the subject includes writings by F.S. Gill[10], P.O. Fanger[11], D.M.K. Kerslake[12] and M.A. Humphreys[13], the last describing work carried out by the U.K. Building Research Station.

Personal protective clothing and devices have often to be worn by construction workers to protect eyes, hearing, hands and sometimes the respiratory system (see Chapter 2.8). It is often difficult enough to ensure that this can be worn in comfort in a common environmental temperature range of 10 to 18 °C. Outside this temperature range the problem of combining it with clothing to suit the prevailing weather becomes even more complicated.

2.6.3 Work in cramped and unnatural positions

HSE call attention in *Construction, Health and Safety, 1981–82*[1] to the high incidence of 'musculo-skeletal and connective tissue diseases' among carpenters and joiners, bricklayers, steel fixers, tunnellers and labourers. These diseases (sometimes referred to as injuries) include the slipped disc, arthritis, frozen shoulder (capsulitis), beat hand (cellulitis) and tenosynovitis (inflammation of the tendons and their sheaths). Such diseases (or injuries) usually result from heavy physical exertion in unnatural or awkward body postures, caused frequently by limited space and difficulties of access. They are also most frequent when the worker is cold through exposure to the elements.

Whilst ergonomics, the science of 'fitting the job to the man' has been applied successfully to the factory floor, it is far more difficult in construction where usually the man has to adapt to the job.

2.6.3.1 The slipped disc and its causes[14,15]

Probably the most common of the musculo-skeletal injuries or diseases is the 'slipped disc', often known under other names such as 'disc lesion', 'backstrain', 'pulled muscle', 'sciatica', 'spinal arthritis', 'lumbago', and 'muscular rheumatism'. Since over 80% of all adults are estimated to suffer at one time or another from one or more disc lesions, most readers will be familiar with their painful symptoms. The subject was treated by the author in some detail in an earlier book[14] and only a bare outline is given here.

Figure 2.6.4 shows front and side views of the spine starting with the coccyx and sacrum at the base. Above these are the lumbar vertebrae, the thoracic vertebrae and the cervical vertebrae, all strong bones. Between

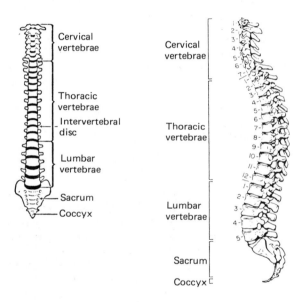

Figure 2.6.4 Front and end views of the human spine.

them lie the invertebral discs which are the shock absorbers of the body. Each consists of a ring of fibrous cartilage, which is attached to the edge of the bone and surrounded by a reinforced capsule which keeps the disc in position. The spinal cord runs in a thick sheath in a canal behind the joint. The spinal cord carries nerves which pass out in pairs opposite the invertebral discs and carry signals to and from the brain to all parts of the body. The spine is a weak part of the human anatomy and easily overloaded. As a result of this, part of a disc can be squeezed out beyond the gap in the vertebrae, where it may protrude into the spinal canal and transmit pressure onto one of the nerve roots (*Figure 2.6.5*). Now the disc has 'slipped'.

The first symptoms are usually slight, a dull ache in the lower back, between the shoulder blades or at the back of the neck. Sometimes the

Work in cramped and unnatural positions 297

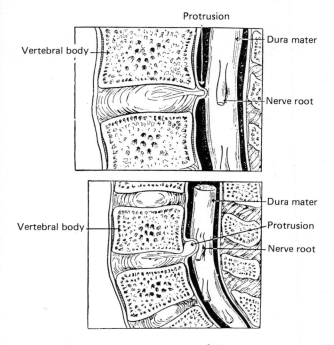

Figure 2.6.5 Pressure transmitted to dural tube and nerve root.

pain appears to come from a limb, calf, thigh, foot, forearm or even fingers, depending on where the nerve which is under pressure leads to. One day, probably when the victim is bending forwards, the joint is opened more than usual at the back and the bulging disc is pressed harder against a nerve root. The pain may become intense and other symptoms, such as numbness, pins and needles in fingers or toes, and a need to urinate frequently or difficulty in urinating, occur. The patient probably thinks that the disc broke when the intense pain began and does not connect it with the early symptoms.

Key points in avoiding slipped discs are:

- maintaining a good back posture, at all times, lying in bed, sitting, standing, walking, bending, lifting, and carrying, and in games and exercises. This is simply illustrated in *Figure 2.6.6* which is a posture chart given to patients by St. Thomas's Hospital, London.
- avoiding activities which impose heavy or sudden impact loads on the back, such as lifting heavy weights and falling heavily on one's heels, and particularly those which cause the body to bend forward at the back (such as holding a heavy object at arm's length in front of one).

When lifting an object from the floor, the knees should be bent, not the back; if necessary the body should be supported by a hand on the thigh or any convenient object. Lifting a large or fairly heavy object should be done by placing one foot on either side of it and slightly behind it (one foot

Figure 2.6.6 St. Thomas's Hospital posture chart showing how to avoid redisplacement in lower lumbar disc trouble.

slightly in front of the other), bending the knees and hip joint, still without bending the back. When moving an object too heavy to lift, it should be pushed forwards rather than pulled. The weights of objects lifted have an obvious bearing on the incidence of slipped discs, although it is difficult to lay down absolute limits. Anybody who has already suffered a back injury is more at risk than other people.

The position in which the weight has to be handled is also most important. Thus a bricklayer who has to handle a concrete block in one hand weighing 15 kg at arms length while stretching forward will have a higher bending moment imposed on his spine than someone lifting a box weighing 50 kg close to his body. The ILO recommendations on this in their 1972 Code of Practice state:

"35.1.5 No adult male worker should lift or carry a load exceeding 50 kg in weight.
35.1.6 No male young person should lift or carry a load exceeding 20 kg in weight.
35.1.7 No female young person should lift or carry a reasonably compact load exceeding 15 kg in weight.
35.1.8 No adult woman should lift or carry a load exceeding 20 kg in weight.
35.1.9 No pregnant woman and no young person under the age of 16 should lift or carry loads."

Another fertile cause of slipped discs is whole body vibration. This is often suffered by drivers of construction plant, such as dumpers, due to the repeated compression strains on the back. Many driver's seats, as well as ordinary chairs, fail to provide proper support in the small of the back where it is most needed for good body posture. If a good backrest cannot be provided when the vehicle is purchased, it should be modified, at the very least by fastening to it a wedge-shaped foam rubber cushion to support the small of the back when the driver is sitting upright.

Once a person is found to be suffering from a slipped disc, the trouble may take years to clear up and change in occupation may be necessary to prevent a repetition. Some exercises, such as swimming, are very helpful, whilst others, such as diving from high boards, squash and hurdling, which are hard on the spine, should be avoided.

Owners, managers and safety specialists of construction firms are recommended to study the books referred to on the subject, and to have access to a good physiotherapist to whom they can turn for help and advice.

2.6.3.2 Beat conditions [16]

'Beat conditions', which go under the medical terms of bursitis and cellulitis include many time-honoured occupational diseases with descriptive names, such as housemaid's knee, hod carrier's shoulder, weaver's bottom, tailor's ankle and Billingsgate hump. All are caused by pressure, friction or repeated blows over a bursa. A bursa is a closed space containing a natural lubricating fluid between two moving surfaces of the body. Most bursae are permanent features of the body, but such is nature's ingenuity, bursae can be developed 'on demand' in tissues which have been irritated by friction. Most bursae with which we are concerned here lie between the covering skin and bony projections, such as the point of the elbow and the knee-cap, but they are also found between tendons and parts of the body which they cross.

In bursitis an excess of fluid is secreted by the walls of the bursa, over-irritated by pressure or friction, so that the bursa becomes distended and tense or it may be punctured following a blow. It generally responds to firm bandaging and draining of surplus fluid, as from an abscess. Cellulitis, manifested by redness, tenderness and swelling, generally results from a wound, punctured bursa, boil, insect bite or sting, turning septic and must be treated to stop the infection spreading. Rest, warmth, draining of pus and treatment with appropriate ointments are needed.

Three beat conditions are recognized as prescribed diseases in the U.K. under the Social Security (Industrial Injuries) (Prescribed Diseases) Regulations (1980). These are Beat Hand (No. 31), Beat Knee (No. 32) and Beat Elbow (No. 33). None are notifiable diseases (which would require compulsory notification by a doctor treating it to the HSE).

'Beat hand' generally starts with an inflamed bursa of the palm which then becomes infected as a result of dirt, cement dust, sawdust, etc. penetrating through abrasions of the skin. It may affect workers using hand tools, particularly picks, hammers and shovels. Among construction workers, navvies and ground and tunnel workers, as well as joiners, are most prone to this disease. Its prevention lies partly in mechanizing the job and using power tools in place of hand tools, partly in using tools with resilient handle coverings, as well as hand protection (leather gloves), and partly in good personal hygiene which requires a reasonable standard of washing facilities being available for the worker.

'Beat knee' is a similar condition of the knee which affects workers who habitually have to kneel at their work, such as floor tilers and other floor workers, roof workers and asphalters and carpenters. Many of these workers use improvised knee pads made of old rubber outer tyres, sometimes attached to the legs of their trousers. No doubt these spread the pressure, but long and close contact between the skin and the rubber of a tyre may lead to other problems as most rubbers contain additives which may permeate the skin and cause toxaemia. Leather knee pads are probably safer, but there seems to be a need for purpose-made knee pads manufactured entirely from sterile and non-toxic materials for these workers. They also need to change and clean their trousers frequently, as well as taking showers or baths or washing their legs and knees.

'Beat elbow' is a similar condition to beat knee, which usually seems to result from one or more isolated blows or injuries to the elbow.

Besides these three prescribed beat conditions construction workers are quite liable to beat conditions on other parts of the body, such as the shoulder or bottom. 'Hod carriers' are less common today in industrialized countries where mechanical hoists are used to raise bricks and building blocks on bricklayers' scaffolds. However in developing countries many building materials are still carried on the head or shoulder and proper resilient padding is essential. Drivers of vehicles with hard and badly sprung seats are liable to the modern equivalent of 'weaver's bottom'.

Health and safety specialists and supervisors in the construction industry should be alert to the physical conditions of work which give rise to bursitis and cellulitis and take all possible steps to alleviate them. They can lead to much lost time while the worker is undergoing treatment and recovering, and once recovered, the worker should not be exposed again to the conditions which gave rise to the trouble.

2.6.3.3 Tenosynovitis[16]

Tenosynovitis, which is also a prescribed but non-notifiable industrial disease in the U.K., is one which affects the wrists and forearms of those involved in rapid repetitive work where twisting and gripping movements are common. It is found most frequently in new employees, on return to work after a holiday or other absence, or following the introduction of a new process or tool which causes unusual muscle strain. The symptoms are local aching, pain, tenderness, swelling and grating sounds from the wrists and forearms. Once these symptoms appear, a change of occupation is usually necessary, although deep massage appears to alleviate the symptoms and may enable the worker to 'keep going' at the same job. Formerly it was common in severe cases to immobilize the wrist and/or forearm, sometimes in a plaster splint.

Tenosynovitis is probably commoner among workers on assembly lines and press operations than in construction, although workers in a number of construction occupations, particularly those involving frequent interaction with a power tool or woodworking machine, are liable to be affected.

2.6.4 Living conditions and welfare facilities[17,18]

Even in industrialized countries the living conditions and welfare facilities of construction workers, particularly when they are working far from their homes and living in a construction workers' camp, often fall short of what is needed to maintain them in good health. The situation is usually far worse in developing countries, where often the only provision is for the contractor's permanent staff and key workers. The large number of migrant and temporary workers, men, women and children, mainly labourers doing work which in an industrial country would be largely done by machines, are simply left to fend for themselves, living in improvised shelters, tents, sometimes inside large drain pipes, or in the partially completed buildings they are working on. This seems to apply most to building sites in urban areas. On remote sites with few inhabitants where dams, bridges and railways have to be built, the need for workers' living accommodation with kitchens, canteens, sleeping quarters and toilets is obviously greater, and here the contractor will often provide simple bachelor-type accommodation for the work force.

2.6.4.1 Construction camps

It is impossible to provide a single detailed and universal standard for workers' accommodation on building sites anywhere in the world. The differences in climate, cultures, habits, tastes and economic circumstances are far too great. In some cold climates accommodation has to be strongly constructed, be wind, rain and snow resistant and have good heat insulation walls and ceilings and be provided with heating. On the other hand in many hot and dry climates men prefer to sleep outdoors, provided they are protected from snakes, disease carrying and biting insects and animals of prey. Beds are not universal and many habitually sleep in hammocks or on floor mats. Similarly, some prefer to sit on a chair, others to squat.

Thus great caution is needed when trying to generalize about the welfare facilities needed for construction workers' camps. People's ideas on necessary welfare facilities vary greatly from country to country. Most large building sites today are provided with a supply of electricity for power and lighting, so electricity can be and generally is provided to the workers' accommodation for lighting, heating, hot water and probably cooking if it is the best option.

Any medium to large sized construction camp should be properly designed and planned with all its services by an engineer trained in the best modern practices for such installations. Wherever possible this exercise should include consultation at an early stage with the representatives of the workers who will use them. Most camps need a camp manager and staff and for projects lasting more than a few months, a workers' club, recreational facilities, first aid room, with nurse and beds, as well as a public address system.

Consideration will have to be given to the security arrangements for the camp. Again this will depend on the population of the camp, ethnic mixes, and its location. While the major concern will be perimeter control, the need for internal law and order should not be overlooked.

The presence of a large number of construction workers of different skills and backgrounds living in a camp provides a good opportunity for instruction in health and safety, for making up for educational deficiencies of the work force (e.g. literacy and hygiene classes), and for craft training.

These opportunities should be borne in mind when the camp and canteen are designed.

2.6.4.2 Food (with special reference to canteens)

Special care is needed in the choice, preparation, storage, cooking, serving and eating of food in canteens, and the person in charge of catering should be trained in the profession, with a proper knowledge of dietetics, hygiene (including local germs and parasites), food storage, preparation and cooking. He or she should also be familiar with local markets, what foods are in season at any time of the year and also with the eating habits of the workers on the site who may be of different religions and have come considerable distances to work on the site. The training qualifications and background of the caterer are therefore very important. Whilst it is very tempting, as well as common, to award the catering to a contractor by competitive tender, this is dangerous. It usually leads to rotten food, ill-nourished and frequently sick workers, and poor morale so that job performance suffers and accidents follow. Thus the illnesses which can result from bad canteen meals should be properly included as occupational diseases.

It pays, therefore, to ensure that workers enjoy a wholesome and adequate diet, properly balanced in carbohydrates, fats, proteins, vitamins and minerals, free from germs, microbes and parasites, and cleanly served on stainless steel or uncracked glazed ceramic plates or special disposable paper plates. Care is needed to ensure that the food preparation and cooking be planned and timed to coincide with the meal breaks. The prices of canteen meals should generally be subsidized to the point where workers find it more economic to take them than try to fend for themselves. This should ensure that they are properly fed.

Workers should have adequate facilities for washing and drying their hands and faces before eating and should be persuaded to do so by their supervisors if this is not their normal habit.

The key person in all this is the caterer, who should be provided with a good wage or salary, all necessary staff and a properly equipped kitchen with facilities for storing (including cold storage and/or refrigeration), preparing, cooking and serving food, and the necessary authority or finance to purchase the total food requirements for the camp. He or she should be under no compulsion or temptation to try to make an illicit profit on the canteen by shady deals, resulting in suspect or low quality food or serving meals which are unbalanced dietically or inadequate in calories. The canteen and the meals served should be inspected regularly and it is even better if the site manager eats in the canteen with the workers.

Care must be taken to cater for differences in dietary habits of the workers based on religion and local backgrounds.

2.6.4.3 Drinking water

A proper supply of filtered and chlorinated drinking water from an uncontaminated source must be arranged. If there is no public supply and a new one has to be provided, the wells, pumps, pipes, storage tanks, treatment vessels, valves, instruments and taps are best obtained and installed through a reputable and trained contractor and should be in accordance with appropriate standards and codes of practice for drinking water supply. Piped drinking water should be supplied to the kitchen and canteen and to drinking water fountains in or adjacent to the workers' sleeping and living quarters and on the site itself. Precautions are given in Reference 19 for cases where drinking water has to be transported and stored.

It is an advantage if sufficient water of good drinking quality can be provided to serve the toilet and washing facilities as well so that no separate supply of lower quality washing water has to be provided.

A competent person should be in charge of the drinking water supply and treatment plant and installations, and he should be provided with facilities, reagents, equipment and training to test the quality of the treated water periodically.

2.6.4.4 Sanitary conveniences[19]

Proper toilet facilities should be provided of a type or types which conform to the social and religious habits of the workers.

If at all possible the toilet facilities should flush, be provided with a proper piped water supply and discharge to a public sewer, septic tank or other means of disposal (e.g. by pipe to the sea) in accordance with approved sanitary codes and practices. Precautions for privies (earth closets) are given in Reference 19.

Plans for the complete toilet facilities should be approved by the appropriate local authority or in the absence of one by a qualified sanitary consultant or inspector who is familiar with the local conditions.

Wash basins with hot and cold water, soap and hot air hand driers should be provided adjacent to the toilets and between the toilets and the canteen and living quarters. There should be sufficient toilets of the various types as well as wash basins en route to the canteen and living quarters, for all workers to be able to use them in comfort and without having to queue for more than two or three minutes. Written notices and/or suitably illustrated posters should be displayed in the sanitary block or building to emphasize the importance of personal hygiene. Special and additional washing and shower arrangements should be available for workers whose hands, faces, hair, body or clothing have been in contact or otherwise contaminated with harmful substances (e.g. cement, paints, asbestos, tar, oils) at the work site.

A supply of suitable barrier creams in suitable dispensers should be available in the washrooms and so placed and arranged that they can be conveniently used by the workers on leaving the site and before entering the camp and canteen.

2.6.4.5 Facilities for clothing[17,18,19]

It is difficult to generalize about the facilities needed for construction workers' clothing. At one extreme there could be the plumber or roof tiler making a brief visit of one or two hours for maintenance, and at the other there is the worker living for months at a time in a construction camp serving a large project in a remote area. The first man is probably self employed and provides and looks after his own working clothing. The second may provide most of his own clothing but should be, and in most cases is, provided with a hard hat and safety shoes, overalls, a rainproof jacket, trousers and gloves. In any case he needs somewhere safe, i.e. a personal locker, in which to keep his own clothing and any clothing issued by his employer, and he generally needs somewhere to dry them. If he is exposed to a toxic dust, such as lead oxide, he should certainly be provided with an impervious outer garment to wear over his own clothing while so exposed. When he removes the garment, if it is not to be cleaned immediately, he needs somewhere safe to store it where any lead dust on it will not contaminate his other clothing, i.e. a second special locker. Another question arises if the worker is employed by a subcontractor. Who should then provide the locker or lockers, drying facilities and perhaps laundering, dry cleaning and repair facilities?

The main point to be made here is that all these questions should be considered, planned for and budgeted for at the contract stage, appropriate action taken and instructions and information given.

The most generally needed facilities are for:

1. Safe-keeping and drying garments which employees shed before starting work.
2. Safe-keeping, drying, cleaning and maintaining employees working clothing and special protective equipment while they are away from work.
3. Changing before and after work, washing and perhaps taking a shower.

Someone has to be given the responsibility of planning and organizing these facilities, and to start with it will help him if he surveys the needs of different workers or groups of workers using a simple questionnaire such as the following:

1. What working clothing, protective clothing and protective devices does the worker need for this job?
2. Is the clothing or the worker liable to be contaminated by any of the following: dust, tar, oil, other liquids, toxic substances, odorous materials, body perspiration?
3. How frequently will the clothing or equipment need cleaning and by what method? (Laundering, dry cleaning.)
4. How frequently will the clothing or equipment need to be inspected, maintained or given other treatment such as dressing, fire-proofing or polishing?
5. What spare clothing or equipment needs to be provided to allow for the above needs?

Once a systematic survey has been made along these lines a plan can be drawn up and discussed with the site manager or whoever else is responsible, and details worked out including the cost of the measures to be taken and how they are to be charged.

2.6.4.6 Temporary accommodation

Temporary accommodation is sometimes required for construction workers at remote sites or urban areas in excess of that which is available within reasonable distance of the work to be carried out. Whilst it is difficult to lay down hard and fast rules as to what will be needed, the main point is that such needs should not be overlooked or left to chance. These needs should be assessed during a preliminary visit to the site by the official responsible for sending the men to work there and he should ensure that they are satisfied by the time the men arrive on the site, and that funds have been allocated to pay for them. Often mobile transportable accommodation provides the most economical and satisfactory answer, using a suitable motor or trailer caravan. Firms whose work frequently involves sending construction workers to work for short periods in remote areas (such as road building) should consider investing in suitable mobile accommodation.

2.6.5 Work related mental stresses[20]

We are all familiar with mental stress and the tense facial expression with which it is associated. It is sometimes described, and perhaps more easily understood, as 'degree of arousal'. Each person has his limits ('I can't take any more') beyond which something snaps. Response to stress is often considered as a personal or national characteristic ('the ...... break before they bend').

Over-stress is recognized as the cause of many accidents. Some accidents attributed to personal failure have happened not so much because of some flaw in his or her make-up but because he or she was 'overstressed'. On the other hand, many accidents occur through sheer boredom, when the person was half asleep. This brings us to an important point. *Not all stress is harmful*. Workers need to be kept at a certain level of arousal, particularly when doing boring, routine and monotonous jobs. This is the raison-d'être of piped music in many factories. A job, on the other hand, requiring intense mental concentration, such as solving the causation of a major accident, is generally best done without background music. There is an optimum stress level for doing most jobs. Trouble can arise both when a person is overstressed and understressed.

Overstress may arise from a variety of causes:

- a feeling of being personally threatened;
- too much to do at once (e.g. eating or holding a serious conversation with one's boss while driving a car);
- high noise levels and discordant noise;
- too high or too low body temperature;
- loss of sleep;
- influence of drugs;
- emotional problems arising out of the home and family, or grievances at work;
- shock on seeing an unexpected and serious accident or on hearing of a serious loss to the person or someone close to him;

- ill health;
- age.

The human brain is not altogether dissimilar to a computer which can break down or go beserk if overstressed or fed with a series of incompatible instructions. Training and job experience are important factors in enabling workers to carry on under stress, but one must beware the danger of trying to turn workers into robots or automata, with reflexes programmed to perform a limited range of tasks and blind to all else.

Safety professionals and supervisors in construction need to give thought and understanding to the problems of mental stress in their workers. For this they require some flair and knowledge of human psychology. There are several facets to this:

- recognition of early warning signs of excessive stress in themselves or others;
- classification of jobs where some personal failure through overstress could cause a serious accident with loss of several lives (e.g. crane operators, scaffolders, drivers of heavy vehicles);
- inclusion of tests to assess the response of applicants under stress in the selection procedure for these jobs;
- thorough job training and trainee testing and certification for those jobs with high accident potential;
- regular discussion with workers of their problems to discover and eliminate causes of stress;
- consideration of motivation at work, as to whether the job and terms of employment provide motivation compatible with safe performance (all too often the opposite seems to be the case);
- alcoholism and drug addiction;

2.6.5.1 Warning signs of overstress

Most people in modern society from time to time fall victim to overstress. Early warning signs are difficulty in getting off to sleep, early waking and excessive anxiety. Further symptoms are changes in weight and sleeping habits, compulsive behaviour and drug or alcohol addiction. Sympathy, keen observation and willingness to listen to workers' problems are needed both to recognize signs of overstress and help lower the tension. In this way a lot may be learnt of the causes of overstress and altered methods of working evolved which reduce the stress element and usually the hazards of the job as well.

2.6.5.2 Classification of jobs with high hazard potential to others

This is usually straightforward, although one must not forget to include 'the bosses'. A crane driver may be able to kill ten men if, through error, he drops a load on them. A boss who gives instructions which fail to ensure the safety of an excavation or work at heights may be responsible for the deaths of many more. If the crane operator has to undergo rigorous training and pass a stringent practical test of his ability before receiving a certificate of competence, why not the boss as well?

Classification of jobs according to their hazard potential to others should not be done or introduced without discussion, where possible involving the workers as well as supervisors. This not only makes for better understanding but it may even prevent mistakes in classification.

2.6.5.3 Selection tests for jobs with high hazard potential

Rational selection of applicants for particular jobs first requires the job itself to be properly defined. This is discussed broadly by Boydell in *A guide to job analysis*[21].

Written tests involving chooing one of several answers have been developed by industrial psychiatrists[22] to enable personalities to be evaluated in terms of percentiles for qualities such as:

emotional stability
self sufficiency
extroversion
dominance
confidence
sociability

There is often also a need to evaluate blind spots, 'hang-ups' and quirks in individuals' personalities which may point to inherent internal stresses which affect the way they may behave when under external stresses. This is properly the task of a psychiatrist, though useful insight for the layman as well as entertaining reading is given by Bernes in *Games People Play*[23].

Two types of test are used to measure the fitness of applicants for particular tasks—aptitude tests and achievement tests. Aptitude tests are designed to show how well the person may be expected to do the job after proper training, while achievement tests show how well a person is performing in a job he is already doing.

The characteristics measured by aptitude tests fall into three groups:

Mental Ability
Spacial Perception
Motor Ability

Mental ability tests are well known and need not be discussed further, except to note that they should be related to the standard required for the job in question. Neglect of this aspect has led to many 'square pegs in round holes'.

Spacial perception is not just good vision, but includes the ability to perceive spacial patterns accurately and compare them rapidly so as to be able to recognize small differences in otherwise identical objects.

Motor abilities are a combination of sensory and muscular abilities and include a number of qualities such as control precision, multi-limb coordination, reaction time, dexterity of hand and fingers, steadiness of hand and arm.

Considerable expertise, which lies outside the scope of this book, is needed in designing and conducting such tests and in evaluating their results. There is, however, a general need for safety specialists to get together with recruiting and training departments to discuss their needs and identify sources of information and assistance on whom they can call. The very small construction enterprise has as usual to 'play it by ear'. The very smallness of the organization and the limited number of workers in it should, however, make for better communications and enable individuals to know and understand each other much more closely.

2.6.5.4 Training and trainee testing and certificates for jobs with high hazard potential

In the U.K. specialist training courses for crane and plant operators, drivers of site vehicles, scaffolders, welders and divers are run by government sponsored Industrial Training Boards. A list of training centres and courses is given in Appendix A. Some large construction companies run their own training courses. These cannot just be left to the foremen or any experienced tradesmen who happen to be around. Training is a highly professional job and the trainer should be well versed in methods of construction, as well as knowledge of the particular job under instruction. Trainers should attempt to break down a particular job into its critical skills and then test their trainees in each of these, later attempting to increase their skills to the level needed by the job.

Critical skills required by all crane operators and most drivers of site vehicles include:

- memory for a sequence of operations
- recognition and correction of mistakes
- accuracy
- sensitivity to controls
- smooth coordination of hand, eye and foot
- knowledge of safety factors
- ability to judge height, distance and angle

A good instructor will involve the trainee from the start and make him think his way through the problems as they arise. Testing must be carried out professionally and certification taken seriously, and the necessity of having a valid certificate of competence for particular jobs impressed on all employees. Retesting and recertification should be required after stated intervals or before radically different equipment is introduced to site.

2.6.5.5 Discussion of workers' problems

An unfortunate feature of construction is the short duration which most workers spend on any particular site, so that they barely get time to know each other and their idiosyncrasies and problems. Many large manufacturing organizations, such as I.C.I. and Shell, realizing that personal problems shared are half-way to being solved, have introduced 'counsellors' in whom workers are encouraged to confide their worries and anxieties, and to discuss their motivation. Managers of large construction companies should, at least familiarize themselves with what is being done in this field, how it operates, and what scope there is for such ideas in their own organization. A useful account of how a counselling scheme operates in a large manufacturing enterprise is given by Lightbody in a chapter in Reference 20.

2.6.5.6 Motivation[24]

Motivation is a complex subject, and usually comes from some source deep in the personality and may be little affected by such commonly thought of motivators as improving work conditions, raising salaries, or shuffling jobs around.

These may merely remove sources of dissatisfaction, but that is not the same as real motivation which only applies when a person 'wants to do it' and needs no outside stimulation. Motivation depends primarily on job satisfaction, for which the five following factors are especially important

- achievement
- recognition
- the work itself
- responsibility
- advancement

Like most factors connected with stress, motivation opens up entirely new fields of interest which lie outside the scope of this book. Some original and fruitful ideas on the subject are given by Herzberg[24].

2.6.5.7 Alcoholism and drug addiction

Excessive use of alcohol is both a cause of and a means of escape from stress, so we quickly find ourselves in a vicious circle, in which there are problems of addiction and body chemistry as well as psychology. The stresses of construction work all too commonly lead to alcoholism, a factor which has almost certainly never been fully appreciated in considering accident statistics in the industry. One main problem is to identify the addict at an early stage, when it is most difficult, since he would not regard himself as one. Early signs of alcohol addiction are:

- extended lunch break
- lack of concentration in afternoons
- lateness or absenteeism on Mondays
- deterioration in work
- deterioration in personal appearance and habits

Alcoholics are frequently sheltered by their work mates out of a mistaken sense of loyalty, for this usually allows the condition to deteriorate further when cure becomes more difficult.

It is unfortunately necessary to be ruthless with alcoholics on construction sites and dismissal is usually the only remedy, if havoc and disasters are to be avoided. Managers and safety specialists have too many other important and urgent matters to attend to than act as nurse, shielder, keeper and confidant of alcoholics. A useful booklet[25] on alcoholics at work is available from the HSE.

References

1. HEALTH AND SAFETY EXECUTIVE, *Construction, Health and Safety, 1981–82*, HMSO, London.
2. PATTERSON, J.H., *The Maneaters of the Tsavo*, Fontana (reprint, 1973).
3. MOURET, A., *Les morsures de serpentes*, Cahiers des comités de prévention du batiment et des travaux publics', Sept. Oct. Nos. 5, p.217–222 (1981), Paris.
4. PRICE, A., LE SERVE, A. and PARKER, D., *Biological Hazards, the Hidden Threat*, Thomas Nelson, Walton on Thames, Surrey (1981).
5. TYRER, F.H. and LEE, K., *A Synopsis of Occupational Medicine*, John Wright, Bristol (1979).
6. NATIONAL SAFETY COUNCIL, *Accident Prevention Manual for Industrial Operations*, N.S.C., Chicago (see Threshold Limit Values for Physical Agents).
7. WALDRON, H.A. and HARRINGTON, J.M., *Occupational Hygiene*, Blackwell Scientific Publications, Oxford (1980) (see Chapter 9 'Heat' by F.S. Gill).
8. EDHOLM, O.G. *The Biology of Work*, Weidenfeld and Nicolson (1967).
9. KINNERSLY, P., *The Hazards of Work: How to Fight Them*, Pluto Press, 6th imp., 89 (1980).
10. GILL, F.S., 'Heat' in *Occupational Hygiene*, edited by H.A. Waldron and U.M. Harrington, Blackwell Scientific, London (1980).
11. FANGER, P.O., 'Thermal Comfort' in *Analysis and Application in Environmental Engineering*, McGraw Hill, Maidenhead.
12. KERSLAKE, D.M.K., *The Stress of Hot Environments*, Cambridge University Press (1972).
13. HUMPHREYS, M.A., *Field Studies of Thermal Comfort Compared and Applied*, Building Research Station CP76/75, Dept. of Environment, London (1975).
14. KING, R. and MAGID, J., *Industrial Hazard and Safety Handbook*, 3rd imp., Butterworths (1982).
15. CYRIAX, J., *The slipped disc*, 2nd ed., Gower Press, Epping (1967).
16. HEALTH AND SAFETY EXECUTIVE, *Guidance Note MS10, Beat Conditions, Tenosynovitis*, HMSO, London.
17. ILO, *Safety and Health in Building and Civil Engineering Work*, International Labour Office, Geneva (1970).
18. ILO, *Building Work. A Compendium of Occupational Safety and Health Practice*, International Labour Office, Geneva (1979).
19. ILO, *Safety and Health in Building and Civil Engineering Work*, Geneva (1972).
20. MARSHALL, J. and COOPER, C.L. (editors), *Coping with Stress at Work*, Gower Publishing, Aldershot (1981).
21. BOYDELL, T.H., *A Guide to Job Analysis*, British Association for Commercial and Industrial Education, London (1970).
22. Useful information available from The Vocational Guidance Association, London.
23. BERNE, E., *Games People Play*, Penguin, London (reprinted many times).
24. HERZBERG, J., *Work and the Nature of Man*, Crosby Lockwood, St Albans (1975).
25. HEALTH AND SAFETY EXECUTIVE, *The Problem Drinker at Work*, Occasional Paper Series OP1, HMSO (1981).

2.7
Hazard monitoring

The adoption of 'threshold limit values' or 'occupational exposure limits' for maximum permissible concentrations or intensities of various health hazards in the working environment has taken place side-by-side with the development of methods for monitoring them. A variety of equipment is now available for monitoring chemical, physical, biological and climatic hazards to which workers are exposed. Many of them are portable and designed for use on the spot, without having to take samples to a laboratory for analysis. Speed and simplicity of measurement are more important than absolute accuracy. Variations in the concentration of a toxic air contaminant are usually very wide and an accuracy of $\pm 20\%$ is often sufficient to show whether a hazardous concentration exists.

Both static and personal exposure monitoring can be carried out. Examples of the former are a noise level meter placed 10 metres from a pile driver, and a carbon monoxide analyser placed at a fixed point in a tunnel. The measurements may be made continuously by automatic instruments, intermittently, or on special request, e.g. to certify that the atmosphere in a pit, vessel or other confined space is safe for persons to enter without breathing apparatus.

Personal monitoring generally takes the form of automatic dose meters worn by the worker, which monitor the hazard continuously over his working day and give a cumulative measure of the hazard to which he was exposed. Such dose meters are now available to measure personal exposure to ionizing radiations, noise and a number of toxic airborne contaminants such as lead compounds, asbestos fibres and various gases and vapours. Extremely compact apparatus has had to be developed which can be worn in comfort by the worker without interfering with his movements. For airborne contaminants, an air sample is drawn continuously through a sampling head which is usually attached to the lapel of his jacket or overall.

A third type of monitoring is that which attempts to measure the rate of emission of the hazard at the source itself, or close to the source, such as measurements of noise power levels of a particular machine, or the rate at which nitrous fumes are emitted in the exhaust of a vehicle.

A fourth method is to monitor some body function of the worker himself, such as concentrations of particular chemical compounds in his urine, blood or breath, or the quality of his hearing and possible loss of it. Such monitoring is mostly too specialized to be discussed in this book.

Monitoring of the outdoor environment for toxic air contaminants to which construction workers are frequently exposed is difficult because of the frequent changes in conditions and the wide variations possible. Continuous monitoring of the atmosphere in tunnels and especially in compressed air workings is, however, generally needed. Gas tests are a necessity in confined spaces, including excavations, to check the levels of oxygen, flammable gases and toxic contaminants such as hydrogen sulphide before men are allowed to enter. Tests should also be carried out to measure the concentrations of asbestos dust in air when asbestos insulation is removed from buildings.

Although rarely done at present, monitoring of the air breathed by painters (especially indoors) for various solvent vapours and by welders for toxic metal fumes, should be practised more widely. The noise to which many construction workers are exposed also needs to be monitored more often.

The various equipment and techniques used are discussed under appropriate categories of health hazards.

2.7.1 Airborne chemical hazards[1,2]

Airborne chemical health hazards may be encountered as gases, vapours, fine liquid droplets (fogs, mists, sprays) or finely divided solids (dusts, fumes, smokes).

Most analytical methods suitable for use on the spot involve drawing a known volume of air through an apparatus which may contain a solid or liquid reagent, an adsorbent (such as charcoal or silica gel), or a filter (used mainly for solids). Where reagents are used, the total quantity of one or more specific contaminants are determined by the extent of a colour change in the reagent. On apparatus using filters and adsorbents, the quantities and composition of the trapped contaminants are determined by removing the filter or adsorbent tube, and analysing it (usually in a laboratory). This type of equipment consists of a sampling head, a pump and a means of measuring the volume of air drawn through the apparatus.

Sampling of air for gas and vapour contaminants is fairly straightforward, since all gases and vapours once mixed in air are uniformly distributed and their concentrations are not altered when the air is drawn through a sampling tube. On the other hand, the sampling of air for finely divided solid and liquid droplets is much more tricky, as these tend to be deposited in the sampling tube. Filters for solid contaminants are, therefore, incorporated in the sampling head itself.

'TLVs' and 'OELs' for many airborne solids are quoted on two bases—'total dust' and 'respirable dust'. The sampling apparatus for 'respirable dust' incorporates a special elutriator or miniature cyclone in the sampling head before the filter. The cyclone separates the larger particles (which would generally be stopped in the nose and upper respiratory tract) and only allows the smaller particles through which would normally reach the lungs. Apparatus approved in the U.K. for determining respirable dust is available from Casella London Ltd and Rotheroe and Mitchell Ltd, two firms who are predominant in the supply of equipment for measuring toxic air contaminants.

Before discussing the range of apparatus now available, and their principles of operation, some methods described in U.K. Government publications in response to specific legislatory requirements are first described.

2.7.1.1 Official methods for the detection of toxic substances in air

The need for these methods arose initially from three pieces of U.K. legislation:

- under section 30 of the Factories Act, 1961, before a person without breathing apparatus may enter any confined spaces where fumes are liable to be present, the space must be tested and certified safe by a responsible person. This requirement is reinforced by Section 63(1) of the same Act.

    Similar precautions related to specific chemical processes are required under Regulation 7 of the Chemical Works Regulations, 1922. This requires a person appointed to examine the vessel or confined space and to certify in writing whether the place is free or not free from danger.
- The Shipbuilding and Ship Repairing Regulations, 1960 have three regulations requiring air to be sampled and tested. These need to be studied in detail, and only the main points are given here.

Regulation 50 requires confined spaces in vessels to be adequately ventilated and tested by a responsible person and certified safe for entry without breathing apparatus before anyone may enter or remain in the confined space.

Regulation 59 requires the atmosphere in any parts of tankers after carrying oil to be tested by a competent analyst and a 'naked light certificate' issued before any naked lights, fires, lamps (other than approved safety lamps) or heated rivets may be used there.

Regulation 60 requires oil tanks on board vessels after carrying oil to be tested by an analyst and an entry certificate issued before any person is allowed to enter without wearing breathing apparatus.

To assist employers in meeting these obligations, a number of relatively simple analytical methods were devised by the Department of Employment—HM Factory Inspectorate and published as booklets by HM Stationary Office[1]. For the most part these are based on a known volume of air contacting a liquid reagent in a glass bubbler. The liquid reagent, usually on subsequent treatment with other chemicals, undergoes a colour change, the degree of which provides a fair measure of the amount of toxic substance present in the sample. A known air volume is drawn through the apparatus by the use of a rubber bulb hand aspirator (usually with a capacity of 120 ml) as supplied by Siebe Gorman & Co Ltd. The tests are not claimed to give a high order of accuracy, but rather a rapid indication of whether or not the atmosphere is dangerous. Even so, they generally

TABLE 2.7.1. Official methods for the detection of toxic substances in air

Booklet number	Toxic substance
1	Hydrogen sulphide
2	Hydrogen cyanide vapour
3	Sulphur dioxide
4	Benzene, toluene and xylene, styrene
5	Nitrous fumes
6	Carbon disulphide vapour
7	Carbon monoxide
8	Phosgene
9	Arsine
10	Chlorine
11	Aniline vapour
12	Organic halogen compounds
13	Mercury and compounds of mercury
14	Lead and compounds of lead
15	Trichloroethylene
16	Acrylonitrile
17	Chromic acid mist
18	Ozone in the presence of nitrous fumes
19	Hydrogen fluoride and other inorganic fluorides
20	Aromatic isocyanates
21	Iron oxide fume
22	Copper fume and dust
23	Acetone
24	Isophorone
25	Zinc oxide fume
26	Cyclohexanone and methyl cyclohexanone

need to be carried out by a trained analyst and are quite unsuitable for personal sampling. A list of them is given in *Table 2.7.1*.

2.7.1.2 *HSE Method of airborne lead monitoring*[3]

Regulation 15 of the Control of Lead at Work Regulations, 1980 states:

"*Air Monitoring*. Every employer shall
(a) have adequate monitoring procedures to measure the concentrations of lead in air to which his employees are exposed unless the exposure is not significant.
(b) measure the concentration of lead in air in accordance with those procedures."

Regulation 3 places similar duties on employers in respect of persons at work on the same premises but who are not his employees.

An Approved Code of Practice for the Control of Lead at Work Regulations provides in paragraph 13, a Table indicating the types of work with examples of industries and processes where significant exposure to lead will occur unless adequate controls are provided. Those in brackets are the authors' thoughts and do not appear in the Code of Practice.

Air sampling techniques and strategies required by the Approved Code of Practice are discussed in the HSE's *Guidance Note EH 28*[3]. An important feature is that 'the measurement of airborne lead should be carried out using personal air sampling techniques.' Paragraphs 22 to 28

Type of work	Examples of industries and processes
Lead dust and fumes	
Lead burning, welding and cutting	of lead coated and painted plant and surfaces in demolition work
Work with lead compounds which give rise to lead dust in air	(extensive use of cartridge operated fixing tools indoors as most detonators are based on lead)
Abrasion of lead giving rise to lead dust in air, e.g. dry discing, grinding, cutting by power tools	(use of angle grinders and disc cutters on lead and lead painted materials)
Spraying of lead paint and lead compounds other than paint conforming to BS 4310/68 and low solubility lead compounds	(spray painting of ships, jetties, bridges and buildings with lead based paints and primers)
Lead alkyls	
Inspection, cleaning and maintenance work inside tanks which have contained leaded gasoline	(modification or demolition of petrol tanks at filling stations)

describe the method to be used by HSE for sampling in order to determine the 8 hour time weighted average concentration of lead in air to which workers are exposed. The apparatus described below applies only to lead dusts and fumes, and a different sampling head containing an adsorbent would be required for lead alkyls.

The apparatus consists of an orifice-type sampling head, illustrated in *Figure 2.7.1*, and a pump of a size capable of being worn by persons while carrying out their normal work and running continuously for 8 hours at 2 litres/min. The sampling head should be attached to a position high on the employee's lapel, shoulder or otherwise positioned in the employee's breathing zone and orientated to ensure that the filter is held close to the vertical plane.

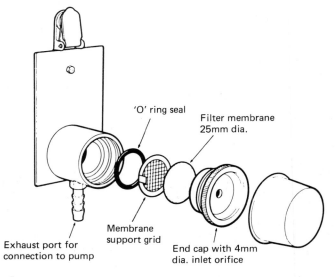

Figure 2.7.1 Orifice type sampling head used for airborne lead (HSE Guidance Note EH 28).

The pump must be capable of providing a constant flow of air through the filter throughout an 8 hour period, and needs to be calibrated frequently against an external flow meter. A clean filter membrane is loaded into the sampling head at the start of the working day. At the end of the working day the filter membrane is removed and its lead content determined by one of several standard analytical methods.

"The air sampling strategy should be developed and executed by a person or persons who have adequate information, instruction and training for the task so that they can carry it out effectively." Even though the loading and unloading of the filter head and the setting-up and calibration of the pump might be done outside a laboratory, the determination of the lead content of the filter can only be done in a laboratory.

Another HSE *Guidance Note, EH 29*[4], deals with the control of lead for outside workers. The following extracts apply particularly to construction workers:

"3. Although the work with lead may be transitory, there will be instances, e.g. welding/cutting operations on lead painted or lead containing materials, when the exposure could be to very high concentrations of lead. The fact that such work is carried out in the open air cannot be taken to be indicative of a low level of exposure to lead because it is possible that the work will be carried out close to the operator's breathing zone and therefore high concentrations of the dust or fume may be inhaled before it has had a chance to disperse.

In addition, there will be circumstances where nominally outside workers, e.g. demolition contractors' employees, may have to carry out lead burning within the confines of a building or other structure where any dust or fume has little or no chance to disperse, thus exposing such workers to even higher concentrations of airborne lead. It is essential that such exposure be assessed as significant and identified as liable to exceed the lead-in-air standard. In these cases the basic step in arriving at the assessment lies in identifying that lead paint or other lead containing material is present. This should be borne in mind whenever welding or cutting operations are to be undertaken and where adequate information is not available, possibly due to the age of the structure, a sample of the paint or material should be analysed to determine the lead content."

After discussing medical surveillance of exposed workers, the note gives the following advice on air monitoring:

"6. In cases where there is insufficient information on the lead-in-air concentrations from a particular type of work, air monitoring will need to be carried out to determine these concentrations. As much of the work will be transitory, air sampling on each job is not a practicable proposition but for each type of job the order of magnitude of the likely lead-in-air concentrations should be known. The results from the sample survey should be used to indicate the significance of the levels of lead exposure at jobs of similar types but allowances would have to be made for varying factors such as partial enclosure of the workplace, any awkward working position which could bring the operator's face closer than normal to the lead source, composition of the lead source, condition of the lead bearing material and the time taken for the job."

2.7.1.3 HSE method of airborne asbestos dust monitoring

The duties of employers in the construction industry to protect employees who are liable to be exposed to asbestos dust is made clear in paragraph 2 of Regulation 5 of the Asbestos Regulations, 1969. Paragraphs 53, 56, 57 and 58 of the Approved Code of Practice *Work with Asbestos Insulation and Asbestos Coating*, drawn up in support of the Regulations, gives the following advice on air monitoring:

> "53. Monitoring to measure the concentration of asbestos in air should be carried out:
> (a) during the work, as often as is necessary to ensure the effectiveness of the control measures in preventing the escape of asbestos from the working area (see paras 3–5),
> (b) at the end of the work to assess the suitability of the site for normal occupation."
> 56. Monitoring and subsequent analytical techniques should be in accordance with established methods set out in guidance notes issued by the Health and Safety Executive."
> 57. Routine monitoring during the course of the work is not necessary in areas where respiratory protective equipment and protective clothing are needed."
> 58. Suitable records should be kept of asbestos in air measurements."

HSE *Guidance Note EH 10*[5], provides general information on the sampling of air liable to be contaminated with asbestos fibres, and on the analysis of the samples, besides giving the names and addresses of suppliers of suitable sampling kits. Due partly to the complexity of the problem, and the need for specialized laboratory equipment and highly trained staff, it does not attempt to give any detailed or routine procedure for sampling and analysis.

In future, however, following a proposal by the Health and Safety Executive that the work of removing asbestos insulation be carried out only by licensed contractors, we would expect those undertaking this work to have available themselves, or ready access to, the necessary monitoring and analytical facilities.

Sampling is considered under three headings:
bulk sampling
environmental sampling
sampling instruments

Bulk Sampling

Airborne asbestos fibres which are encountered today by construction or demolition workers usually arise during the removal of old insulation. Before the work of removal starts and appropriate precautions are taken to protect workers, it is necessary to know what types of asbestos are present and in what concentrations. The analysis of the samples requires a laboratory equipped with specialized and sophisticated physical instruments ranging from simple optical microscopes to x-ray diffraction equipment and electron microscopes. Prior to the removal work starting, assistance on sampling and analysing the insulation and identifying the types of asbestos present can be sought from the Health and Safety Executive or consultants recommended by the Executive.

Environmental Sampling

The concentration of asbestos dust in the atmosphere may be determined either by static sampling or by personal sampling, with a sampling device attached to the worker's clothing and within 300 mm of his nose. In both cases, a known volume of air is sucked through a special membrane filter by a hand or battery operated pump. Apparatus from various suppliers may be used, and no universal standard sampling procedure is described in the Guidance Note. The counting of the asbestos fibres on the filter, however, requires more specialized techniques and equipment. The method recommended by the HSE is that given in Technical Note No. 1 published by the Asbestosis Research Council. This requires the use of an optical microscope fitted with a 4 m/40 × objective at a magnification of 450–500 using transmitted light and phase contrast techniques. Most construction firms will require assistance in making these determinations. Pumps and filter heads and complete sampling kits may be obtained from the firms whose names and addresses are given in Section 2.7.1.7.

2.7.1.4 Physical methods of monitoring chemical hazards

Monitoring instruments which are based on physical principles do not depend on measurement of the volume of air passed through the instrument. Some have detecting heads which are simply exposed to the air, without the need for any pump at all. When the air inside confined spaces is to be tested by physical principles, a portable instrument is used which contains the measuring device. This gives a reading on a dial on the top or side of the case. A long sample probe is connected by a flexible tube to one side of the instrument and to the other side a rubber squeeze bulb is attached by which air is drawn from points in the confined space into the instrument.

Examples of the use of physical principles include:

Electrochemical measurement of oxygen, carbon monoxide and hydrogen sulphide concentrations in air

The instrument contains a semi-permeable membrane through which the gas in question can diffuse, and the air is passed over it. Depending on the membrane selected, oxygen, carbon monoxide or hydrogen sulphide pass through it into an electrochemical cell on the other side. This gives an electrical response (a voltage) which is proportional to the partial pressure (hence concentration) of oxygen, carbon monoxide or hydrogen sulphide in the air.

Flammable gas or vapour monitors

These depend on the rise in temperature when air containing a flammable gas or vapour passes over an electrically heated catalyst on which the flammable gas or vapour is completely oxidized to water and carbon dioxide. These give the concentration of flammable gas or vapour as a percentage of the lower explosive limit (LEL). They should, however, be calibrated for the flammable gas or vapour in question. Users should be aware that since they depend on the presence of atmospheric oxygen as well as the flammable gas or vapour, they do not function at high concentrations of flammable vapours when there is not enough oxygen for their complete combustion.

Instruments which depend on the absorption by the air contaminant of light, generally infra-red or ultra-violet, of a particular wavelength
These are widely used for monitoring industrial processes, although they tend to be rather too complicated for monitoring most toxic air contaminants. Examples of their use are for mercury vapour in air and certain types of automatic fire alarms.

Most monitoring instruments which depend on a physical principle and produce an electrical response will operate continuously, have no moving parts, are relatively trouble-free and can be supplied with audible alarms which operate when the concentration of oxygen or a particular contaminant reaches a pre-determined level.

Although not much training is needed to set up and use these instruments, they can go wrong and need to be calibrated and adjusted by trained personnel who understand their principles of operation. It is also difficult to prepare test mixtures of air with different concentrations of contaminants for calibration of the instruments without access to a laboratory. Typical faults that may occur with apparatus for testing the air in a confined space are:

- leak on connection to sample probe so that little or none of the air sucked into the instrument is drawn from the confined space at all.
- the catalyst of a flammable gas detector is poisoned by some vapour (such as leaded petrol) which causes it to give a false reading.
- flat instrument battery.

2.7.1.5 *Apparatus and methods involving a known air volume*

The use of portable equipment involving the passage of a known volume of contaminated air through a liquid or solid reagent, a filter or an adsorbent was referred to earlier. All the official methods described in Sections 2.7.1.1, 2.7.1.2 and 2.7.1.3 depend on this feature. The growing emphasis on personal sampling devices has led to the development of compact lightweight equipment, particularly low flow rate pumps weighing less than 0.5 kg which can be carried in the worker's pocket. These are electrically operated by small batteries and give reasonably constant flow rates of 20 to 200 ml per minute over an 8 hour period. Most of these pumps employ a diaphragm or a piston, or work on the peristaltic principle.

For static sampling, on a 'one-off' basis, bellows-type hand pumps and rubber bulb aspirators which enable a fixed volume (usually about 100 ml) to be drawn through the apparatus are most commonly used.

Where a given volume of air has to be collected and taken to a laboratory for analysis, evacuated containers, plastic bags and gas-tight syringes can be used.

For personal sampling, a wide range of standard tubes containing a solid reagent which reacts with a specific air contaminant to produce a colour change are now available and preferred to the use of liquid reagents in bubblers.

'Short-term' tubes are available for spot samples taken with a hand pump, and larger long-term tubes are available for use with a personal sampling pump over a period of 8 hours.

Among the air contaminants for which solid reagent tubes are available are:

Acetone	Methylene chloride
Ammonia	Nitrous fumes
Benzene	Oxygen
Carbon dioxide	Ozone
Carbon disulphide	n-Pentane
Carbon monoxide	Per-chloro ethylene
Carbon tetrachloride	Petroleum hydrocarbons
Chlorine	Phenol
Cyclohexane	Phosgene
Ethyl acetate	Phosphine
Formaldehyde	Styrene monomer
n-Hexane	Sulphur dioxide
Hydrochloric acid	Toluene
Hydrocyanic acid	Toluene di-isocyanate
Hydrogen	Trichloro-ethane
Hydrogen fluoride	Trichloro-ethylene
Hydrogen sulphide	Vinyl chloride
Methyl bromide	

For gases and vapours for which no suitable reagent tubes are available, standard adsorbent tubes filled with charcoal or silica gel have been developed, largely by the National Institute of Occupational Safety and Health in the U.S.A. These are known as 'NIOSH' tubes and one is illustrated in *Figure 2.7.2*. The adsorbent is microporous and has a large area of active surface to which the molecules of many air contaminant vapours are attracted and adhere.

The tubes consist of two portions. The first and larger charcoal portion faces the air stream and is separated from the second by a piece of plastic foam. Each portion is analyzed separately and if compounds of interest are detected in the second, it is clear that the first portion has become saturated. The sampling is then repeated using a smaller air volume. Before use, the flame-sealed ends of the tube are removed to allow air to

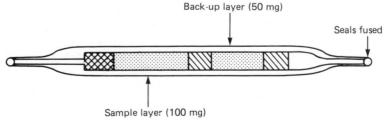

Figure 2.7.2 NIOSH charcoal tube (from *Occupational Hygiene*, edited by H.F. Waldron and J.F. Harrington, Blackwell Scientific, 1980).

pass through. The tube is then fitted so as to make an air-tight seal inside a special holder which is clipped to the worker's clothing. The holder is connected by flexible tubing to a low flow sample pump which may be carried in a breast pocket or a small leather case worn on the belt. When sampling is complete, the tube is capped and returned to a laboratory for analysis.

Personal monitoring of particulate solids in the air requires the use of a filter head such as that described for lead in Section 2.7.1.2, together with a cyclone or elutriator if 'respirable dust' only is to be measured. A variety of different filter discs made from materials such as cellulose acetate, PVC, sintered silver, glass and paper are available. The choice of filter depends both on the nature of the solid or liquid contaminant in the air and on the method of analysis used in the laboratory.

2.7.1.6 *Personal air monitoring kits relying on diffusion*

In spite of the compactness of some modern low flow air pumps used in personal sampling dosemeters, they still have to be carried in a pocket and connected by flexible tubing to the sampling head, as well as maintained and calibrated. Any compact personal dosemeters for toxic air contaminants which dispense with the pump are more likely to be accepted by workers. Some firms, Draeger in particular, offer very small personal dosemeters which can be used to measure a worker's daily exposure to the vapour of a number of volatile organic liquids, including some paint solvents. These operate on the principle of diffusion and adsorption. An adsorbent, normally charcoal, is contained in the middle of a very short glass tube which has plugs or membranes of porous material sealed in one or both ends. Vapours present in the air diffuse through the plug or membrane at a rate proportional to their concentration in air, and are trapped by the adsorbent. The tubes are sealed by caps which are removed immediately before use. One of these tubes can be fitted to a small holder which is clipped on to the breast pocket. After being worn throughout a working day the tube is removed and its ends again capped. The adsorbent in the tube is then analyzed for specific toxic organic compounds. This method was primarily developed for workers in the chemical industry, but it has considerable potential for checking the exposure of painters to organic solvents.

2.7.1.7 *Firms offering portable air monitoring kits and equipment*

Some firms in the U.K. offering portable air monitoring kits operating on the principles described in Section 2.7.1 are listed in *Table 2.7.2*.

2.7.2 Monitoring physical hazards

The physical hazards discussed in Chapter 2.5 were noise, vibration, ultra-violet radiation, visible light, infra-red radiation, lasers, ionizing

TABLE 2.7.2. Some U.K. firms offering air monitoring kits and equipment

Firm	Address
Chubb Fire Safety (Gas Monitoring Division)	St Clements Road, Parkstone, Poole, Dorset
Casella London Ltd	Regent House, Britannia Walk, London N1 7ND
Draeger Safety	Draeger House, Sunnyside Road, Chesham, Bucks, HP5 2AR
MDA Scientific (UK) Ltd	No 1 Haviland Road (Unit 6), Ferndown Industrial Estate, Wimborne, Dorset
Rotheroe & Mitchell Ltd	Victoria Road, Ruislip, Middlesex HA4 0YL
Siebe Gorman & Co Ltd	Avondale Way, Cumbran, Gwent, Wales
Vinten Instruments Ltd	Jessamy Road, Weybridge, Surrey, KT13 8LE
W & G Instruments Ltd	Progress House, 412 Greenford Road, Greenford, Middlesex, UB6 9AH

radiation and electricity. Some of these, vibration, bad illumination and electricity, are as much a safety as a health hazard (i.e. they cause accidents and sudden injuries rather than gradual loss of health). Means of monitoring all of these hazards exist, but of those to which construction workers are exposed, noise monitoring is perhaps best developed and most important. Vibration monitoring is more advanced in relation to machinery than to workers exposed to it.

Protection of workers from other physical hazards—UV light from welding, lasers, etc.—depends largely on the use of standard welding and laser equipment whose hazards are known and on the use of standard eye and face shields etc. which are appropriate to the hazards. Welding inspectors who use radioactive sources should wear radiation dosemeters which give a record of the radiation to which they are exposed over a given period. But provided normal precautions over the use of radioactive sources are observed, there should be no need to monitor the exposure of other workers.

The quality of illumination on construction sites, discussed in Section 2.5.3.2 needs to be checked to prevent accidents. A more detailed discussion was given in the writer's earlier book[6], but its quantitative monitoring is too complex and specialized a subject to be discussed here. The monitoring of electrical hazards (see Section 2.5.4) is also a specialized subject with which all electricians should be familiar, so is not discussed further here.

2.7.2.1 Noise and hearing measurement, training requirements

The basic features of noise and acoustics were given in Section 2.5.1. Whilst modern instruments with in-built computing and integrating circuits have greatly simplified the tasks of taking noise and hearing measurements, the subject remains inherently complicated. A good theoretical and practical training is still needed in order to make reliable measurements and understand what they mean. Guidance Note EH 14[7], issued by the HSE, details the training required by technicians and safety officers making noise surveys.

The standard of training required is given in Section 5.1.4 of the 'Code of Practice for reducing the exposure of employed persons to noise'[8]. The theoretical course in basic acoustics and the effects of noise in man is similar to the treatment given in Section 2.5.1. Practical noise measurement covers the following subjects:

- sound level meters, uses and limitations. British and International standards for meters.
- integrators; transportable types and personal-wear dosemeters.
- tape recorders; AM and FM types; uses and limitations
- graphic level recorders
- other instruments; statistical analysers; oscilloscopes, etc; and their applications and limitations
- maintenance of instruments; daily checks and periodic calibration
- surveys; procedures and records. The problem of people who move about
- impulse and impact noise
- noise tests for machines

In addition, the course covers the principles and practice of Noise Control and Ear Protection, both of which have many facets which require careful study and practical demonstration[8,9].

The minimum suggested duration of the course, which is applicable only to those with quite advanced training in mechanical or electrical engineering or physics, is 5 days. By comparison, the course for the Institute of Acoustics' Diploma in Acoustics and Noise Control[10] which can be taken at various Technical Colleges, requires 330 hours of theoretical and practical training (i.e. 8 to 10 weeks).

Here we discuss only three instruments used for noise and audiometry measurements on building sites, and their use:

- Sound level meters
- Personal noise dosemeters
- Audiometers for checking the hearing of noise-exposed employees.

2.7.2.2 Sound level meters [11,12,13,14]

Lightweight industrial grade sound level meters are normally used in fixed positions, preferably supported on a light tripod. If hand held, they should be used at arm's length, with the microphone as far as possible from the body to minimize errors caused by reflection of obstruction of sound waves by the body. To minimize reflection from their cases, the front ends of industrial sound level meters are usually conical. Some instruments are provided with additional extension rods or cables between the microphone and the case to reduce interference by the case and the body of the user still further.

The main components of sound level meters are:

- microphone
- amplifiers
- sound filters of four types:
    A, B, C or D weighting levels

octave band
constant bandwidth
constant percentage
- attenuator, adjustable in 10 dB steps
- meter circuits of two types:
root mean square (RMS) signal (standard requirement)
controlled response time (fast and slow speeds)
- output, consisting of the microphone signal or an amplified output as input to tape recorders or readout devices.

A sound level meter, of course, contains a battery. Some contain built-in clocks and integrators so that they can be used continuously over a specified period to give an average noise level over the period.

The sound level in decibels shown by the meter is normally a special kind of average known as root mean square (RMS) which is related to the amount of sound energy.

Meters are fitted with a fast and slow response switch. For steady sound levels, the fast position is used, but for fluctuating sound levels the slow position is used to damp down the movements of the meter needle. Many meters are fitted with a peak rectifier and special 'hold circuits' for the measurement of impulse and peak sound levels. Some standards require the peak sound level to be measured, whilst others require an RMS value of the impulse sound over a short defined time interval.

When measuring the sound from individual machines or sound sources, account has to be taken of other noise sources, and also of whether the walls, ground and other objects surrounding the machine absorb or reflect sound waves. Several type of microphone are used:

- free field microphones are used with the plane of the microphone diaphragm facing the sound source;
- pressure microphones are used with the plane of the diaphragm in the direction of sound travel;
- random microphones are designed to respond uniformly to signals arriving from all angles.

To eliminate the noise caused by wind blowing across a microphone, a special windscreen consisting of a ball of porous polyurethane sponge needs to be placed over the microphone when working outdoors. This is also effective in shielding the microphone from rain, snow and hail.

Sound level meters are primarily used in construction for the following purposes:

- measurement of the sound level at the periphery of a construction site, particularly in front of the facades of nearby residential property. This is to check compliance with sections of the Control of Pollution, 1974, dealing with construction noise.
- noise surveys over the noisiest areas of construction sites to identify areas where special protection for employees' hearing is needed.

The results are usually presented as a noise map or topograph showing the positions of major noise sources, with contour lines of equal sound level. Zones where ear protectors should be obligatory (e.g. dB 'A' values in excess of 90) may be marked in red.

The constantly changing position and use of noisy plant and other noise sources on construction sites makes such noise maps quickly out of date, although they have a certain educative value in showing what to expect when particular noise sources are present. They are of more permanent value in areas such as wood working machine shops where the positions of the main noise sources are more permanent.
- investigations of the noise power levels of noisy plant and machinery, identification of noise sources within the machine, and the effects of barriers, silencers and modifications to the machinery to reduce noise levels. This work which was discussed in Sections 2.5.1.4 and 2.5.1.7 may require additional sound monitoring equipment.

It is important when using sound level meters that the results be properly documented. Reports should contain the following information:

1. A dimensioned scaled sketch of the measuring site showing the location of the microphone and the main noise source or sources.
2. Type and serial number of instrument.
3. Method of calibration.
4. Weighting networks and meter responses used.
5. Type of sound (impulse, continuous, tones, etc.).
6. Background noise level.
7. Meteorological data and date.
8. Data on noise sources investigated (machine type, load, speed, etc.).

2.7.2.3 Personal noise dosemeters[12,15]

Personal noise dosemeters are designed to be carried in a worker's pocket with a microphone which may be clipped on to his hat, ear muffs or coat collar. They contain many of the features of sound level meters, plus an electronic clock and integrator and they measure the cumulative noise dose to which the worker has been subjected over the course of a working day. A typical noise dosemeter weighs 250 g including the battery, and measures about 125 mm × 75 mm × 30 mm, excluding microphone and clip. Results are usually displayed as a percentage of the maximum allowable daily exposure which is equivalent to 90 dB 'A' for 8 hours. Most, in addition, incorporate a Peak Excess Detector which warns of exposure to dangerously high noise peak levels (e.g. above a preselected peak level between 135 and 145 dB 'A' over very short periods of as little as 100 microseconds). Such detectors, when activated, usually display a 'P' or other sign until the meter is read at the end of the working day (or any shorter period), thus warning of the existence of the hazard after it has passed.

Noise dosemeters provide the only reliable means of assessing the noise exposure of construction workers, whose working positions and exposure to noise are constantly changing. Their use is expected to increase considerably, particularly if the EEC proposed directive[16] on noise exposure of workers is accepted by the U.K. and other member countries and its measures become law.

326   Hazard monitoring

Health and Safety professionals who attempt to initiate hearing protection programmes using personal noise dosemeters need a number of skills. First in persuading managements to buy them, secondly in persuading workers to try them, thirdly in calibrating and maintaining them, fourthly in recording and reporting their results, and last but not least in instigating and implementing measures to protect the hearing of workers exposed to excessive noise. Further developments and miniaturization of noise dosemeters are progressing, and it is rumoured that one the size of a pocket pencil or even an ear pendant may be round the corner.

2.7.2.4   Audiometry [11,16]

The need to check the hearing of workers before employment and at regular intervals subsequently to determine changes in their hearing threshold was stressed in Chapter 2.5. Screening audiometers for such use must be simple for the operator to use and the test procedure must be easy for the subjects to understand and perform. Automatic recording audiometers which satisfy these requirements are now available. Test signals are directed to each of the subject's ears in turn via a set of close-fitting headphones. The subject controls the signal level by a hand held switch, and his response is automatically recorded on a preprinted chart. He is simply instructed to 'press the button when you can hear the sound and release it when you can't'. Such audiometers are best kept in the company's (or site's) medical centre and operated by a trained nurse. Like other sound instruments, they require periodic calibration against special equipment (artificial ears and precision sound level meters).

Regular checking of employees' hearing forms an essential part of any hearing conservation programme within a company. It enables reductions in hearing performance to be noticed quickly and prompt corrective action to be taken, i.e. introducing measures to reduce ambient noise levels, or moving particularly noise sensitive employees to quieter work. Great stress is laid in the EEC proposed directive[16] on the greater use of audiometry within industry, and before long this may become compulsory in law.

2.7.2.5   Vibration monitoring [12,18]

Monitoring of machine vibration is widely used in providing early warning of failure and in diagnosing faults, but monitoring the vibration experienced by workers is in its early stages. However, since vibration is nearly all transmitted by oscillating or out-of-balance machinery, the problem is best tackled at source by concentrating on the machine, and the transmission paths between it and the worker. This is as much the work of mechanical engineers as of health and safety specialists.

Many different techniques are used to measure vibration, using one of the three following parameters:

displacement
velocity
acceleration

Tentative standards for human exposure to vibration (see Section 2.5.2.4) are based on acceleration and frequency. Measurements of acceleration are thus easier to relate to these standards than measurements of displacement or velocity.

Accelerometers measure acceleration. Most accelerometers are piezoelectric devices which produce a small alternating voltage proportional to the acceleration experienced, and of the same frequency as the vibration. The accelerometer, sometimes referred to as a 'pick-up', may be attached to the vibrating surface (whether man or machine) by adhesive, wax, threaded rod or a magnetic shoe, or it may be hand-held in contact with the surface. Errors are introduced through poor contact between the vibrating surface and the pick-up, particularly at high frequencies. The accelerometer is connected by a short length of special cable to an amplifier, analyser and recorder or dial instrument.

Some sound level meters may be used in conjunction with a piezoelectric accelerometer to measure vibration, but unless the meter was designed as a dual purpose instrument with frequency ranges extended below 20 Hz, it will be limited to the measurement of vibration in the audio frequency range. Specially designed human response vibration meters are available for measuring three categories of vibration:

- whole body vibration, in the frequency range 1 to 80 Hertz, in relation to draft British Standard DD 32, 1974. For this a special triaxial seat accelerometer positioned under the buttocks of the seated person may be used.
- hand arm vibration in the frequency range 4 to 2000 Hertz, in relation to draft British Standard DD 43, 1975. Here the pick-up is usually attached to the handle of the vibrating tool.
- whole body vibration at very low frequencies which causes 'motion sickness' and which is dealt with in an addendum to the 'whole body' standard.

2.7.3 Monitoring other health hazards

The other health hazards discussed in Chapter 2.6 are grouped under the headings of biological, weather, work in cramped and unnatural positions, living conditions and welfare and mental stresses. The monitoring of the first two of these hazards is discussed here. For the remainder, little can usefully be added to the discussions on them in Chapter 2.6.

2.7.3.1 Biological hazards[19]

These tend to be as much an environmental as an occupational health problem, with which the local public health authority is usually as much concerned as the building contractor. Employers should discuss these problems with the local health authority. Hazards of wild animals, parasitic insects and poisonous plants tend to be local and records of them are usually available. Local inhabitants and authorities have usually evolved their own methods, albeit sometimes primitive, of monitoring and dealing with them. Contractors working abroad, especially on remote sites, should ask their local agent to make full inquiries about such hazards before sending personnel to work there. These hazards, when present, are usually too specific for general advice on monitoring to be given, and must be treated on a case-by-case basis.

The first warning the employer usually gets of bacteriological and virus hazards is an outbreak of illness among his workers—sometimes just one or two isolated cases. Medical records and surveys are important in pinpointing these hazards. Management usually needs to develop systems which encourage workers to report these problems. Many will be revealed by asking such simple questions as "Do you suffer from rashes, skin problems, boils and septic cuts? Do you have chest or breathing difficulties?" In some, but by no means all cases, these are caused by bacteria or viruses.

Screening tests of individual workers should include:

- chest x-rays for lung infections
- stool samples for the presence of food poisoning organisms or parasitic worms
- urine specimens and blood samples for infectious organisms.

The atmosphere breathed by workers may be sampled for airborne bacteria using the same personal sampler with air pump and filter head as used for lead, asbestos and other hazardous dust particles (see Section 2.7.1). In this case the particles are trapped into slides previously coated with a tacky jelly and examined by a trained technician under the microscope. Special samplers are available for sampling the atmosphere for bacteria (and spores), for example the 'Slit Sampler' or 'Airborne Bacteria Sampler' and the 'Hirst Spore Trap'. The bacteria are trapped in a growing medium which is incubated and examined microscopically when the number and type of micro-organisms present can be determined. This work can only be done by trained technicians.

2.7.3.2 Weather hazards [20]

Weather conditions are of the utmost importance in construction and affect it in numerous ways. Most construction contracts have clauses to allow for abnormally bad weather, so that contractors are not unfairly penalized, although an 'average number' of hours of heavy rain, strong winds, fog, frost, snow, low and high temperatures and humidities have to be allowed for by the contractor. This information may be needed in any part of the country or abroad, and at any time of the year.

The adverse effects of particular weather conditions on various building operations are shown in *Table 2.7.3*.

When claiming under adverse weather clauses in a contract the contractor has to produce evidence to support his case. He is thus much dependent on reliable weather records (to provide the 'norm' for the time and place), and forecasts (to help him plan his work). In industrialized countries, most of these records and forecasts are provided by the national meteorological service. Besides having numerous weather monitoring and recording stations throughout the country, they have access to more sophisticated equipment such as weather satellites as well as to world wide and up-to-the-minute data. As well as providing regular weather reports

TABLE 2.7.3. Weather adverse to building work - from Meteorological Office pamphlet, 'Weather Services for Builders'

Building operation	Weather condition						
	Strong wind	Rain	Low temp.	Wet ground	High humidity	Fog	Snow
Transport, site movement				×		×	×
Earth moving, excavating, use of plant			×	×			×
Bricklaying, concreting, asphalting		×	×				×
Scaffolding, steel erection	×		×				×
Tall cranes, cradles	×						
Wall sheeting, roofing	×						×
Painting and plastering (external)		×	×		×		×
Drying out			×		×		
Partly completed work	×		×				
Materials on site	×	×	×				×
Site hazards increased	×		×				×

and forecasts by radio, these meteorological offices provide consultancy services and reports by telephone for particular construction sites.

The contractor can also, and in some cases may be obliged to, provide his own weather station, records and forecasts. The instruments used in such stations, their housing and location are generally standardized, and at the minimum would consist of:

4 thermometers—wet and dry bulb, maximum and minimum
1 barometer
1 rain gauge
1 wind direction indicator
1 wind speed indicator
1 sunshine recorder

These instruments are normally read and reset at 09.00 hours every morning.

To provide a constant record throughout the day automatically recording instruments are used. Provided these are readily accessible and frequently consulted, such instruments can provide warning to supervisors on construction sites of the onset of hazardous weather conditions.

In the U.K., as *Table 2.7.3* shows, the main weather hazards (to the health and safety of building workers) are considered to be strong winds, low temperatures and snow. However, all of the adverse weather conditions listed in the table introduce some degree of risk.

In other countries, different hazardous weather conditions may be more common, such as high temperatures (specially combined with high humidity), large hailstones, dust storms and lightning.

Of the various weather hazards in the U.K., high winds are the most dangerous and one manufacturer at least, R.W. Munro Ltd, supplies wind speed and direction recorders with alarm systems (audible and visual) which operate when wind gusts exceeding a preselected speed are experienced. Most anemometers are of the rotating cup type, and for comparison with other wind measuring instruments elsewhere they are usually mounted at a standard height of 10 metres above open ground, well clear of obstacles.

330 Hazard monitoring

Whilst high winds are a serious hazard to most outside construction activities (as discussed in Section 2.6.2.2), cranes and their operation are particularly vulnerable to such conditions. According to BS 2573[21] cranes are designed to be operable up to wind pressures of 240 N/m^2, which corresponds to a wind gust speed of 17.2 m/s. With the jib free to rotate, cranes are designed to withstand a load of up to 1630 N/m^2, corresponding to a wind speed of 44.3 m/s. When handling light and bulky loads, however, such as lightweight panels or shuttering, cranes or at least their loads may be unmanageable at lower wind gust speeds, i.e. at about 12 m/s or even less. Wind gust speeds are, of course, higher, depending on various factors including the surrounding terrain and neighbouring buildings, than the mean windspeed—sometimes as much as 100% higher. The Critical Windspeed Alarm System is, therefore, particularly important for monitoring the safety of crane operations. Since abnormally high local windspeeds are commonly found on building sites on some sides and corners of buildings, a hand held anemometer (also supplied by R.W. Munro Ltd) is also useful in evaluating prevailing conditions at particularly windy spots.

Three other useful parameters for assessing hot and cold weather hazards as discussed in Section 2.6.2.1 are wet bulb globe temperature

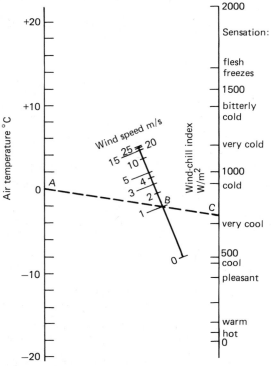

Figure 2.7.3 Nomogram for calculating wind chill index. To use the nomogram, lay a straight-edge through the temperature and wind speed scales and read the index from the right-hand scale, e.g. the line ABC through 0°C and 1 m/s gives an index of 740 W/m^2 (from R.E. Lacey, *Climate and Building in Britain*, HMSO).

(WBGT), the wind chill index (WCI) and the air cooling power, measured by the Kata thermometer. A simple assembly of three thermometers for determining WBGT was illustrated in *Figure 2.6.1*. Each thermometer has to be read and the WBGT then calculated. An instrument which measures these three temperatures, computes the WBGT and records it continuously, would doubtless be useful on large construction sites in hot climates, particularly if combined with a high WBGT alarm.

The Wind Chill Index, which is the rate of heat loss from a dry surface at a temperature of 33 °C (corresponding to dry skin temperature), is a function only of the air temperature (dry bulb) and windspeed. Knowing these two parameters, the Wind Chill Index may be calculated using the nomogram in *Figure 2.7.3*. It is understood that a Wind Chill Index recorder is under development by one U.K. instrument manufacturer. This should be valuable on large construction sites in the U.K. and in cold and temperate climates, particularly in warning workers when it is time to add or shed a sweater or other article of clothing.

The Kata thermometer[22] is used by heating and ventilating engineers, public health officials and HM inspectors concerned with human comfort in mines, tunnels, factories, etc. It is an alcohol-filled thermometer with a large bulb and with upper and lower graduations on the stem with a 3 °C interval between them. The bulb is first warmed, then suspended in the air, the time taken for the temperature to drop over the 3 °C interval is then measured with a stopwatch. Wet and dry Kata thermometers are used. Details of its use are given in Reference 22.

References

1. *Methods for the Detection of Toxic Substances in Air*, Booklets 1 to 26 (see *Table 2.7.1*), HMSO, London.
2. G.L. LEE, 'Sampling: Principles, Methods, Apparatus, Surveys', Chapter 2 in *Occupational Hygiene*, edited H.A. Waldron and J.M. Harrington, Blackwell Scientific, London (1980).
3. HEALTH AND SAFETY EXECUTIVE *Guidance Note EH 28, Control of Lead: Air sampling techniques and strategies*, HMSO, London.
4. HEALTH AND SAFETY EXECUTIVE *Guidance Note EH 29, Control of Lead: outside workers*, HMSO, London.
5. HEALTH AND SAFETY EXECUTIVE *Guidance Note EH 10, Asbestos—hygiene standards and measurement of airborne dust concentrations*, HMSO, London.
6. KING, R.W. and MAGID, J., *Industrial Hazard and Safety Handbook*, Butterworths, London (1982).
7. HEALTH AND SAFETY EXECUTIVE, *Guidance Note EH 14, Level of training for technicians making noise surveys*, HMSO, London.
8. DEPARTMENT OF EMPLOYMENT, *Code of Practice for Reducing the Exposure of Employed Persons to Noise*, HMSO, London (1972).
9. British Standards:
   BS 5228:1975. *Code of Practice for noise control on construction and demolition sites*.
   BS 5330:1976. *Method of test for estimating the risk of hearing handicap due to noise exposure*.
   BS 4142:1967. *Method of rating industrial noise affecting residential and industrial areas*.
   BS 4813:1972. *Method of measuring noise from machine tools excluding tests in anechoic chambers*.
10. INSTITUTE OF ACOUSTICS, *Syllabus for Diploma in Acoustics and Noise Control*, Institute of Acoustics, London.
11. BURNS, W., *Noise and Man*, 2nd ed., John Murray, London (1973).
12. *Handbook of Noise and Vibration Control*, 3rd ed., Trade and Technical Press, Morden, Surrey.

Hazard monitoring

13. BRÜEL and KJAER, *Measuring Sound*, B & K, Naerum, Denmark.
14. BS 3489:1962, *Sound level meters (industrial grade)*, British Standards Institution, London.
15. BRÜEL and KJAER, *Product Data Sheet*, Personal (ISO), Noise Dose Meter, type 4428, B & K, Denmark.
16. BRÜEL and KJAER, *Audiometry and Hearing Aid Testing*, B & K, Denmark.
17. EEC Proposed Directive, COM (82) 645 final (15.10.82) Brussels.
18. BRÜEL and KJAER, *Measuring Vibration*, B & K, Denmark.
19. PRICE, A., LE SERVE, A. and PARKER, D. *Biological Hazards, the Hidden Threat*, Thomas Nelson, Walton on Thames, Surrey (1981).
20. LACY, R.E., *Climate and Building in Britain*, HMSO, London (1977).
21. BS 2573, *Specification for possible stresses in cranes and design rules, Part 1 structures, 1977, Part 2 mechanisms, 1980*, British Standards Institution, London.
22. HUMPHREYS, M.A., *Field studies of thermal comfort compared and applied*, Building Research Station, CP 76/75, Department of the Environment (1975).

2.8

Personal protective clothing and equipment

This chapter deals with special protective clothing and equipment used by construction workers and follows the same pattern the author's earlier treatment[1] of the subject for industrial workers. It covers:

1. The philosophy behind the use of such clothing and equipment and some of the problems of adopting them.
2. British legal requirements regarding their use.
3. Clothing and equipment available and guidelines to their selection.

2.8.1 Philosophy and problems

There is no clear-cut line between the special protective clothing and equipment used to protect against particular hazards—hard hats for falling objects, PVC jackets and trousers against corrosive liquids and heavy rain—and the normal working clothing used to conserve body heat, protect the skin against abrasion and provide padding. None the less the clothing and equipment developed to protect workers against particular man-made hazards are mostly 'special extras', cause some inconvenience to the wearer, and restrict his movements or sensory perception. Work should, therefore, be planned to eliminate these hazards so that special personal protection is unnecessary. Unfortunately this is not always possible and even when the hazard has been largely, but not totally, eliminated, personal protective clothing and equipment may still be required for emergencies and as a last line of defence. The general requirements for protective clothing and equipment are listed in *Table 2.8.1*.

Sometimes the need for personal protection is obvious and readily accepted, e.g. the need for eye protection by welders. In others, e.g. the use of hard hats by steel erectors, the probability of injury is less evident and it is more difficult to persuade workers to use the protection provided. The problem thus is a threefold one:

- to identify situations where the hazard justifies the use of personal protective equipment;

334   Personal protective clothing and equipment

TABLE 2.8.1. General requirements of personal protective clothing and equipment

1. Adequate protection against specific hazard for which intended.
2. Minimum weight and minimum discomfort compatible with efficient protection.
3. Flexible but effective attachment to the body.
4. Weight carried on part of body well able to support it.
5. Durability.
6. Parts accessible for ease of inspection and maintenance.
7. No additional hazards introduced by use of item.
8. Wearer not restricted on movements or sensory perception essential to the job.
9. Construction in accordance with accepted standards.
10. Should not make worker 'look stupid' and should, as far as possible, be attractive.

- to select a satisfactory item or range of items;
- to convince the worker of the need to wear such items and to ensure that they are properly used and looked after.

When a persistent hazard exists in a given task which cannot be entirely eliminated and where accident and medical records demonstrate the need for a particular form of protection, it should be worn. It is recommended that its wearing for the task in question be made a condition of employment.

Interestingly, the use of protective items by workers varies considerably even in industrialized countrie (*Figure 2.8.1*). It would be expected that the higher education and social standards of such countries would have resulted in widespread use of such equipment. The reason for such variance is not easy to define. For instance in the U.S.A. safety helmets are

Figure 2.8.1 '... the use of protective items by workers varies considerably ...'

a badge of office among construction workers, while in the U.K. the general consensus is that it is not part of the industry's 'macho' image, or perhaps the wearability of the helmets is not as well developed in the U.K. So the use of protective clothing may also involve psychological factors which will only be changed by re-education and retraining and revising the image of the industry. In such cases the use of penalties may be of little benefit, especially in view of the difficulties of uniform application and inconsistencies with individual management who are often loathe to force the issue.

Managers and visitors to areas where the hazard exists and who may only be exposed to it temporarily should be issued with and obliged to wear protective equipment (e.g. hard hats and eye shields). It is an important pychological point that if managers do not take the hazard seriously for their own safety and health, workers are likely to take the same attitude. Precise instructions are, however, needed on when and where the protection is to be worn, and whether it must be used all the time or just while a particular job is being done. This can be made crystal clear by written job descriptions and the message can be reinforced by posters.

Personal protection should generally be allocated for the individual use of one worker only, and it should be marked with his name and/or number. He should be responsible for its care and, for items not subject to legal requirements, he should contribute to the cost of another if he loses it. Management must, however, provide a locker or equivalent where the worker can keep the item securely on the site and should also arrange for its regular cleaning, disinfection, testing, maintenance and replacement.

The needs of site visitors, e.g. hard hats and eye shields, must also be catered for. Someone should be made responsible for issuing them on a loan basis when the visitor arrives, collecting them when he or she leaves, and cleaning and disinfecting them.

Personal protective clothing and equipment may be classified according to the part of the body which they protect and the hazard they protect against. *Table 2.8.2* sets these out in grid form as a rough check list when considering the needs for such protection. A three star rating of importance is used where:

*** = Serious risk if protection not provided and used
** = Protection often needed
* = Protection occasionally needed.

While such classification is based on single hazard to a particular part of the body, the protective equipment identified as necessary may provide protection for other parts of the body also. For example, protection for the eyes and respiratory system can be achieved by a full-face mask respirator which protects the whole face as well. Similar examples abound.

2.8.2 *Legal requirements for protective clothing and equipment in the U.K. construction industry*

Protective clothing and equipment are prescribed in U.K. statutes and regulations, some dating back to 1910 and earlier, for various tasks in particular industries. Only those which apply directly to construction and

TABLE 2.8.2. Parts of body and hazards against which protection may be needed

Part of body	Gravity		Toxic and harmful substances	Noise	Physical hazards Vibration	Radiation UV, IR and visible	Burns and Electrocution	Cuts, abrasions, moving machinery	Weather heat, cold wind, etc.	Biological insects, bacteria & viruses
	Falls	Falling objects								
1. Eyes	*		***			***	**	**	*	
2. Respiratory system			**						*	**
3. Face	*		**			**	**	*	**	*
4. Head	*	**		*					***	
5. Ears	*			**					*	
6. Hands	*		**		*	**	**	**	**	*
7. Feet	**	**								
8. Body, including neck, arms, legs	**		*		*	*	**	*	**	**

closely related industries, such as shipbuilding, are included here. Usually the employer or occupant of the premises is obliged to provide the specified items as well as arranging for them to be washed, disinfected and maintained and providing a proper place to keep them. The detail with which the particular equipment is specified varies considerably; some Regulations merely stipulate that 'suitable' clothing etc. shall be provided. Most regulations also require particular workers to wear the equipment stipulated when engaged in particular tasks, but most welfare orders merely require the employer or occupier to provide it whilst leaving it to the employee whether he uses it. A list of these legal requirements which apply to construction is given in *Table 2.8.3*. As the conditions attached to these requirements vary considerably employers and workers subject to them should study the Regulations carefully.

Diving equipment and breathing apparatus, rescue and resuscitation equipment, and fireproof clothing are not included owing to their highly specialized application.

General requirement to provide appropriate protection for workers against risks to health and safety are contained in Part 1 of the Health and Safety at Work, etc. Act, 1974. The provision of hard hats is not obligatory under any Regulations, although protection for operatives on sites where there is a foreseeable risk of head injuries being sustained is required under Rule 23 of the National Working Rules for the Building Industry[2]. These Rules form part of the Contract of Employment of building operatives, non-compliance with which can be dealt with by disciplinary measures. It is understood that the HSE now proposes to enforce this Rule under their powers granted by Sections 2 and 7 of the Health and Safety at Work, etc. Act, 1974, and by way of guidance has specified the following circumstances where safety helmets should be worn:

1. Working under persons who are at a higher level
2. Working near to lifting appliances
3. Working in excavations over 4 ft (1.22 m) deep
4. Working near the loading and unloading of vehicles
5. Working on a site where any piling is taking place
6. Working on a site where demolition is taking place

In most Regulations there is provision whereby the Chief Inspector of Factories can grant exemption from providing and wearing the protection specified in cases where the employer or occupier can demonstrate that he has eliminated the hazard for which protection is required out of his plant, etc.

Eye protection receives very special attention in the Protection of Eyes Regulations, 1974, which, although coming under the Factories Act, 1961, also apply to all construction operations. Schedule 1 of these Regulations contains a list of specified processes for which approved eye protectors are required and Schedule 2 contains a list of processes where protection is required for persons at risk but not actually employed in the processes. Most processes listed are qualified in the Regulations by words such as 'where there is a reasonably foreseeable risk of injury to the eyes of any person engaged on the work'.

TABLE 2.8.3. British regulations and orders prescribing protective clothing and personal protective equipment which apply in the construction industry

Title of regulation or order	Reference no.	No. of prescribing section or regulation	Task or persons for whom protection is prescribed	Protection	Party liable to provide & maintain protection	Is wearing obligatory?	Under which regulation?	Remarks
The Construction (Health & Welfare) Regulations 1966	SI 1966 No. 95	15	All persons required to work in open air during rain, snow sleet or hail	Adequate and suitable protective clothing	Contractor	No		Suitable clothing described in BS 4170
The Factories Act 1961 The Protection of Eyes Regulations 1974	Part IV SI 1974 No. 1681 amended by SI 1975 No. 303	65	See Table 2.8.4	Eye protectors, meaning goggles, visors, spectacles and face screens	Employer	Yes	11	Section 65 of the FA 1961 is an enabling section under which these and other regulations were made. These regulations make the provision and wearing of eye protector obligatory on employer and employee alike over a range of processes unless exempted in writing by the Chief Inspector
The Factories Act 1961		30	Work inside any chamber, tank, vat, pit, pipe, flue or similar confined space in which dangerous fumes are present or in which the	Breathing apparatus of a type approved by the Chief Inspector i.e. a) self contained breathing apparatus approved under the	Occupier	Yes	30	

Regulation	SI/Reference	Section	Circumstances	Equipment Required	Responsible	Approved	No.	Notes
			proportion of oxygen in the air is liable to have been substantially reduced	Chemical Works Reg 1922 or b) A properly fitted helmet or face piece with necessary connections by means of which a person breathes ordinary air				
The Shipbuilding & Ship Repairing Regs 1960	SI 1960 No. 1932 as amended by SI 1969 No. 690 & SI 1974 No. 1681	50	Workers in confined spaces where there may be fumes or shortage of oxygen	Approved breathing apparatus	Every employer	Yes	50	
			All persons employed when using gas cutting or welding inc. in machine caulking or machine rivetting or in transporting, stacking or handling plates at machines	Adequate hand protecion inc. suitable gauntlets to protect the hands and forearms from hot metal and rays likely to be injurious	Every employer	No		
		76	Scaling, scurfing or cleaning boilers combustion chambers or smoke boxes where injurious dust occurs	Breathing apparatus of a type approved for the purpose of this regulation	Every employer	Yes	76	
The Asbestos Regs 1969	SI 1969 No. 690	2	Work where asbestos dust may be present in the atmosphere where it cannot be removed by exhaust ventilation	Approved respiratory protective equipment and protective clothing	Employers & occupiers of factories	Yes	3	Gives no protection to the public during demolition of buildings containing asbestos insulation
The Electricity Regs 1908	SR & O 1908 No. 1312 as amended by SR & O 1944 No. 739	24	Where necessary	Insulating boots and gloves	The occupier	Yes	24	
The Control of Lead at Work Regs 1980		7	Any person exposed to airborne lead in excess of the lead in air standard	Approved respiratory protective equipment	Employer	Yes	13	
		8	Employees whose exposure to lead is significant	Adequate protective clothing	Employer	Yes	13	

TABLE 2.8.4. Summary of specified processes in construction for which eye protection is required under Protection of Eyes Regulations, 1974 (Section 65 of Factories Act, 1961)

Schedule & Part	Specified processes

Schedule 1
Part 1. Processes for which approved eye protectors are required
	1. Shot blasting of concrete.
	2. Shot cleaning of buildings or structures
	3. Cleaning by high pressure water jets
	4. Striking masonry nails (by hand or power tool)
	5. All work with hand held cartridge tools
	6. All work on metal involving the use of a chisel, punch or similar tool by means of a hammer or power tool
	7. The chipping or scarifying of paint, scale, slag, rust or corrosion from metal and other hard surfaces by a hand or power tool
	8. The use of power driven high speed metal cutting saws, abrasive cutting-off wheels or discs
	11. Any work on plant which contains or has contained acids, alkalis, corrosive substances or substances harmful to the eyes unless the plant has been treated, designed or constructed to prevent risk of eye injury
	13. Driving in or on of bolts, pins or collars to a structure or plant by a hammer, chisel, punch or portable hand tool
	14. Injection by pressure of liquids into tall buildings or structures which could result in eye injury
	16. Breaking, cutting, dressing, carving, or drilling by a hand or portable tool any of the following:
	   (a) glass, hard plastics, concrete, fired clay, plaster, slag or stone or similar materials or articles consisting wholly or partly of them
	   (b) bricks, tiles or blocks of brickwork, stonework, blockwork (except wooden blocks)
	17. Use of compressed air to remove swarf, dust, dirt or other particles
	22. Cutting wire or metal straps under tension

Part II. Processes for which approved shields or approved fixed shields are required	24. Processes using exposed electric arcs or exposed arc plasma streams
Part III. Processes in which approved eye protectors or approved shields or approved fixed shields are required	25. Oxy-gas metal cutting 27. Hot cutting, boring, cleaning, surface conditioning or spraying of metal by an air-gas or oxy-gas burner 28. Instruments such as lasers which produce light radiation which can cause eye injury
Part IV. Processes in which approved eye protectors or approved shields or fixed shields are required	29. Trueing or dressing abrasive wheels 31. Dry grinding of materials by applying them by hand to a wheel, disc or band or by applying a power driven portable grinding tool to them 34. Machining of metals, including any dry grinding process not elsewhere specified 35. Electrical resistance and submerged arc welding of metals
Schedule 2 Cases in which protection is required for persons at risk but not employed in the specified processes	1. Item 6 of Schedule 1 2. Item 24 of Schedule 1 5. Item 28 of Schedule 1

So many of the processes in the Schedules apply to construction that they are summarized in *Table 2.8.4*.

The Regulations interpret certain expressions used in them as follows:

'approved' means for the time being for the purposes of these Regulations by certificate of the Chief Inspector.

'eye protectors' means any of the following (being equipment to be worn by a person) that is to say, goggles, visors, spectacles and face screens.

'factory' includes any premises and place to which these Regulations apply.

'fixed shield' means a screen which is free standing or which is or is made to be attached to machinery, plant or other equipment or to a building or structure.

'shield' means a helmet or hand shield, being equipment made to be worn or held by a person.

Many processes used in construction are specified in the Regulations as requiring eye protection, and safety specialists, managers and supervisors should take careful note of them.

Under Regulation 11 of the Protection of Eyes Regulations, the employee is responsible for looking after and wearing eye protectors provided to him, and for reporting any defects in them to his employer. At the same time the employer must ensure that the employee can meet these obligations.

2.8.3 Clothing and equipment available and guidelines to their selection. General

Much useful information on U.K. suppliers, clothing and equipment available, and advice on selection is given by the Industrial Safety (Protective Equipment) Manufacturers Association, who publish a biennial reference book[3]. Before purchasing protective clothing and equipment employers are advised to review their requirements carefully, find out what firms supply suitable equipment, and then invite several to send their representatives to discuss and demonstrate their equipment. Finally, the employer is advised where possible to standardize on a single supplier for a particular line of protective equipment, for greater ease in maintenance and supply of replacement parts and accessories.

British Standards which cover personal protective clothing and equipment applicable in construction in the U.K. are listed in *Table 2.8.5*. Where these are identical or equivalent to an International Standard, the ISO number is also given. The standards are listed according to the part of the body protected and referred to under that part in the following.

2.8.4 Eye protection

The loss of sight in one or both eyes is so serious a handicap that special care must be taken to protect the eyes from injuries at work. Evolution has

TABLE 2.8.5. British Standards for protective clothing and equipment used by construction workers

Part of body protected	BS no.	Year	Title	Amendments in years	ISO no.
Eyes	2092	1967	Industrial eye protectors (radiation protection not included)	1970, 1972, 1975, 1977(2), 1979, 1980, 1981	
	679	1959	Filters for use during welding & similar industrial operations	1960, 1961, 1963, 1969, 1977	
	2724	1956	Filters for protection against intense sunglare	1961	
Eye & face	1542	1960	Equipment for eye, face & neck protection against radiation arising during welding and similar operations	1962, 1963, 1966, 1969, 1974, 1977, 1981	
Respiratory system	4275	1974	Recommendations for the selection, use and maintenance of respiratory protective equipment		
	4667		Breathing apparatus		
	Part 1	1974	Closed-circuit breathing apparatus	1974	
	Part 2	1974	Open-circuit breathing apparatus	1977	
	Part 3	1974	Fresh air hose and compressed air line breathing apparatus	1977	
	Part 4	1975	Escape breathing apparatus		
	2091	1969	Respirators for protection against harmful dust, gases & agricultural chemicals	1970, 1978	
	4555	1970	High efficiency dust respirators		
	4558	1970	Positive pressure powered dust respirators		
Respiratory system — Head & body	4771	1971	Positive pressure, powered dust hoods and blouses		
Head	5240	1975	General purpose industrial safety helmets		
Ears	5108	1974	Method of measurement of attenuation of hearing protectors at threshold	1976	4869
Hands	1651	1966	Industrial gloves	1970	
	697	1977	Specification for rubber gloves for electrical purposes	1978, 1983	
Feet	1870		Safety footwear		
	Part 1	1979	Specification for safety footwear other than all rubber and all plastic moulded types	1980	
	Part 2	1976	Specification for rubber lined safety boots (with toe caps and protective midsoles or toe caps only, protection against shock not covered)	1978, 1979	
	Part 3	1981	Specification for polyvinyl chloride moulded safety footwear		
	5145	1975	Lined industrial rubber boots	1975	± 2023
	5462	1977	Specification for lined rubber boots with protective (penetration-resistant) midsoles		

	6159		Polyvinyl chloride boots	
	Part 1	1981	Specification for general industrial lined or unlined boots	
	5451	1977	Specification for electrically conducting and anti-static rubber footwear	
Body	4679	1971	Protective suits for construction workers and others in similar arduous activities	1971
	4170	1967	Waterproof protective clothing (for building workers, etc)	
	4171	1981	Specification for donkey jackets	
	2653	1955	Protective clothing for welders	1960, 1964, 1978
	3314	1982	Specification for protective aprons for wet work	
	4610	1970	Colours for high visibility clothing	
	5064	1974	Optical performance of reflective agents for use in high visibility garments and accessories	
	4724	1971	Method of test for resistance of air-impermeable clothing materials to penetration by harmful liquids	
Fall protection	3713	1982	Safety nets	
	1397	1979	Specification for industrial safety belts, harnesses and safety lanyards	1981(2)
	5062	1973	Self-locking safety anchorages for industrial use	1976, 1979
	5845	1980	Specification for permanent anchors for industrial safety belts and harnesses	

provided the human eye with various built-in protective devices against nature's hazards:

- the bone structure shields the eyes against large objects;
- the surrounding muscles absorb the shock of blows;
- eyebrows divert water running down the forehead from entering the eyes;
- eyelids and lashes close rapidly by reflex action to trap small particles or insects or shut out sudden bright light;
- the long optic nerve allows some displacement of the eye ball without permanent damage;
- the tear ducts try to wash away foreign bodies which have entered the eyes.

In spite of these defences, the eyes are very vulnerable to various man-made hazards, as the numerous scheduled processes shown in *Table 2.8.4* clearly indicate. These hazards for which eye protection is needed come under the following broad categories:

- impact from flying particles
- splashes of molten metal or other liquids
- dust
- toxic and irritating gases
- chemicals
- radiation
- combinations of two or more of the above

The hazard to which construction workers are most exposed is impact from flying particles, as *Table 2.8.4* shows. A proportion of workers are also exposed to splashes from harmful chemicals, and many are exposed to dust. Few other than welders and gas cutters and those working near them are seriously exposed to radiation and drops of molten metal. Those few who are exposed to toxic and irritating gases generally require combined eye and respiratory protection—often skin protection as well.

Two British Standards cover the definition, choice, construction and performance of eye protectors for all these hazards in detail and should be consulted carefully before standardizing on and ordering eye protectors for groups of construction workers. BS 2092:1967 with its numerous later amendments covers design and construction of eye protectors against the first four hazards listed, but not against harmful radiation. However, by incorporating suitable light filters in the lenses specified under BS 2092 (e.g. filters covered by BS 679), protection is provided against the radiation to which most construction workers are exposed.

Eye protection for welders and gas cutters is covered by BS 1542:1982. This Standard not only specifies the protection requirements for the eyes but also for the face and neck. Protection of workers other than those actually involved in welding and gas cutting is discussed here first.

Spectacles and goggles of various types cater for the majority of situations. Unless they can be worn for long periods without discomfort, there is a risk that eye protectors will not be worn when most needed.

2.8.4.1 Spectacles

Individually sized and fitted spectacle frames, with suitable lenses and generally with side pieces, whilst tending to be more expensive than goggles, are more comfortable and more readily accepted. Many workers in any case wear spectacles with prescription lenses for sight correction, which provide them with some measure of protection against flying particles.

Where the protection offered by such normally worn spectacles is judged to be inadequate, there is normally little difficulty in persuading a worker to wear a pair with a strong plastic or plastic covered frame, prescription impact-resisting lenses and side protectors perforated for ventilation which comply with BS 2092. (Metal frames can be uncomfortable in hot or cold weather.) The majority of workers today even with perfect eyesight will, at some time, have worn sun glasses, often badly fitting ones, and do not find the wearing of properly sized and fitted safety spectacles with plain lenses stressful.

In view of the high risks of eye injuries from flying particles faced by construction workers, it seems desirable that all construction workers be provided with personally fitted safety spectacles with side protectors and prescription lenses where needed.

For persons such as supervisors and steel erectors who are incidentally exposed to radiation from welding, lightly tinted lenses which absorb IR and UV radiation without significantly impairing general vision are recommended. For those seldom if ever exposed to welding radiation, or very strong sunlight, untinted lenses are preferred. Clip-on and hinged flip-up tinted eye shades are available for attachment to spectacles with clear lenses for protection against radiation, although they tend to be cumbersome, and easily damaged or displaced. Special lenses are available today for use against sunlight and other forms of radiation, the tint of which automatically adjusts itself to the light intensity.

The main disadvantage of most spectacles which applies to prescription as well as safety ones, is that their side arms make it impossible to achieve an air-tight seal when respiratory protection is worn. A similar problem exists when ear muffs are worn. Safety spectacles with special frames can, however, be obtained which obviate this problem. Because they only rest on the bridge of the nose spectacles offer less protection against impact than goggles which fit the face more closely.

2.8.4.2 Goggles

Goggles which enclose the eyes completely, with or without ventilation, and which are kept in position by an elastic band round the back of the head, provide more complete eye protection than spectacles but are generally far less comfortable to wear. Goggles may be considered for the following operations and hazards for which spectacles may not provide adequate protection:

- gas welding and cutting (see BS 1542:1982)
- work in high concentrations of dust

- work where highly injurious chemicals (such as hydrofluoric acid or caustic soda) are used or where other less injurious chemicals (such as hydrochloric acid) are sprayed or may be present as fine sprays or mists
- work with toxic and irritating gases
- work where impact from large particles is possible

Special lightweight disposable plastic eye shields which are sometimes classified as goggles are often issued to visitors to building sites and factories where they may encounter one or other of the eye hazards mentioned above. They afford a fair degree of all-round protection, and may be worn over spectacles with prescription lenses.

Goggles are classified in BS 2092 either as 'box type' with one-piece protective lenses, or 'cup type' with individual protective lenses which enclose each eye. Goggles are also classified as 'gas tight', 'dust' and 'chemical' goggles.

Of the applications mentioned above, the use of tinted goggles for gas welding and cutting is perhaps the most important. The other three applications generally require respiratory and often skin protection as well. Whilst goggles and some types of gas mask or respirator can be worn together, a wider range of hazards can be covered by the use of breathing apparatus with a full face mask or even a complete protective suit.

Before purchasing goggles, care must be taken to ensure that the type chosen is the most suitable one for the hazard involved. For protection against gases and vapours that attack the eyes, gas tight goggles are needed. These of course readily mist up inside. To alleviate this problem the lenses should be treated with a hydrophilic coating. Most other goggles have limited ventilation, although the tendency to mist up remains.

2.8.4.3 Other types of eye protection

Face shields, hand shields, helmets and fixed shields are types of combined eye and face protection needed for class 3 electric arc welding and cutting and class 4 gas shielded arc cutting and welding. These requirements are discussed in detail in BS 1542.

A number of different types of protection is available. Some can be supplied with compressed air via an air line both to provide a cleaner atmosphere for the wearer to breathe and to cool his face. Neither spectacles nor goggles alone provide sufficient protection for these operations. Owing to the multi-functional nature of this equipment, it is discussed subsequently in Section 2.8.6 under 'Face Protection' (which also includes protection of the eyes).

2.8.5 Protection of the respiratory system

BS 4275 gives a guide to the selection of a bewilderingly wide variety of protective equipment. Respiratory protective equipment falls into two broad categories:

respirators
breathing apparatus

Protection of the respiratory system 347

Respirators utilize the surrounding air and remove impurities from it by different media before it is breathed by the wearer. Breathing apparatus on the other hand relies on the use of an independent air or oxygen supply. Both types have to prevent the surrounding contaminated air from entering the nose or mouth of the wearer. This may be done in five ways by the use of:

- ori-nasal (half mask) respirators which cover the nose and mouth only;
- full face masks, which cover the nose, mouth and eyes, and which incorporate their own lenses or eye protectors;
- hoods, which cover the whole head;
- combined hoods and blouses, which cover the upper part of the body;
- by the use of full protective suits, which cover the entire body.

Most respirators which depend on the lungs of the wearer to suck incoming air through purifying canisters or cartridges are restricted to half masks and full face masks. However, respirators which include a portable air fan, motor and battery to reduce the breathing effort of the wearer are not so restricted and are available as hoods and other types.

Two points of special importance apply to all types of respiratory protection, but particularly to masks.

- Mainly because of the difficulty in providing an air-tight seal round the wearer's nose and mouth, all types of respiratory protection allow a small leak of contaminated air from the surroundings to the wearer's breathing zone. Figures for typical leakage of contaminant past various types of respiratory protection are quoted from about one in fifty to one in two thousand. These assume a good fit between the face of the wearer and the mask. This can only be achieved if the correct size of face mask has been chosen, and usually only if the wearer's neck, cheeks and temples are clean shaven. Sideboards, whiskers and beards often make it impossible to achieve a proper seal between the face and face mask (full or half), thus leaving the wearer with a false feeling of security.
- In the case of respirators, the canisters, cartridges and air filters used have only a limited working life before they become saturated with the contaminant and cease to protect. Manufacturers instructions on how long a cartridge, canister or filter may be used should be followed carefully. In general, a canister or cartridge should be treated as spent after any use and disposed of, and replaced by a new one.

Respiratory protection so restricts the activities and impairs communications of the wearer that every effort needs to be made to eliminate respiratory hazards or protect the worker from them in other ways (e.g. by wet methods of work and improved ventilation) before they are specified. In cases where respiratory equipment is needed, an effort should be made to monitor the air for harmful impurities as discussed in Chapter 2.7.

Particular construction activities where respiratory protection is or may be needed, include:

- paint spraying to protect against solvent vapours and fine paint particles
- removal of old asbestos insulation, to protect against asbestos fibres
- shot blasting
- all work producing fine silica and chrome dust, e.g. work in tunnels, grinding granite and laying high silica or chrome furnace bricks
- gas and electric welding and cutting of steels containing or coated with toxic elements (e.g. lead, cadmium, beryllium)
- laying floor tiles (in poorly ventilated rooms) with adhesives containing organic solvents
- painting and paint removal by powerful solvents
- the use of highly toxic chemicals for wood treatment, insecticides and herbicides and work which produces fine dust from objects so treated
- entry into sewers, pits and confined spaces where atmospheres cannot be made safe.

Since particular types of breathing apparatus have more general application than respirators, provide better and more continuous protection, and generally provide less strain on the wearer, they are discussed first.

2.8.5.1 Breathing apparatus

Two main types and five sub-types of breathing apparatus are covered by British specifications, as shown below:

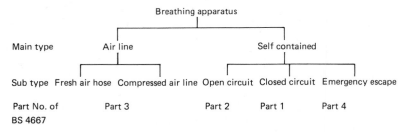

Those used in actual construction operations fall almost entirely in the first main type. Self-contained breathing apparatus as used in various forms by firemen and divers, and for escape from tunnels and fires, are so little used in actual construction work that they will be only briefly discussed here.

Both air line types of breathing apparatus place some restrictions on the wearer's movements, since like a diver he is connected to the end of a flexible air line.

Fresh Air Hose Breathing Apparatus
With fresh air hose breathing apparatus, a wide diameter (approximately 1 inch) non-kinking air hose is used with the air at substantially atmospheric

Protection of the respiratory system 349

pressure. Uncontaminated air is drawn into the hose from a safe place and enters the face piece of the wearer via a non-return valve. The air intake end of the hose must be secured so that it is not drawn into the contaminated atmosphere. The face piece has a similar non-return valve to discharge exhaled air to the atmosphere. This type of apparatus may be used with or without a portable hand or motor operated air blower. Without a blower, the air is drawn through the air hose when the wearer inhales. This restricts the length of the air hose to a maximum of 9 m (30 ft). With a portable blower (hand or motor operated) the length of the hose may be increased up to 36 m (120 ft).

When this type of apparatus is used, the wearer should also wear a safety belt attached to a safety rope held by at least one other person, located at the air intake end in a safe atmosphere (see Section 2.8.12.2).

Compressed Air Line Breathing Apparatus
The compressed air line type of breathing apparatus (*Figure 2.8.2*) is better suited to regular and continuous work, such as paint spraying and shot blasting, than the atmospheric pressure hose line. Here a much smaller diameter air line replaces the 1 inch hose, and since its length is less restricted, it allows the wearer a much greater freedom of movement (working range). The face mask is a little more complicated, since it incorporates a pressure reducing valve on the air line as well as the other

Figure 2.8.2 Compressed-air line breathing apparatus.

automatic valves. Compressed air line breathing apparatus is available only with face masks, helmets or completely impervious protective suits.

Two types of compressed air line breathing apparatus are available, constant air flow and demand valve types. The first provides a constant flow of clean air at a pressure slightly above that of the surrounding atmosphere through the face piece and past the wearer's mouth and nose. Thus danger of contaminated air leaking into the face mask is minimal, while the wearer does not have to exert himself while inhaling. With the demand type, the inlet non-return valve on the face mask only opens when the wearer breathes in, and when the pressure in the face mask has fallen slightly below that of the surrounding atmosphere. The main advantage of the demand type is that it uses far less air than the constant air flow type.

Compressed air line breathing apparatus is critically dependent on a constant supply of clean and pure compressed air. The air from many compressors used in construction is often contaminated with fine droplets of lubricating oil and sometimes with toxic decomposition products of the oil—particularly if the compressor is running hot. Compressed air of guaranteed purity for use with compressed air line breathing apparatus can be bought in cylinders which is more economical for occasional use than a special adapted compressor. Where compressed air cylinders are used, a cylinder regulator with two pressure gauges is needed to reduce the pressure in the cylinder to that required in the air line. The contents of the cylinder are shown by the pressure gauge on the inlet of the regulator.

Special non-lubricated air compressors of the diaphragm or carbon ring type can generally be safely used for providing a clean and safe compressed air supply. The air compressor should be fitted with an after cooler, a water separator and an air reservoir on the discharge side. These prevent water condensing in the air supply line and ensure that there is sufficient compressed air to allow the wearer to leave the contaminated atmosphere and reach a safe place before removing his face mask.

When working in tanks, pits and enclosed spaces where exposure to the atmosphere could cause collapse or unconsciousness, the compressed air cylinders or air compressor should be stationed outside in clean air, with another man in attendance. A safety line should be attached to the belt (see Section 2.8.12.2) of the wearer of the breathing apparatus, so that he can be pulled out in an emergency. The man on duty outside the tank, etc. must keep a close check on the compressed air supply. If a single compressed air cylinder is used, he must signal to the man working in the hazardous atmosphere to leave as soon as the pressure falls to some pre-set value. This should be such as to ensure an adequate reserve is available to enable the wearer to leave the hazardous atmosphere area safely.

Compressed air line breathing apparatus is critically dependent on the proper functioning of the various pressure and flow control valves on the apparatus. They must therefore be regularly inspected and maintained. Because of these complications, compressed air line apparatus has not usually been recommended for work in atmospheres which are dangerous to life. Ultimately, however, the safe use of such equipment depends largely on the reliability and state of maintenance of the apparatus, on the training of its wearer and the attendant on the outside. Provided these are all of a high standard, there is no reason why this type of apparatus should be any less safe than a fresh air hose apparatus or a completely self-contained breathing apparatus.

Self-Contained Breathing Apparatus

Of the various types of self-contained breathing apparatus available, only the open circuit type with portable compressed air cylinders is likely to be needed on most construction sites, and then only for rescue purposes. An example is the rescue of a worker who has been trapped or incapacitated in a confined space where the atmosphere is hazardous. The only real difference between this type and the compressed air line apparatus is that the wearer carries a small compressed air cylinder on his back instead of being connected to a large cylinder via a flexible compressed air line.

Open circuit apparatus made to BS 4667, part 1, comprise a portable compressed air cylinder, a body harness, a face mask, a means of speech transmission and various short connecting hoses and valves. The maximum time for which the apparatus can be used continuously with the single cylinder carried varies between 20 and 40 minutes. This type of apparatus should only be used by men in good physical condition who have been thoroughly trained in its use.

Closed circuit breathing apparatus which uses oxygen and a gas purifier (to remove carbon dioxide) in place of air, though lighter than open circuit apparatus, is considerably more complicated. Its use would be limited on construction sites to very special circumstances, such as rescue from tunnels.

Escape breathing apparatus is covered by BS 4667, Part 4 and may be of open or closed circuit type. These differ from general use equipment in being smaller and lighter, with smaller compressed air or oxygen cylinders. Its main use is in enabling people trapped in a tunnel or coal mine to escape through zones or passages where the atmosphere is very hazardous or will not support life.

2.8.5.2 *Respirators*

Respirators which depend on removing specific contaminants from the air breathed by the worker are of many types. The canisters, cartridges and filters must, therefore, be specially selected to remove the contaminant in question. Air contaminants are removed in three ways:

- by filtering the air through a fine filter medium (such as paper, felt and impregnated pads); this removes dusts, mists and fumes but not gases and vapours;
- by passing the air through a granular solid which reacts chemically with the impurity (such as soda lime to remove hydrogen chloride and other acid gases);
- by passing the air through a granular porous solid which adsorbs gases and vapours on its large surface area (such as active charcoal to remove the vapours of chlorinated hydrocarbons).

These various types of respirator with the numbers of the relevant British Standards are shown diagrammatically below:

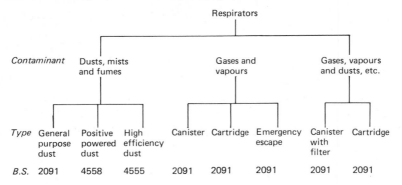

Respirators used by construction workers are almost exclusively of the first type for protection against dusts, mists and fumes. Workers exposed for any length of time to hazardous gases and vapours are better protected by breathing apparatus.

General purpose dust respirators have many applications against fine dusts which cannot be suppressed or avoided, e.g. from dry grinding, handling dry cement, and dressing silica based stone and bricks. In some cases a simple dust mask which covers the nose and mouth only may be sufficient. This, however, needs careful checking since very 'simple' dust masks which effectively trap large particles may allow the more dangerous fine particles to pass through.

For demolition and other workers involved in removing old asbestos insulation, the general purpose respirator covered by BS 2091 is unlikely to offer adequate protection. Unless air line breathing apparatus is used, a positive powered dust respirator (BS 4558) or a high efficiency dust respirator (BS 4555) is generally required. Positive powered respirators combined with a helmet are now available with a small battery and blower incorporated into the helmet itself.

2.8.6 Face protection

Face protection is always combined with protection for the eyes or respiratory system or both, depending on which of these more sensitive body organs require protection from the hazard in question. Electric arc welding, particularly of the gas shielded type, creates intense UV radiation which can damage the unprotected skin, as well as the eyes. Protection is usually offered by a visor, face shield or welding helmet. These must comply with BS 1542:1982. Visors and face shields may be hand held or supported from a 'hard hat' worn for head protection. In the latter case these are best purchased from the same supplier as the hard hat to ensure compatibility. Clear face shields are also sometimes required when there is a hazard of flying particles which may injure the face as well as the eyes, e.g. certain grinding and machining operations.

Clean-air helmets are used with compressed air line breathing apparatus to protect the whole face, head, eyes and respiratory system and also cool the head. Even where only the respiratory system strictly needs protection, they may be preferred to face masks because of the problems of air leakage mentioned with them in Section 2.5.8.2.

2.8.7 Head protection

By head protection we generally refer to some sort of hat to protect the top of, rather than the whole, head. BS 5240:1975 covers General Purpose Industrial Safety Helmets, which protect against four main hazards:

falling objects
abnormal heat
sideways crushing
chemical splash

Of these, the first is usually the main concern of construction workers.

Safety helmets or hard hats consist of a strong outer shell made of a thermoplastic, such as ABS, glass reinforced plastic, phenolic resin laminate or aluminium, which is supported clear of the head by a harness and adjustable headband to ensure a sound fit. Harnesses are available in plastic, fabric and leather.

Safety helmets may have a peak in front only or a rim all round (for protection against chemical splashes and heavy rain) but the rim can be a menace in confined and cramped spaces. A helmet with only a peak is generally more practical for construction workers.

The specification does not restrict the materials used so long as the helmet passes various obligatory tests. These are shock absorption under various conditions, flammability and penetration resistance. In addition, the customer may require tests for electrical insulation, lateral rigidity and shock absorption and penetration resistance at a low temperature (-20 °C). Helmets may have various extras, e.g. chin strap, neck protection, nape strap, inner (winter) liner, draw lace and attachment devices for a lamp and cable, ear muffs, air line breathing apparatus, face shield and welding visor. The weight of a safety helmet (shell plus harness) must be shown on a label attached to it if it exceeds 400 g.

Aluminium helmets, whilst generally lighter and offering better radiant heat protection than most others, are electrically conducting and are best avoided on construction sites where there are temporary overhead electric cables.

It is obviously important for a construction worker that his helmet should not fall off if he bends over, or be blown off by the wind, and also that his head, neck, ears and forehead should be protected from the cold in winter. Thus purchasers should insist that helmets bought for construction workers be supplied with chin straps (and/or nape straps) as well as warm liners and neck, ear and forehead protectors (best supplied integrally with the liner). Further, since these are optional extras and not covered by the specification, their effectiveness, particularly that of the chin strap (and/or nape strap), should be carefully checked, preferably by field trials.

When purchasing helmets it is also as well to ensure that face shields or visors and ear muffs can be fitted to them, and to buy the helmets from a firm which also makes and supplies these items as well.

354 Personal protective clothing and equipment

In addition to these general purpose safety helmets, the following other head protection is often needed:

- welders' helmets, to which visors or face shields are attached, which protect the head from welding sparks and heat, as well as offering some protection against impact from falling objects. These are generally lighter than the general purpose safety helmets.
- acoustic helmets, for those exposed to extremely high noise levels (i.e. over 130 dB 'A') for which even ear muffs do not offer sufficient protection (because the bones of the skull also transmit sound)
- bump hats and scalp protectors (BS 4033) for protection against bruising and abrasion by those working in confined spaces such as garage service bays and tunnels
- caps and hairnets for machine workers (e.g. wood working machinists) to prevent hair catching in moving machinery
- cape hoods and 'sou'westers' as part of 'foul weather clothing'
- chemical proof hoods and helmets which may form part of an internally ventilated suit with air line breathing apparatus
- shot blast helmets
- 'air stream' helmets. These contain an electric blower, powered by a battery worn on a belt, which forces air through a filter in the roof of the helmet and blows clean air over the face of the wearer. A face visor attached to the helmet gives eye protection as well. These are really special respiratory devices, whose main use is in welding and in working with toxic dusts.

These appear to cover most of the requirements of construction workers, but even this list is not complete when we consider helmets for firemen, motor cyclists, anti-riot police, divers, and others. Construction companies need to consider the head protection requirements of their workers rather carefully to avoid landing up with a mass of assorted and incompatible head gear from different suppliers.

2.8.8. Ear protection

Whilst we are mainly concerned here with noise and hearing protection, ears are also very vulnerable to low temperatures. Fur hats worn in Russia and China in winter time have retractable ear flaps, which our own hat makers might do well to copy. There are still no British Standards for hearing protectors, although a subjective standard, BS 5108:1974, has been published for measuring their effectiveness.

Hearing protectors range from simple ear plugs through more efficient ear muffs to noise helmets for the noisiest situations.

Most types of ear muffs and noise helmets can be provided with some type of audio communication system which incorporates receivers in the ear cups.

2.8.8.1 Ear plugs

These are mainly for workers continuously exposed to noise levels of 90 to 100 dB'A'. Several types are available:

1. Disposable ear plugs of wax impregnated cotton wool or glass down which are roughly shaped with the fingers and inserted into the ear canal.
2. Semi-disposable types made of foamed plastic which are compressed with the fingers and inserted into the ear canal where they expand, fitting it closely.
3. Moulded pre-shaped plugs of rubber or plastic.
4. Individually moulded plugs made by inserting a synthetic rubber compound in paste form into the ear canal and allowing it to cure and set there in situ. The shaped plugs are then removed and inserted when needed.
5. Pre-moulded hollow rubber or plastic plugs with a valve system which absorbs high noise levels whilst allowing speech sounds to pass.

The attenuation provided by typical plugs and ear muffs was shown in *Figure 2.5.6*.

One disadvantage noted with ear plugs is that it is not easy for supervisors to check whether they are being used. Another is ensuring that they are fitted hygienically, without risk of infection, when a worker about to start a noisy job decides to fit ear plugs when his hands are dirty.

2.8.8.2   Ear muffs

Ear muffs provide more protection than ear plugs and are generally suitable for noise levels up to 120 dB'A'.

There are two types, circumaural and superaural. Circumaural muffs are more common. They enclose the ears and seal against the side of the head. Typically they consist of two semi-rigid cups containing sound absorbent liners, each with a soft sealing ring in contact with the head. The cups are attached to a springy metal or elastic headband which holds the muffs in close contact with the head. Alternately, the muffs may be attached by springy arms to safety helmets designed to carry them. Unfortunately, it is difficult for a person wearing spectacles with normal side arms and ear bows to wear these muffs in comfort and enjoy the full benefit of the noise reduction possible with them.

Superaural ear muffs are smaller than the circumaural and seal against the ears themselves. Whilst lighter and probably less effective than the circumaural type, they are less affected by spectacle frames.

2.8.8.3   Noise helmets

These enclose the neck and skull and are used by persons exposed to very high noise levels, e.g. above 115 dB'A'. They insulate the bones of the skull and neck which can transmit high noise intensities to the ears.

2.8.9   Hand protection

By hand protection we generally think of a close fitting covering of fabric, leather, rubber or plastic in the form of gloves, gauntlets, mittens, hand guards (to protect the palm only), finger and thumb stalls, and may possibly include barrier and cleansing creams. All these are considered here, and the term 'gloves' is used in its broadest sense to include gauntlets, mittens and other forms of hand covering.

2.8.9.1 Gloves

Gloves are used to protect against numerous hazards which make varied requirements on them. It is unfortunately impossible to find a single material which offers protection against all hazards, so that a wide range of gloves are made in different materials. Added to this, even when gloves fit well and are light and flexible, a certain manual dexterity and delicacy of touch are lost, so that they are only worn when the need is very obvious, e.g. in very cold weather or when handling hot objects. Also, where there is but a single external hazard, there is often a need for different types of gloves for tasks which provide different exposure to the hazard. This is well illustrated by the totally different gloves worn by cricket batsmen and wicket keepers to counter the same hazard—impact from a hard leather ball. Very close familiarity with the hazards and with the tasks performed is needed before appropriate gloves can be chosen. The great diversity in size and shape of hands and fingers must also be catered for.

BS 1651, though somewhat dated, gives details of a wide range of glove materials and types, with guidance for their selection for various tasks. BS 697 gives specifications for rubber gloves for electrical purposes (i.e. work on live circuits).

Hand hazards among different occupations in construction which may require the use of gloves are shown in *Table 2.8.6*, together with an indication of the most suitable material. This table can only be used as a general guide, and manufacturer's advice and chemical resistance charts should be consulted before ordering.

The need for gloves varies considerably throughout the occupations listed. Those with a special need are asphalters, electricians (when working on live circuits), pneumatic tool operators, ground workers, painters, glaziers and brick and stone cleaners. Most outdoor workers need gloves in cold and rainy weather, and those chosen should be appropriate to their work. Only five out of a large number of possible glove materials are shown under 'material' in the table. Leather and cotton in various forms are generally most comfortable, and are used for protection whilst handling hot and cold objects, or against cuts, splinters, abrasion, bruises and pressure. They are not, however, suitable for wet work or solvent resistance unless specially treated with rubber, PVC or other plastics.

Natural rubber has good abrasion resistance and can be used in wet conditions, but its solvent resistance is limited. Synthetic rubbers, particularly neoprene and nitrile rubbers have much better general solvent and chemical resistance, although they cannot be used in all cases. A range of PVC and PVC impregnated gloves are available with resistance to abrasion and many solvents and chemicals. But to make PVC flexible it contains a 'plasticiser' which is leached out by hydrocarbon and chlorinated hydrocarbon solvents, thus making the gloves hard and brittle. Some rubber and leather gloves contain compounds (chrome used in tanning,

357

| Occupation | \multicolumn{18}{c}{Hand hazards} | \multicolumn{5}{c}{Protection suggested – Type} | \multicolumn{5}{c}{Material} | \multicolumn{3}{c}{Weight} |

Occupation	Cold	Heat & burns	Rain & wet	Cuts	Pricks & splinters	Abrasion	Bruising	Pressure	Vibration	Hot asphalt	Solvents	Paints	Adhesives	Acids & alkalis	Cement and lime	Other chemicals	Dirt	Electric shock	Gauntlet	Glove	Mitten	Hand guard	Finger or thumb stall	Leather	Cotton & moleskin	Natural rubber	Synthetic rubber	PVC	Heavy	Medium	Light
Bricklayers	×		×			×									×					×						×		×	×		
Stone masons	×		×			×									×					×						×		×	×		
Carpenters & joiners, outside work	×		×	×	×	×	×													×						×		×		×	
Carpenters & joiners, indoor work				×	×		×													×										×	
Roof workers, tilers & slaters	×		×			×							×							×					×			×		×	
Asphalters (roof & road surfacing)	×	×								×									×				×		×				×		
Concrete workers	×		×			×	×								×	×				×								×		×	
Bar benders & steel fixers	×		×		×	×	×													×						×		×	×		
Scaffolders	×		×			×	×	×												×		×		×	×			×	×		
Steel rigger-erectors	×					×	×	×												×				×	×			×	×		
Electricians (installation & linesmen)																		×		×						×†					
Heating, ventilating & ductwork fitters																		×		×								×	×		
Welders		×																	×	×				×							
Plumbers		×	×																	×				×						×	
Crane erectors & operators	×				×	×	×	×												×				×						×	
Thermal insulation fitters (laggers)					×	×	×	×											×			×		×						×	
Pneumatic tool operators						×			×											×										×	
Ground workers	×		×			×			×							×	×			×						×		×		×	
Painters & decorators	×		×									×	×			×	×			×						×	×	×		×	
Glaziers	×		×	×	×									×					×	×				×	×						×
Tilers (indoors)													×							×						×		×		×	
Flooring layers											×	×	×													×	×	×		×	
Demolition workers	×		×			×											×			×				×				×	×	×	
Brick & stone cleaners	×													×		×	×		×	×							×	×	×	×	

†For work on live electric circuits. Gloves should comply with BS 697:1977

accelerators and antioxidants in rubber) which can actually cause dermatitis in sensitive persons.

Other materials used in gloves include Kevlar (a modern plastic), asbestos, wool, chain mail, polyvinyl alcohol, butyl and Viton rubbers.

Some materials, such as rubber and PVC, do not allow the skin to breathe, and the hands can become wet and clammy when wearing them. This can create problems with wearer acceptability which must be borne in mind when selecting these type of gloves. An inner lining may help reduce this problem, but this tends to be used more for warmth than sweat absorption.

The question of whether gloves are needed on particular jobs should be decided jointly between employers and workers' representatives. Where it is agreed that gloves should be worn, HASWA, Section 2, obliges employers to provide them, and Section 7 obliges workers to wear them. Special care is needed in keeping gloves clean and in good condition, both inside and outside. In many cases the effectiveness of gloves against chemicals and solvents can be destroyed by their being removed and either intentionally or accidentally turned inside out. Thus the contamination comes into contact with the hands negating the protection for which they were supplied.

2.8.9.2 Barrier and other hand creams

Barrier creams once enjoyed a certain vogue as a substitute for gloves to protect the hands against oils and greases, solvents and various chemicals. This followed a Ministry of Labour pamphlet in 1946. Subsequently they came in for critical examination by various members of the medical profession about 1960. From this is appeared that they did little good, and increased rather than decreased the degree of penetration of the skin by various harmful contaminants. One long-maintained claim for barrier creams is that when applied before work they make it much easier to remove tars, oils, chemicals and dirt when the hands are washed after work. This claim has since been largely refuted by medical researchers.

A different type of UV absorbent barrier cream is effective in protecting the skin against strong sunlight and arc welding radiation, and is recommended for workers exposed to these hazards.

There is, however, one real need for skin creams for construction workers. This is to replace natural oils and waxes in the skin lost during work by contact with solvents and chemicals, or during hand cleaning after work using powerful detergents or other hand cleaners. These skin creams should however be applied after washing the hands during and after work, and not before.

2.8.10 Foot protection

All workers today in industrialized countries wear shoes, boots, sandals or clogs, but this is far from the case in many developing countries. In discussing foot protection for construction workers, we are, therefore, talking about additional protection against particular hazards over and above that provided by the employee's own footwear.

Whilst there are no regulations in the U.K. requiring safety shoes to be provided for and worn by workers, comprehensive British Standards on Safety Footwear exist, particularly BS 1870, Parts 1, 2 and 3 and several manufacturers produce safety footwear conforming to these standards. One common feature to all such safety footwear is an in-built steel toe protector which covers all toes except the little toe and protects them from being injured by falling objects. This is a hazard faced by most construction workers.

As safety boots and shoes with toe protection are as comfortable, attractive and economic as most other footwear worn by construction workers, it is sound safety policy to encourage workers to buy and wear them in preference to other types.

Safety footwear (which may be of leather, rubber or PVC and includes boots and shoes), may incorporate the following three optional features:

- protective mid soles
- oil resistance
- electrical conductance or resistance, including protection against the hazard of igniting sensitive materials by sparks arising from the build-up of static electricity on the body.

Of these, protective mid soles protect the wearer from upstanding nails and other sharp objects. This protection is valuable to ground workers, but unfortunately protective mid soles make shoes too rigid for climbing and should not be worn by steel erectors, scaffolders, roof workers and others who are constantly using ladders.

Some safety boots and shoes can be provided with an armoured flap which protects the front of the foot against falling and crushing hazards, although this is rarely done. The one weak spot of safety boots and shoes is that they provide no protection against objects piercing them through the side.

The most common footwear on construction sites is the rubber or PVC ankle boot, which offers good protection from puddles and wet ground, and from toxic or hazardous chemicals, etc. which may be underfoot. However, employers can encourage employees to buy their own safety footwear through the 'Golden Shoe Club'. This arranges for a representative of the manufacturer to visit them at work, fit them with safety boots or shoes, and arrange for their payment by a small weekly deduction from their wages.

2.8.11 Body protection

It is difficult to draw a hard and fast line between the everyday working clothes used by building workers, whose main object is to provide thermal insulation to conserve body heat, and provide some padding against pressure and abrasion, and special clothing and protective equipment required for particular hazards encountered in the industry. Often in the past there was simply no suitable and durable clothing which the construction worker could afford and he had to manage without.

The commonest protective clothing used by craftsmen and labourers until well into this century were aprons and padding[4]. Head caps and 'knots'—cushions resting on the back of the neck supported by a strap over the forehead—were commonly used in the U.K. 150 years ago and may be seen today in parts of India. Shoulder pads are commonly used by dustmen, and hip pads were used till recently in brickworks. Leather aprons and paper hats (to keep the hair clean) were used by builders and other tradesmen in the nineteenth century[4].

The only statutory requirement in the U.K. for the provision of personal protective clothing for construction workers is contained in Regulation 15 of the Construction (Health and Welfare) Regulations, 1966, which states:

> "Every contractor shall provide adequate and suitable protective clothing for any person so employed who by reason of the nature of his work is required to continue working in the open air during rain, snow, sleet or hail."

Special clothing is also required to protect the body from toxic and corrosive chemicals which can penetrate or injure the skin as well as to protect normal clothing from damage or contamination.

Since most construction workers are exposed at various times to moving machinery, their clothing should have no loose ends which may get entangled in it.

Another hazard against which special body protection is required is that of falling, for which safety lines, harnesses and belts are available. This is discussed in the next Section 2.8.12.

Much of the clothing available is covered by British Standards which are discussed below.

2.8.11.1 Normal working clothing for building workers

BS 4679:1971 gives a specification for suits (generally two piece) to provide protection from the elements throughout the year for those working outdoors on building sites. The Standard does not specify style, material or weave of cloth, but it provides for insulation to cover an ambient temperature range from -10 °C to $+16$ °C by adding or removing liners. It allows no loose shirts, cuffs, belts or projecting fastenings. Plastic coated fabrics may be used but are not obligatory. Specifications for abrasion resistance, air permeability, cold crack temperature, colour fastness to light and rubbing, flex cracking resistance, hazardous liquid resistance, tear resistance, breaking load and water penetration are all given.

Seams and pockets must be waterproof. Fastenings should not be able to catch on projecting equipment, but should be able to be manipulated with one hand, and certain specifications are made for ventilation of the garments.

The clothing covered by this specification is of course only intended for temperate and cold climates. For work in hot climates, much lighter clothing is needed—principally a pair of old shorts, plus any necessary protection against abrasion, pressure, insects, dust storms and strong sunlight.

Another garment commonly used by construction workers is the 'Donkey jacket' covered by BS 4171:1981 which, like BS 4170, originated from requests by the building industry and local authorities. A donkey jacket is simply a strong jacket with leather shoulders and patch pockets which will stand a considerable amount of abrasion, keep the wearer warm and protect his shoulders from pressure when he carries ladders, parts of scaffolds, hods of bricks, etc.

Details of welders' clothing are given in BS 2653:1955 which covers welders' caps and berets, capes, sleeves, aprons, leggings, gaiters and spats made from leather, asbestos fabric or textile materials suitable for flameproof industrial clothing. The standard has three sections:

1. Materials (requirements for leather of various types and asbestos)
2. General Requirements (seams, stitching, edge finish and fasteners)
3. Specifications for equipment referred to above—caps, berets, capes, etc.

Protective aprons for wet work are specified in BS 3314:1982. This deals with materials, dimensions, making-up and labelling of such aprons. Whilst not specifically intended for construction workers, there are a number of activities in construction—painting and decorating, stone and brick cleaning, plastering—where aprons complying with this Standard may provide the most suitable protection.

2.8.11.2 *Foul weather clothing*

BS 4170, prepared at the request of the Builders' Plant and Equipment Industry and Local Authorities, gives specifications for Waterproof Protective Clothing made from coated fabrics to cover coats, trousers, leggings and sou'westers. The material for the garments is a PVC coated fabric as specified in BS 3546.

The standard covers garment sizes and measurements, make up, style and manufacture.

A few details are given below, but for full information the standard should be consulted.

Coats are single breasted with fly front, buttoned to the neck, the top button buttoned through. Three types of coat are considered, long coats, three-quarter length coats and jackets. Seams are stitched and welded over. Patch pockets with a 2 inch flap may be provided on each hip, and a self hanger is required. For ventilation at least 2 metal eyelets are required under each arm.

Leggings are made in pairs, each cut in one piece with a double thickness strap at one side with two button holes.

Trousers are unlined with an elasticated waist and side vents for access to pockets. The seat is in one piece with a reinforced crutch.

Sou'westers are pilot shape, with welded crown seams and chin straps.

However well designed such foul weather clothing may be, a good thermal balance is hard to achieve when wearing it, and the rate at which sweat can evaporate is very limited. A person wearing such clothing tends

Figure 2.8.3 'In hotter climates ...'

to get too hot when engaged in strenuous physical work, but when he stops to rest he soon gets too cold.

In hotter climates, the scantily clad worker may feel no need for rain protection and will welcome the cooling feel of monsoon rain on his bare skin, providing it is not mixed with hailstones (*Figure 2.8.3*).

2.8.11.3 High visibility clothing

Construction workers working in bad weather and poor visibility outdoors are sometimes at risk of being run down by plant operators and drivers of site vehicles. Similarly slingers and banksmen working in conjunction with crane operators need to be sure that they and the arm signals they make are clearly visible. The visibility of clothing has been radically improved by the introduction of fluorescent pigments into PVC coated fabrics, which give off a bright glow under the action of UV light.

Garments treated in this way include:

- bib-type waistcoats of the 'sandwich board' style, tied at each side by tapes;
- elasticated sleeves used for signalling;
- foul weather jackets, trousers and sou'westers covered by BS 4170;
- sleeveless jerkins

Colours for high visibility clothing are covered in BS 4610:1970 *Specification for colours for high visibility clothing* and reflective agents are specified in BS 5064 *Optical performance of reflective agents for use in high visibility garments and accessories.*

2.8.11.4 Chemical protection

PVC coated fabrics with a nylon or terylene base provide good protection against most chemicals used in construction, with the exception of some organic solvents and hydrofluoric acid. For these, fabrics coated with neoprene rubber, polyurethane or chlorobutyl rubber are more resistant. It is, however, important to inform clothing suppliers of the particular chemicals or harmful substances against which protection is needed and to secure their help and advice. British Standard BS 4724:1971, *Method of Test for Resistance of Air-Impermeable Clothing Materials to Penetration by Harmful Liquids* arose mainly from the needs to protect agricultural workers from toxic crop sprays. The fact, however, that a fabric meets all the requirements of this test offers little guarantee that it will resist all the various toxic and harmful chemicals encountered in construction.

The Industrial Safety (Protective Equipment) Manufacturers Association[3] divide chemical resistant clothing into three groups, light duty, medium duty and heavy duty.

Light duty chemical protective clothing generally consists of uncoated cotton or synthetic fabrics with a water repellant finish, in the form of overalls, laboratory coats or smocks. It is recommended for workers who are at some slight risk from relatively innocuous chemicals such as very dilute acids and alkalis.

Medium duty chemical protective clothing consists of an apron of one of several impervious and resistant materials such as natural rubber, nylon fabric coated with PVC, neoprene, etc., and leather. They are useful for workers who use and are exposed to a wide range of chemicals, oils and hydrocarbons, and include many machine tool operators.

Heavy duty chemical resistant clothing is based mainly on PVC or PVC coated synthetic fabrics in the form of boiler suits, long surgical coats, bib and brace overalls, leggings and three-quarter length coats.

Sometimes the foul weather clothing specified in BS 4170 proves suitable for this purpose. Contaminated heavy duty chemical protective clothing should always be washed or hosed down before the wearer removes it, and proper provision must be made for storing, inspecting and maintaining it.

In addition to these three types, some operations, such as paint spraying, may require the use of a completely impervious outer suit with an air line breathing apparatus and hood. Many suits of this type allow the air supplied to be used for body cooling and evaporation of sweat as well as for breathing. These are specially useful in hot climates where it is very uncomfortable to wear heavy duty protective clothing.

2.8.12 Fall protection

Most construction workers whose work carries a risk of falling are, or should be protected by the provision of proper scaffolding complete with guard rails and toe boards and safe means of access. It is not, however, practical to protect all workers in this way. Scaffolders themselves, steel erectors and steeplejacks are particularly at risk, for reasons discussed, in Chapter 1.5. For them, other means of fall protection, safety harnesses, safety belts and safety nets are available although not used to the extent needed to make an impact on the fatal injury statistics of these professions.

Safety belts and harnesses have been available for many years, but have not been widely used by steel erectors and others who most need them.

Figure 2.8.4 Anchorage line greatly extends the zone in which a worker can operate. The line can be used in conjunction with a full arrest device as shown here (Barrow Hepburn Equipment Ltd).

Two reasons for this have been:

- insufficient safe anchorage points for attachment of the belt or harness;
- severe restrictions on the movement of the worker when attached to any one anchorage point, and the constant need to change from one anchor point to another.

Greater flexibility in the use of safety belts and harnesses has been achieved in recent years with the development of anchorage lines suspended from a permanent anchorage at a considerably higher elevation, and to which the safety belt or harness may be attached (*Figure 2.8.4*).

These greatly extend the zone in which a worker can operate whilst attached to a single fixed anchorage point.

2.8.12.1 Safety nets

Safety nets are sometimes used to catch workers and debris falling from elevated places, and are covered by BS 3913:1982. Apart from their familiar use in circuses, their main industrial use has been in bridge construction. BS 3913 gives requirements for two types of safety nets, one suitable for duty heights up to 6 m and one for a maximum duty height of 1 m. The duty height is the maximum vertical distance between the working level and the level at which the safety net is to be placed in use. An earlier version of the standard included requirements for safety nets at duty heights up to 12 m. This has been dropped on safety reasons, including the danger of bounce after impact.

Safety nets are made from various man-made fibres with the exception of polyethylene. The fibres should contain an ultraviolet stabilizer and unnecessary exposure of the nets to sunlight should be avoided. The minimum nominal size of industrial safety nets is 4 m × 3 m. The area covered by a net is somewhat smaller than the net size since it sags in the middle. Nets to catch human beings have a square or diamond mesh, with a maximum length of mesh side of 10 cm.

Overlay nets of smaller mesh may be placed over a safety net to catch falling tools and debris. Safety nets can only be used where there are strong structural members to support them at the sides and corners.

The Standard contains appendices which give details of drop tests, inspection and care.

Two problems have hindered the wider use of safety nets on construction sites:

- there are often too many solid objects between the employees' working level and the position of the net; and
- there are often insufficient or inadequate supports or anchorages for the safety net at the required height and position.

2.8.12.2 Safety belts, harnesses and lanyards

BS 1397:1979 *Specifications for industrial safety belts, harnesses and safety lanyards* gives details of five types of belts and harnesses with guidance on their selection:

Type A Pole Belts are intended for use in a fully loaded condition by linesmen working on poles, and could be better classified as 'working belts' rather than 'safety belts'. They are intended for situations with a maximum drop of 60 cm. They consist of an adjustable body belt and adjustable pole strap whose length can be altered whilst it is still attached to the body.

Type B General Purpose Safety Belts are used with safety lanyards (1.2 m long) which can be attached to anchorage points. They are intended for situations where only limited mobility is required and where the drop can be limited to a maximum of 60 cm.

Type C Chest Harnesses incorporate a chest belt with shoulder straps linked with fabric to support the torso, and are attached by safety lanyards to anchorage points. They usually incorporate extensible webbing to break the fall. These should only be used where the possible drop is limited to 2 m allowing for the position of the anchorage, the length of the lanyard, the attachment point on the harness and the extension of the webbing.

Type D General Purpose Harnesses incorporate thigh straps and shoulder straps and may be built into lightweight suits. They are attached to anchorage points by safety lanyards. Like Type C, they should be used so that the maximum drop cannot exceed 2 m.

Type E Safety Rescue Harnesses are worn by persons working in confined spaces and are designed to enable the person to be withdrawn by a rescue line attached to the harness in an emergency (*Figure 2.8.5*). They allow for a maximum drop of 60 cm.

BS 1977 gives detailed specifications including performance tests with articulated dummies for all five types of safety belts and harnesses.

Safety harnesses are generally recommended rather than safety belts because of the much greater support given to the body.

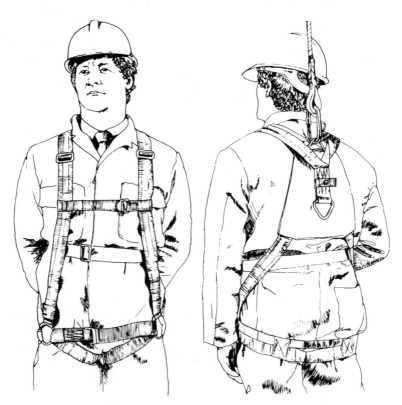

Figure 2.8.5 Safety and rescue harness designed to allow comfort and freedom over long periods (Barrow Hepburn Equipment Ltd).

A number of points need to be checked carefully when using safety belts and harnesses. These include:

- anchoring points for belts should be above the head to limit the free fall to 60 cm.
- anchoring points for harnesses should never be below waist height in order to limit the fall to 2 m.
- safety lanyards should never be looped round sharp edges
- users of safety belts and harnesses should be fully familiar with the maker's instructions
- equipment must be inspected, maintained and stored in accordance with the maker's instructions
- equipment must be attached only to points and structures which are strong enough to hold the wearer in the event of a fall.

2.8.12.3 Fall arrest devices

Two types of fall arrest device are covered by BS 5062:1973 for use in conjunction with safety belts and harnesses. Type 1 runs on a fixed anchorage line and has the safety belt or harness attached to it. Type 2 is attached to a fixed anchorage point and contains a special reel with a spring which winds or unwinds an extendable anchorage line which is attached to the safety belt or harness of the wearer.

A sudden tug on the device causes it to lock automatically. It must be able to halt the fall of a 136 kg weight released from a point 1 metre above it.

These devices, particularly the second type, enable a safety harness to be used by a scaffolder or steel erector with the minimum interference to his movements.

2.8.12.4 Fixed anchorages for industrial safety belts and harnesses and fall arrest devices

BS 5845 advises on selection of anchors best suited to particular structures, and gives specifications for the following types:

Eyebolts
Cast-in sockets in reinforced concrete
Expanding and other types of sockets
Through type anchors for solid walls
Anchors for cavity walls
Anchors for steel structures
Chemically bonded eyelets

The standard gives detailed specifications for the design, dimensions, materials used, heat treatment and hardness of various types of permanent anchors for incorporation into new and existing buildings and structures. It also makes recommendations on their location and methods of fixing.

References

1. KING, R.W. and MAGID, J., *Industrial Hazard and Safety Handbook*, Butterworths, London, 3rd ed. (1982).

2. NATIONAL JOINT COUNCIL FOR THE BUILDING INDUSTRY, *Working Rule Agreement, National Working Rule 23*, 18 Mansfield Street, London W1M 9FG.
3. INDUSTRIAL SAFETY (PROTECTIVE EQUIPMENT) MANUFACTURERS ASSOCIATION, Reference Book of Protective Equipment, 6th ed., ISPEMA, London 1982).
4. CUNNINGTON, P. and LUCAS, C. *Occupational Costume in England from 11th Century to 1914*, A & C Black, London (1967).

2.9

Portable ladders

2.9.1 General

Portable ladders are widely used in construction for ascent or descent to working places and sometimes as working places themselves. They may be in use for a matter of a few minutes only, e.g. by window cleaners, or they may provide access to a scaffold which is in use for several months.

Timber ladders, though cheaper than aluminium ladders, are heavier, more subject to damage and require closer inspection and maintenance. Extension ladders incorporate two (or three in the case of aluminium) pieces of plain ladder, with metal hooks and locks which allow one piece to slide over the other and be attached at various positions. The minimum overlap between the two parts of extension ladders when fully extended depends on the closed length of the ladder, as shown below:

Closed length	*Number of rung spacings overlap*
Up to 4.9 m	2
4.9 to 6.1 m	3
6.1 and over	4

Step ladders and trestles are free-standing units made of two parts joined at the top by hinges to enable the ladder or trestle to fold flat for carrying and a strong cord at the bottom to control the angle between the two parts and prevent the legs from spreading.

Roof ladders are types of rung ladders with a metal hook on the top end to secure the ladder over a roof ridge, and usually with wheels with rubber tyres near the top end to enable the ladder to be moved up and down the roof for positioning and removal without damaging the roof weathering.

Falls from ladders have always been a prolific cause of accidents in construction, particularly in building operations. Figures from 1978 to 1982 for fatal and reportable injuries resulting from falls from heights are summarized in *Table 2.9.1*[1,2].

The high percentage of fatal fall accidents from ladders among the self employed should be particularly noted. The same high figures repeat themselves with dismal monotony and no improvement with time is to be observed.

TABLE 2.9.1. Accidental and fatal fall injuries from ladders and stepladders, 1978–1982

| | | 1978 | | | 1979 | | |
		Building operations	Works of engineering construction	Total	Building operations	Works of engineering construction	T
Reportable Accidental Fall Injuries	From ladders	1360	85	1462	1280	76	1
	Total from heights	4510	481	5044	4149	453	4
	Ladders as % of total	30.1	17.7	29.0	30.6	16.8	
Reportable Fatal Fall Injuries	From ladders	5	nil	5	5	nil	
	Total from heights	51	15	66	53	7	
	Ladders as % of total	9.8	—	7.6	9.4	—	
Non-reportable Fatal Fall Injuries (self employed)	From ladders	4	1	5	8	—	
	Total from heights	18	2	20	15	1	
	Ladders as % of total	22	50	25	53	—	

In its 1981–82 Construction Report, the HSE looked at fatal accidents for an activity closely linked to construction which depends much on the use of ladders, 'window cleaning'. Although not subject to the construction regulations or the Factories Act 1961, window cleaning is subject to HASAWA 1974. Nineteen window cleaners were known to have been killed at work between 1977 and 1981, all from falls. Nine of them were self employed. Twelve falls were from window ledges and four from ladders.

The importance of ladders as a cause of accidental falls in the EEC building and public works industry was pinpointed in a study report of November, 1981 which was made available to us through the International Labour Office[3]. It contains the following passages:

"Analysis . . . revealed that using a ladder as a workplace, particularly when the worker needs both hands for his task or if he leans to one side, is the source of the vast majority of accidents by falling from a ladder. On the other hand, accidents resulting from the use of a ladder as a means of access are less frequent, and virtually no accidents are ever caused by a side rail or rung breaking."

"Portable ladders should no longer be generally used as work places but gradually replaced by work platforms (with guard rails) which can easily be moved."

2.9.2 Regulations and standards

Ladders are dealt with in the Construction (Working Places) Regulations, 1966. Regulation 4 (Interpretation) tells us that 'ladder' does not include a folding step-ladder, and Regulation 15 tells us that "scaffolds which can be moved on wheels . . . shall be adequately secured to prevent movement when any person is working upon it or upon any ladder . . . being a ladder supported by the scaffold."

	1980			1981			1982		
ding 'a-	Works of engineering construction	Total	Building opera- tions	Works of engineering construction	Total	Building opera- tions	Works of engineering construction	Total
	60	1398						
	414	4479	not analysed			not analysed		
.0	14.5	31.2						
	nil	9	8	1	9	5	nil	5
	5	68	43	7	50	52	5	57
.3	—	13.3	18.6	14.3	18.0	9.6	—	8.8
	—	7	4	4	4	6	—	6
	—	21	11	—	11	15	—	15
.3		33.3	36.4		36.4	40		40

Regulation 31 deals with the construction and maintenance of ladders and folding step-ladders, and makes the following requirements of them:

- good construction of suitable and sound material of adequate strength for purpose
- proper maintenance
- prohibition on use if one rung or more is missing or defective
- every rung to be properly fixed to the stiles or sides
- no rung to depend purely on nails, spikes or similar fixing for its support
- tenon joints of wooden ladders to be secured by reinforcing ties and not by wedges
- the grain of wooden stiles and rungs to run lengthwise

Regulation 32, which deals with the use of ladders, makes the following requirements for them:

- to be securely fixed at the upper end, or where not practicable, at or near the lower end
- footing to be firm and level (not loose bricks, etc.)
- to be secured as needed to prevent undue swaying or sagging
- to be equally supported on both sides

The Regulation makes the following further requirements:

- ladders should extend at least 3 feet 6 inches above the landing place, or the highest rung to be used by the feet, unless there is other adequate handhold
- there must be sufficient space at each rung to provide adequate foothold
- in all other cases, ladders should be securely suspended, equally on both sides or stiles, and secured to prevent undue swaying or sagging

- ladder runs should not exceed 30 feet vertically without a landing place provided with guard rails and toe boards
- any opening in a platform through which a ladder passes should be as small as reasonably practicable.

A point to be noted about these Regulations is that, whilst they dwell at length on the need for safe working places as provided by proper scaffolds with guard rails and toe boards, they seem curiously permissive in allowing ladders (which afford only a precarious foothold, and have no lateral fall protection) to be also used as working places. This probably merely reflects the general lack of safe alternatives for painters, window cleaners, electricians and others using ladders as temporary working places at the time the Regulations were drafted.

The construction of portable ladders is covered by:

BS 1129 for timber ladders
BS 2037 for aluminium ladders
BS 4211 for steel ladders

Portable ladders may be classified by type, e.g.

plain rung ladders or standing ladders
extension ladders
step ladders and trestles
roof ladders

or by material:

timber
aluminium
glass reinforced plastic
steel

2.9.3 Purchasing

Ladders should be purchased by experienced men who are familiar with their construction, materials used and with the work for which they will be used. They should comply with the relevant national and local standards, and where these are inadequate or insufficient the construction firm buying them should establish its own standards.

2.9.4 Records, inspection, maintenance and storage[4]

Every ladder used by a firm should, on purchase, be given an identification mark, preferably a small disc screwed to the inside of one stile, and all ladders should carry such marks. A register of all the firm's ladders should be kept, with details of inspections, faults and repairs.

A competent person should be made responsible for inspecting all ladders on first receipt, before and after use, after repair and at fixed periods. Timber ladders should be protected by wood preservative and clear vanish (which enables faults to be seen) and not by paint, and new timber ladders should be so treated before entering the store. A list of points to be checked by inspection is given in *Table 2.9.2*

TABLE 2.9.2. Ladder inspection check list

General
Missing steps or rungs
Loose steps or rungs
Cracked, split, worn or broken stiles, braces, steps or rungs
Damaged or worn non-slip bases
Twisted or distorted stiles
Identification disc missing or illegible
Cleanliness and surface condition free from splinters and paint
Decay
Excessive wear
Makeshift repairs

Aluminium ladders
Distortion
Tightness of rungs in stiles
Security of ropes, pulleys and fittings on rope operated extension ladders
Security of rivets and fastenings
Corrosion
Sharp edges on stiles and rungs

Stepladders
Wobble (from side strain)
Loose or bent hinge spreaders
Broken stop on hinge spreader
Broken, split or worn steps
Loose hinges
Worn, broken or missing cords

Extension ladders
Loose, broken or missing extension locks
Defective locks which do not seat properly
Rusted and corroded metal parts
Defective cords and pulleys on rope operated extension ladders
Loose tie rods

Trestles
Loose hinges
Wobble
Loose or bent hinge spreaders
Broken stop on hinge spreader
Extension guide out of alignment
Defective extension lock

Ladders should be stored under cover on edge, clear of the ground in racks or on wall brackets, which support the edge of the ladder to avoid twist and warping. At least three points of support should be provided for ladders over 6.1 m in length.

All defective ladders must be removed from service. Those which can be repaired should be handed over to an authorized person for repair, with precautions to ensure that they cannot be used again until they have been repaired and inspected. Those which are beyond repair must be destroyed.

2.9.5 Carrying ladders

When a ladder is carried by a single person, the front end should always be kept high enough to clear a person's head. Special care is needed when passing through doorways, around corners and when turning.

2.9.6 Erection and lowering of ladders[4]

Before a ladder is erected the ground and top supports to be used should be checked that they are secure and level and that the ladder is long enough. If traffic or people are liable to pass by the base, barriers and warning notices should be placed around it. Doors close to ladders should be locked shut, or secured in the open position with a man on guard. The angle for plain and extension ladders as used in the U.K. should be 75°, with a vertical:horizontal ratio of 4:1. With steeper angles, there is a risk of the person (and ladder if not well secured) falling backwards. With shallower angles, the bending moment on the ladder increases, and there is a risk of it sagging or even breaking when the user is half way up.

In the U.S.A. a somewhat shallower angle than in the U.K. is recommended. "The base should be one fourth the ladder length from the vertical plane of the top support. Where the rails extend above the top landing the ladder length to the top support only is considered." This formula is perhaps a little easier to apply than the U.K. one, since one should know the height of a ladder exactly, whilst not always knowing the height of the building.

Angles greater than 75° may be used for access to scaffold platforms, provided they are secured at the top before anyone ascends them.

Plain and extension ladders with metal ties or reinforcements on one side only of the rungs should be erected with the ties or reinforcements underneath the rung.

Landing places for ladders should be provided at not more than 9.14 m. Where a ladder passes through an opening in a platform, the opening should be as small as practicable.

2.9.6.1 Short plain rung ladders or standing ladders

Short ladders may be erected by one person, by first placing the bottom against the base of a wall or other solid object, lifting the top of the ladder and pushing upwards to a vertical position. Once vertical, it may be transferred to the required position.

2.9.6.2 Long plain rung (standing ladders) and extension ladders

These should be erected by two people. The ladder is first laid on the ground with the foot at the spot where the ladder has to stand. In the case of an extension ladder, the two parts are adjusted and locked together to give the required height. One then lifts the top by the top rung with the wire or metal support under the rung, while the other stands with one foot on the bottom rung. The first then walks towards the anchor-man while holding the ladder by the sides and moving his hands while the second, with a foot on the bottom rung, clasps the third or fourth rung as the ladder is raised and pulls to assist its erection. The two then lift the ladder to the vertical, at which time effort has to be concentrated on steadying the ladder and lowering it to its resting place. For long ladders it may be easier, where possible, to run a rope from the ladder top over some high fixture which can be steadied by one man and assist in controlling the ladder during its erection. On very long ladders more than two persons may be needed to carry out erection safely.

When erecting extension ladders, these should be set to the required length while horizontal and no effort made to erect them closed and then jump/climb them out. Rope operated extensions are the exception as they can be erected closed and then extended from the ground.

A ladder should never be rested against a window sash. The ladder must extend at least 1.07 m above the landing or the highest rung to be stepped on by anyone using it.

The foot of the ladder should be placed on flat, level ground which provides a secure and firm base for both stiles. It should never be placed on boxes, bricks, barrels or other unstable base to give extra height, nor should wedges or loose materials be used to level ladders on uneven ground. A ladder should never be supported by its bottom rung on a plank.

Plain and extension ladders are lowered by reversing the erection procedure. If the ladder is heavy, a rope should, if possible, be tied to the top and passed round some secure object above it to enable it to be lowered by the rope. Otherwise, adequate assistance is needed before starting to lower the ladder.

2.9.6.3 Stepladders and trestles

Before erecting a stepladder or trestle (used mainly for indoor work) a check is needed that the area on which it is to be erected is firm, level and large enough. If not, a suitable platform should be prepared with scaffold boards. The stepladder or trestle should then be positioned with the legs as far apart as the retaining cord allows, and a check should be made that the stepladder or trestle is level, with all four stiles firmly resting on the floor or base before anyone ascends.

2.9.6.4 Roof ladders

A plain or extension ladder should first be erected and secured, and one person should ascend it to eaves level. The roof ladder is then passed or pulled up by a rope round one of the rungs of the erected ladder, and it is passed up the slope of the roof, with the anchor hook uppermost. The man doing this should be secured to the erected ladder with a belt and snap hooks. If possible, the roof ladder should be attached to a line passing right over the ridge and down to the ground on the other side of the roof, where someone can hold the line taut to prevent the roof ladder sliding down. When the roof ladder has passed over the ridge, it is turned over and lowered until the hook engages the anchorage. Proprietary roof ladders incorporate spreader feet on the underside to prevent damage to the weathering. For roof ladders not provided with spreader feet, hessian bags filled with straw, etc. are placed between the roof ladder and the weathering. The two ladders are then lashed together, so that a man can step from the access ladder to the roof ladder.

2.9.7 Securing plain and extension ladders

If the ladder is for access to a scaffold platform, it is best secured by one man climbing it and lashing both stiles to a scaffold ledger, while another foots the ladder. Ladders are generally best secured at the top, but the point to which they are lashed should be strong and secure, and roof gutters or hooks screwed into an eaves or fascia board are not good enough. Serious accidents have resulted from tying ladders to them. If the ladder cannot be secured at the top, upper parts of each stile may be secured with rope to a fixed anchorage in the structure, while the base may be secured by driving a peg into the ground adjacent to one of the stiles and lashing the stile to the peg. For short ladders of less than 3.05 m, used for work of short duration with hard floor and wall surfaces, stiles with rubber friction pads may be used for the base and top. Ladders to be supported at the top by a horizontal rail at constant height from the ground may have hooks attached to the underside of the stiles to engage the rail.

2.9.8 Use of ladders[4]

2.9.8.1 Plain and extension ladders

Only one person should use a ladder at any time. Nothing should be carried in the hands when climbing or descending. Tools and small items can be carried in a belt or shoulder bag, but all other items should be raised or lowered by a suitable line, rope or tackle from a safe position.

Ladders should only be used as working platforms as a last resort, for work of short duration, such as painting and window cleaning, where the erection of any other form of working platform is quite impracticable.

One hand should always be on the ladder when working from plain or extension ladders, except when a safety harness or belt, clipped to the ladder, is used. When painting from a ladder, the paint pot should be attached by a hook to a rung of the ladder. Great care is needed when working from a ladder not to lean or reach too far sideways or upwards or to attempt to reach round the ladder. Men should never climb or attempt to work from the underside of ladders.

Plain and extension ladders should never be used horizontally, nor should planks be supported on the rungs of these ladders.

2.9.8.2 Use of steps and trestles

Steps should, wherever possible, be placed at right angles to the work with either the front or back of the steps facing the work. Reaching, pulling or pushing sideways should be avoided.

The top step or tread should only be stepped on if it is constructed as a platform and there is secure handhold. Where steps or trestles are used to support a working platform, great care is needed to avoid excessive spans.

2.9.8.3 Use of roof ladders

Additional protection is generally required when using roof ladders and this is discussed in Section 1.5.3 under 'Roof Work'.

2.9.8.4 Use of ladders near electric cables

Aluminium and steel ladders should never be used in the vicinity of electric cables, and glass reinforced fibre or wooden ladders are preferred, provided the weather and ladder are dry. Great care is in any case needed when using ladders near live electric cables.

2.9.9 Other precautions

Care should be taken not to drop ladders, and if one is dropped it should be thoroughly inspected before being used again.

Ladders should not be left in position for any longer than they are required for use.

Where ladders have to be left in position overnight, e.g. as access to scaffolding, unless the site is fenced around and guarded, a scaffold board should be lashed securely to the lower part of any ladder to which children or unauthorized persons may have access, to prevent them climbing the ladder.

References

1. HEALTH AND SAFETY EXECUTIVE, *'Construction, Health and Safety', 1979–80*, HMSO, London.
2. HEALTH AND SAFETY EXECUTIVE, *'Construction, Health and Safety', 1981–82*, HMSO, London.
3. Study Report, 560/RE, 'Analysis of the Causes of Accidents due to Falls in the Building and Public Works Industry', Institut National de Recherche et de Securif, 30 Rue Olivier Noyer, F. 75680, Paris Cedex 14.
4. *Safety with Ladders*, W.C. Youngman, Manor Royal, Crawley, Middx.

PART 3

FIRE, EXPLOSION AND ALLIED HAZARDS

Introduction

The construction industry, like all others, faces risks of fire and explosion which destroy property and endanger lives. Compared with most other industries and businesses, construction seems ill prepared for such risks in its own work, although it may be constructing buildings with high built-in standards of fire safety.

Probably the most serious risk of this type in construction comes from the careless use of LPGs (liquefied petroleum gases). These are a particularly convenient form of fuel for construction.

Other fire and explosion risks in construction arise with:

- gas welding and cutting
- burning off paint on woodwork of older buildings
- site storage of timber and other combustible materials
- bonfires on demolition sites
- use and storage of solvents, paints, varnishes and chemicals
- use of explosives in demolition
- use of cartridge-operated fixing tools
- work in compressed air atmospheres
- use of electricity

We deal here first with the hazards of compressed and liquefied fuel gases used in construction and supplied in special containers. The other fire hazards are discussed in Chapter 3.2.

3.1
Compressed and liquefied fuel gases supplied in containers

The gases in this category which are used in construction are mainly confined to LPG (both propane and butane types) and the welding gases acetylene, hydrogen and oxygen. Some non-combustible gases which are sometimes found in construction are considered under 'Chemical Health Hazards' in Chapter 2.4.

3.1.1 General features of the gases

Both types of LPG, propane and butane, are gases at ambient pressure and temperatures above 0 °C. They can be readily liquefied by compression followed by cooling, and handled as liquids under pressure. Propane, which has the lower boiling point, requires the higher pressure for liquefaction and storage, although the pressure (about 15 bar gauge) is only moderate by engineering standards.

Note that pressures may be quoted as 'absolute', when zero pressure refers to a perfect vacuum, or as 'gauge' when zero pressure refers to the pressure of the surrounding atmosphere. The bar is a unit of the SI system and equal to 10^5 newtons per metre^2 (Nm^{-2}) or 14.5 pounds per square inch. The atmospheric pressure at ground level is generally close to 1 bar absolute. Thus, one bar gauge is approximately equal to two bars absolute. Pressures here are quoted as bars gauge, or bar g for short.

Acetylene is used only for welding and gas cutting (and sometimes for illumination). The very high temperature reached in the oxy-acetylene flame allows many steel alloys to be welded and cut readily. Acetylene has a lower boiling point than propane and butane, but it cannot be compressed much beyond one bar g without the risk of exploding spontaneously. It is a very unstable compound. It is, however, quite soluble in some organic solvents, such as acetone. When the solution is impregnated on an inert porous solid such as kieselguhr, it may be safely stored and handled at higher pressures. Acetylene is thus stored in cylinders similar externally to propane cylinders, but containing acetone and inert porous solid at a pressure of 17 bar g.

When the valve on top of an acetylene cylinder is opened, gas comes out of solution in the liquid like soda water, and the acetone and kieselguhr remain in the cylinder. Besides being supplied in cylinders, acetylene is sometimes generated in situ in portable generators from calcium carbide and water.

Hydrogen and oxygen are both 'permanent gases' with very low boiling points. They cannot be liquefied at ambient temperature at any pressure, however high. For small scale use on construction sites, hydrogen and oxygen are supplied at high pressures (over 100 bar) in thick wall steel cylinders which are longer and thinner than LPG and acetylene cylinders. When there is a steady demand for oxygen it is sometimes more economic to supply it as a liquid at very low temperatures in specially insulated containers. The liquid oxygen then has to be vapourised in specially designed equipment on site, and the gas distributed by pipes to the points where it will be used.

Hydrogen, like acetylene, is used as a fuel gas with oxygen in welding and cutting. It does not give as high a temperature as acetylene and is mainly used for metals such as aluminium and lead with lower melting points than steels. Even an oxy-propane flame can be used to a limited extent in welding and cutting metals, e.g. lead.

Oxygen, of course, is not a fuel in itself but a supporter of combustion. It is used in place of air in special burners with one of the fuel gases where temperatures higher than those obtainable with air are needed, and for work under water.

All five gases are supplied for modest demands in steel cylinders with a single valve at the top. The cylinders are identified by painting to a standard international colour code[1] (ISO 448-1977 which closely resembles BS 349). LPG and hydrogen cylinders are painted signal red; acetylene maroon; and oxygen black. The chemical formulae C_2H_4 and O_2 have also to be marked on acetylene and oxygen cylinders. LPG cylinders in the U.K. have to be marked 'Highly Flammable LPG' in accordance with the Highly Flammable Liquids and Liquefied Petroleum Gases Regulations, 1972[2].

The outlet valves fixed to the cylinders terminate with a threaded female socket to which pressure reducing regulators or manifolds are attached. The threads on LPG, hydrogen and acetylene regulators and cylinders are left hand, whilst those on oxygen cylinders are right hand. When in use, LPG and acetylene cylinders have to be upright with the valve on top to ensure that only gas is drawn off. Oxygen and hydrogen cylinders can, however, be used in the horizontal position, since they only contain compressed gas, although it is sometimes recommended that these too be always kept upright during storage and whilst in use.

3.1.2 The LPG gases

LPG gases as used in the U.K. are defined in the Highly Flammable Liquids and Liquefied Petroleum Gases Regulations[2] as 'commercial butane, commercial propane and any mixtures thereof'. They are products of the petroleum, natural gas and petrochemical industries. Specifications for commercial butane and propane are given in BS 4250[3].

Commercial propane is essentially composed of the hydrocarbons propane and propylene which contain 3 carbon atoms per molecule, whereas commercial butane consists essentially of the two butanes and four butenes, hydrocarbons containing 4 carbon atoms per molecule. Small amounts of volatile organic compounds ('stenching agents') are added to LPG. This enables escapes of gas to be detected by smell at concentrations well below those at which the gas–air mixture could catch fire or explode if ignited. The main physical properties of liquefied petroleum gases of interest to users in construction are given in *Table 3.1.1*[4].

TABLE 3.1.1. Properties of liquid petroleum gas as used in the U.K.[3]

Property		Commercial butane	Commercial propane
Density of liquid (relative to water) at 15°C		0.57–0.58	0.50–0.51
Density of vapour (relative to air) at 15°C		1.90–2.10	1.40–1.55
Ratio of gas to liquid volume at 15°C and 1.0159 bar (760 mmHg)		233	274
Boiling point °C approx at atmospheric pressure		−2	−45
Vapour pressure, bar (g) at 20°C		1.0–2.5	7.2–9.0
at 50°C		3.5–7.0	15.2–19.5
Flammability limits in air, % V/V	Lower	1.8	2.2
	Upper	9.0	10.0
Volume of gas/air mixture at lower limit of flammability from 1 volume of liquid at 15°C		12,900	12,450
Heating value gross at 15°C and 1.0159 bar kJ/m³		121,820	93,130
	kJ/kg	49,310	50,010
Latent heat of vaporisation at 15°C	kJ/kg	372	358

The liquids are considerably lighter than water and float on it, boiling vigorously (at atmospheric pressure). Their vapours on the other hand are heavier than air. Gaseous escapes of LPG thus tend to travel along the ground and readily enter pits, gullies, trenches and drains, along which the gas can travel for considerable distances. On meeting a source of ignition the whole trench bursts into flames and the flame 'flashes back' to the source of escape. Escapes of liquid vapourise rapidly on the ground to give a spreading layer of gas close to the ground.

Commercial propane has a considerably higher vapour pressure than commercial butane and vaporises at pressures somewhat above atmospheric, even in the coldest weather. Its specification allows only a small variation in vapour pressure at any temperature.

Due to its lower vapour pressure, commercial butane does not vapourise at all in very cold weather. Because of the wide variations possible in its composition, its vapour pressure at any temperature may show a wider variation than is allowed for commercial propane. Whilst its lower vapour pressure makes its use in many ways safer than that of propane, it cannot be used outdoors in cold weather. In countries with cold winter climates, mixtures of propane and butane are used in place of butane. In northern Europe, commercial propane is far more widely used in construction than butane, although commercial butane would be preferred in tropical climates.

When LPG cylinders are filled, sufficient free volume is left to allow for thermal expansion of the liquid up to a certain temperature, which in the

U.K. is 50 to 55 °C. Above that, the pressure in a full cylinder increases so much as to cause the relief device to open or the cylinder itself to leak or rupture.

Thus due both to vapour pressure and liquid expansion, there is a real danger of LPG cylinders leaking or rupturing when they are exposed to heat—even direct sunlight on a very hot day. The principal danger is when such cylinders are involved in fires. A fire in an LPG store is almost impossible to control.

The valves fitted to most LPG cylinders today include a relief device which releases gas if the pressure or temperature exceeds a set value. In the event of a fire they can therefore add fuel to the flames. However, up to about 1975 such relief devices were not fitted in the U.K. Thus a fire in an LPG store led to a bombardment of bursting cylinders with violent conflagrations as the escaping LPG mixed and burnt in the air, and flying fragments of hot metal were projected sometimes hundreds of metres[5].

Makers of LPG cylinders of course allow a considerable margin of safety in the inherent strength of the materials used, to allow them to withstand any normal temperature likely to be encountered. Temperatures considered abnormal in one country may, however, be common in another. Thus cylinders considered safe for use in northern Europe may be liable to leak or burst in North Africa or India.

From *Table 3.1.1* we see that the upper flammability limits of propane and butane in air are about five times the lower limits. Thus there is a wide range of gas concentrations in which an explosion (resulting from a leak) can occur. The volume of a room or closed space which could be filled with an explosive mixture of LPG and air at the lower explosive limit is 15 000 times the volume of the LPG as a liquid. Thus two litres of liquid LPG escaping from a cylinder into a closed hut of 3 m × 4 m × 2.5 m (i.e. 30 cubic metres) would be sufficient to fill the entire hut with an explosive mixture. In practice, less would cause a disastrous explosion, which could happen when only part of the hut was filled with an explosive mixture. To quote a maxim familiar to yachtsmen, 'It takes only a cup full to petrol to blow up a cabin cruiser'.

The calorific values of LPGs are high, somewhat higher than petrol or kerosene. This gives them premium fuel value, but increases the fire hazard.

The latent heats of vapourisation, whilst not high, must be taken into account when the use of LPG is considered.

3.1.2.1 *Use of LPG in construction*[6]

Most LPG used in construction is supplied as propane in portable refillable steel cylinders ranging in capacity from about 10 to 100 kg. Among its uses in this form are:

- in portable space heaters, often supplied with blowers to dry buildings, accelerate the curing of cement and improve the working environment in cold weather;
- in melting and applying asphalt and pitch for roads, roofs, pavements, etc.;

- in softening and drying old asphalted surfaces before applying fresh material;
- in stripping old paintwork;
- in plumbing, soldering and brazing brass, lead and non-ferrous pipes;
- in welding and cutting some metals (withoxygen);
- as fuel for internal combustion engines.

For these applications the cylinder, fitted with regulator, is brought near to the job. A flexible tube or hose is used to connect the regulator to the burner. Most accidents with LPG arise with such temporary hook-ups.

On large construction sites with canteens, washrooms and other amenities, LPG is often used for cooking, supplying hot water and space heating. Here it is used with fixed pipes designed as a more permanent installation to the same standards as used in hotels, offices, etc. The LPG may be delivered in much larger bulk containers holding a tonne or more, and special handling facilities are provided for them.

As such installations are designed and constructed by experienced engineers and accidents with them appear to be few, no separate discussion on their hazards is given here. Their design in the U.K. is covered by BS 5482[7] and the LPG ITA Code for bulk storage at consumer's premises[8].

A third form in which LPG may be supplied is in small non-returnable thinwall cartridges, to which a blow lamp or other burner may be fitted directly. Whilst popular with 'do-it-yourself' painters and house repairers, they are not much used professionally due to the high cost of purchasing LPG in this way.

The portable LPG cylinders with a single valve on top are used in an upright position so that only gas leaves the container when a valve is opened. The amount of gas present as such in the cylinder is very small and would only last a short time. It can only be replaced by boiling of the liquid in the cylinder. This requires heat, which has to be drawn from the surroundings. The temperature of the cylinder therefore falls below that of its surroundings until the rate at which heat enters the container from the surrounding equals the rate at which gas is withdrawn from the cylinder multiplied by its latent heat of vapourization. The lower temperature of the cylinder is often evident from the formation of a film of water or even ice condensed from the atmosphere on its lower outer surface. At the same time the pressure of gas in the cylinder falls with the temperature to equal the vapour pressure of the LPG at the lower temperature.

When large offtakes of LPG are required, a single cylinder or container may not provide enough surface for the LPG to be vaporized fast enough. This problem can be surmounted in one of three ways:

1. By using a specially-designed and heated LPG vaporizer between the LPG cylinder and the point of application and by withdrawing the LPG as a liquid from the base of the cylinder instead of as a gas. This is used in fixed bulk insulations for kitchens, washrooms, etc., but since it is not portable it has limited application.
2. By connecting several cylinders in parallel to a suitable header at cylinder pressure, so that LPG gas can be withdrawn from several cylinders at once. For most uses of LPG a pressure regulator is fitted

on the outlet of the cylinder to give a constant pressure below that in the cylinder at the valve of the burner. In this case the pressure regulator is fitted between the header and the burner valve.
3. By heating the cylinder. This, however, is very risky since the cylinder may easily become overheated when the offtake of gas is reduced or ceases. A pressure relief device if fitted will then discharge LPG to the surroundings, or if not fitted the cylinder may leak or burst. The application of flames, steam, hot air blast or hot water on the outside of an LPG cylinder should be absolutely forbidden. There may, however, be occasions where it is permissible to move the container from some very cold location to a warmer one without serious risk. Skilled supervision is then needed.

3.1.2.2 The hazards of LPG in portable cylinders

Whilst the main hazards of LPGs are those of fire and explosion, others must not be forgotten. Being heavier than air and able to remain for long periods in pits and cellars, LPG gases present a certain asphyxiation risk. A normal person would detect the presence of LPG by the smell of its stenching agent long before there was any risk of asphyxiation. Unfortunately however, the sense of smell becomes dulled to unpleasant odours by frequent exposure (as well as by head colds and nasal infections). Thus a person continually using LPG or someone with a bad cold may be unable to detect it.

Another hazard of LPG is carbon monoxide poisoning through burning LPG with insufficient air or in poorly maintained burners. Whilst the deaths of sleeping holidaymakers in chalets with flue-less LPG heaters have made headlines[9], the painter using an LPG blow torch to strip old paint in a closed room is also at risk.

Yet another hazard of LPG, particularly of propane, is that of frostburn, should the liquid come into contact with the skin. If liquid propane at room temperature in a container escapes into the atmosphere, part of it immediately evaporates ('flashes') to form gas, whilst the rest of the liquid drops in temperature to its boiling point of -45 °C. Serious injuries leading to loss of fingers, etc. can occur in this way.

The hazards arising from escapes of LPG (fire, explosion, asphyxiation, frostburn) result largely from the following causes:

1. damaged or defective cylinders and valves;
2. leaking or damaged connectors, regulators, pipes and hoses;
3. blow-out of burner, allowing gas to escape from an unlit burner to the air.
4. the attempted use of a cylinder on its side, or upside down, when liquid instead of gas is withdrawn.

To prevent such escapes care is needed in all phases of LPG handling and use. For convenience, these are divided into the following steps:

- transport, unloading and storage of LPG cylinders at site
- use of LPG on site.

3.1.2.3 Transport, unloading and storage of LPG cylinders at site

Something first must be said about the containers themselves. The commonest are portable steel cylinders with capacities ranging from about 10 to 100 kg, with dished ends, a carrying ring, a single valve fixed centrally to the top and a short cylindrical skirt at the base. Most cylinders today also have a safety relief device combined with the cylinder valve. The valve also incorporates a third feature—a shut-off device which closes when the pressure in the cylinder falls below a certain figure, when all liquid has evaporated. This is to prevent air entering the empty cylinder. Most cylinders also have a steel cover which fits over the valve and is screwed onto the top of the cylinder to protect the valve during transport. Cylinders are often transported from distribution centres to site stores in lorries and the valves can easily be damaged or loosened by careless offloading, especially if the protective cover is missing.

The first point needing care then is the unloading of the cylinders themselves. They should be slid or rolled to the ground down a special inclined chute and never dropped. Some LPG suppliers have delivery vehicles fitted with a small crane to facilitate safe loading and unloading. Before being put into store the valves and relief fittings of all cylinders should be checked for leaks. This is easily done with a plastic squeeze bottle filled with detergent solution or soapy water. A leaking cylinder should on no account be put into store. The proper procedure is to take it to a well ventilated open space where there is no ignition hazard, mark or label the cylinder as defective, place it in the upright position and display 'no smoking or naked lights' notices beside it. Barriers should be placed round it to prevent general access and the supplier should be informed at once. If it is decided to empty the cylinder before it is removed this should only be done on a windy day, and not near any pits, trenches or drains. The cylinder valve is then cracked open. When the cylinder is empty the valve should be closed and the cylinder returned to the supplier for repair or scrapping. No attempt should be made to repair it on site.

Precautions for the storage of LPG cylinders on construction sites are given in HSE Guidance Notes CS 6[6] and CS 4[10]. The main points of CS 6 are summarized below.

Preferred conditions of storage
- in open air, in well ventilated area;
- at ground level on firm flat surface;
- at least 3 metres from hollows (excavations, etc.) where vapour could collect;
- good access to area (for delivery vehicles, hand trucks);
- area free from combustible material (wood, packing material, weeds);
- no adjacent sources of heat (boilers, bonfires, etc.);
- any weather protection to be non-combustible with open sides.

Alternate conditions of storage if open air storage impracticable
- in well ventilated storeroom built of non-combustible materials;
- storeroom and its electrical installation to comply with requirements given in Reference 10.

Note: These are very stringent. Storage is only allowed on the ground floor and the building or storage space must be specially designed for the purpose.

Separation from other materials in store
The following materials to be kept at least 3 m from LPG cylinders:
- oxygen cylinders
- other flammable liquids
- oxidizing materials such as sodium chlorate
- toxic or corrosive materials

Condition of cylinders in store
- all cylinders upright, with valves uppermost (whether full or empty)
- all valves closed
- all valve plugs, shrouds and caps securely in position

Separation and quantities of LPG in store
- Minimum separation distances for open air LPG stores as recommended by HSE Guidance Note 6 are given in *Table 3.1.2*.
- Cylinders should be placed in groups or stacks, each containing no more than 1000 kg of LPG;
- Maximum heights of stacks and minimum widths of gangways between them and maximum amounts of LPG per vertical stack are given in *Table 3.1.3*.
- For storing quantities of LPG greater than 4000 kg Table 1 of reference 10 should be consulted.

Electrical equipment for open air stores
- Only explosion-proof electrical equipment should be allowed within the area given by the separation distances in *Table 3.1.2*, except for:
- lighting units which may be located within the area provided they are mounted at least 1.5 m above the uppermost cylinder.

TABLE 3.1.2. Minimum separation distances for open air LPG stores

Total LPG storage, including nominally empty cylinders	Kilogram		
	50–300	300–1000	1000–4000
Min. distance to huts, buildings, site boundaries or fixed sources of ignition from nearest cylinder where no radiation wall provided, metres	1	3	4
Ditto, from radiation wall where provided	nil	1	1

TABLE 3.1.3. Heights of stacks, widths of gangways and maximum amounts of LPG per stack

			Unpalletized stacks	Palletized stacks
Max. height			2.5m	6 pallets
Min. width of gangway, metres			1.5	2.5
Amount of LPG in any cylinder, kg	Up to 6	Max. amount of LPG in any vertical column of stack, kilograms	30	35
	6–15		45	75
	15–20		50	80
	20–55		55	110

Fencing and exits for open air LPG stores
- The storage area should be enclosed by a fence of non-combustible material approximately 2 m high
- The fence should not impede natural ventilation, particularly at ground level and should preferably be made of wire mesh.
- The fence should have at least two gates as far from each other as practicable
- The gates should open outwards and not lock automatically when closed
- Both gates should be unlocked when anyone is inside the storage area.
- No cylinder should be placed nearer than 1.5 m to the fence, unless it is a boundary fence, when *Table 3.1.2* applies
- Notices marked 'Highly Flammable—LPG' and prohibiting smoking and naked lights in or near the store should be displayed on the fence.

Storage of small quantities of LPG in wire cage in open air
- A lockable wire cage (*Figure 3.1.1*) may be allowed when less than 300 kg of LPG has to be stored and neither a fenced open air compound nor a storeroom is practicable.

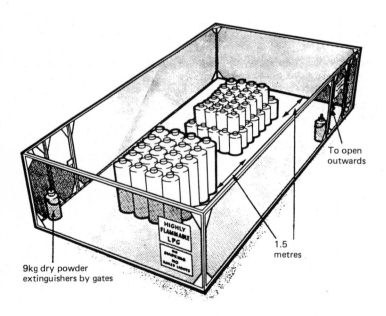

See Table 3.1.2 for minimum distances from any cylinder to buildings, boundaries and sources of ignition

Figure 3.1.1 Typical compound for storing LPG cylinders.

- Only one exit from the cage is required providing there is no danger of anyone being trapped inside it.
- Similar notices should be displayed on the cage as on a fenced compound.

3.1.2.4 Precautions in the use of LPG cylinders on construction sites

The precautions given below are again mostly summarized from References 6 and 10.

Precautions against tampering by children and unauthorized persons
- All cylinders not connected to equipment should be returned to the locked store at the end of the working day.
- Gas supplies to all equipment should be isolated at the end of the working day.
- Special care must be taken to ensure that neither cylinders left connected to equipment nor the equipment itself are tampered with during non-working hours.

Siting and ventilation of cylinders and LPG fired equipment
- Areas where LPG fired equipment is used must be kept free of combustible materials. If combustible materials cannot be removed they should be screened by suitable insulation.
- Cylinders should be placed on a firm horizontal base always in the upright position.
- The position of cylinders and equipment should be such that there is ready access to the cylinder valves and regulators and that the cylinders are not heated by the equipment.
- Burners should always be situated in a well-ventilated space to ensure complete combustion of LPG without formation and build-up of carbon monoxide gas.
- Cylinders should be in a well-ventilated space to ensure that any minor and undetected leaks of LPG are dissipated.
- Commercial propane cylinders should always be sited in the open air.

Handling and connecting of LPG cylinders
- Cylinders should be handled with care, never thrown or dropped, and never lifted or levered by the valve.
- Specially designed trolleys should be used for moving cylinders.
- Before using any cylinder, all nearby fires, naked lights, cigarettes, etc. should be extinguished.
- Cylinders should be examined and checked for leaks before being connected. Faulty cylinders should not be used, nor should attempts be made to repair them (see Section 3.1.2.3).
- All valves on cylinders and equipment should be shut before a cylinder is connected or disconnected (except when an automatic changeover device is fitted).
- All connections and equipment should be complete (with washers where used) and mutually compatible and compatible with the type of LPG being used (propane or butane).

- Hoses, hose fittings and regulators should only be used as recommended by their manufacturers for the type of LPG in question.
- The correct sized tools, particularly spanners, should be used to tighten all connections and care must be taken to avoid overtightening.
- Once equipment has been connected to an LPG cylinder it should be checked for leaks as described in Section 3.1.2.3.
- Where several cylinders are to be connected in parallel by a manifold, this should incorporate non-return valves which allow one cylinder to be removed without shutting off the others.

Regulators, flexible tubing and hoses
- Flexible tubing and hoses should be kept as short as possible and only used for final connection to appliances.
- Only hoses supplied to a nationally recognized standard should be used.
- Where abrasion damage is possible, hoses should be specially protected or steel braid reinforced hoses should be used.
- Flexible tubing should be coloured black and hoses orange for pressures up to 50 millebars gauge in accordance with BS 3212 and 5120.
- Regulators must be suitable for the type of gas, inlet and outlet pressures and gas flow rates to be used.
- Regulators should be checked for leaks using soapy water. Defective regulators should be returned to the makers for servicing.

Lighting up, use and shutdown of LPG burning equipment
- Maker's procedures should always be followed.
- Gas valves on equipment should be checked and be closed before the cylinder or supply valve is opened.
- If unlit gas from a burner has entered the equipment there is a danger of an explosion if an attempt is made to relight the burner before the unlit gas has dispersed. The procedure should include precautions to avoid such explosions and only persons trained in the procedure should be allowed to light LPG burning equipment.
- Only properly constructed and maintained equipment should be used.
- If there is any smell of gas from the equipment the supply should be shut off and the cause investigated.
- If the flame goes out, the burner should be shut down and the main valve closed until unburnt gas has dispersed and the cause has been determined.
- Where burning equipment is used in an enclosed space great care must be taken to ensure adequate ventilation to remove products of combustion, fumes, and excess oxygen supplied to the burner, and to ensure that the air remains respirable. Such work in enclosed spaces should only be allowed under a special permit system which ensures that all necessary precautions have been taken.

- Cylinders used in connection with work in confined spaces should not be taken into the space but located in a safe position outside in the open air.
- After use the supply or cylinder valve should be closed as well as the valve on the equipment.
- When not in use, portable equipment should be stored in a secure place.
- No equipment should be moved whilst it is alight unless it has been designed to be moved in use, e.g. blow torches.
- LPG burners should not be left burning unattended, particularly at night-time, unless fitted with flame failure and/or other appropriate safety devices.

Disconnecting and transport of cylinders
- All valves, including the cylinder valve, must be closed before a cylinder is disconnected, whether empty or not.
- After disconnection, valve protection caps and plastic thread caps or plugs should be replaced.
- Cylinders should always be adequately secured with their valves uppermost during transport.
- Substantial quantities of LPG should always be carried in open vehicles.
- Small quantities may be carried in closed vehicles provided they are well ventilated at low level.

Inspection and maintenance
- All equipment should be regularly inspected for leakage (using detergent solution or soapy water to verify) as well as for damage and missing items such as washers and hose clips.
- Burners need regular cleaning and adjustment.
- Faulty equipment should be taken out of service at once.
- Repairs, modifications and maintenance should only be carried out by trained and competent persons.

Training and instruction
- Many accidents with LPG are due to ignorance.
- All persons using LPG should be instructed in its hazards and the precautions needed.
- Special emphasis should be laid on correct lighting procedures and the need to shut off the supply after use.

Fire precautions and procedures
- Smoking should be prohibited in LPG storage areas and where cylinders are loaded and offloaded.
- Fire extinguishers should be of the dry powder type. An adequate number should be provided and operators should be trained in their use.
- Water should only be used to cool other cylinders threatened by a fire, not to extinguish an LPG fire (except by the fire brigade).
- LPG fires should, where possible, be put out by closing a valve or valves on the line leading to the point of emission.

- To extinguish an LPG flame from a leak whilst the emission continues is to risk an explosion which may be worse than the fire.
- All persons not directly involved in fire fighting should be evacuated from areas of LPG fires.
- Fire precautions in pressurized workings[11] are dealt with in Chapter 3.2.

Use of LPG equipment in site huts and small buildings
- The main hazards have already been noted, i.e. asphyxiation and explosion through leaks and carbon monoxide poisoning through incomplete combustion.
- All cylinders, regulators and high pressure fittings should be located outside the hut or building except where they are an integral part of the equipment. Cylinders thus stored outside the building or hut will normally be propane except in climates with warm winters. Those used inside the hut or building will normally be butane.
- Heaters requiring large quantities of air for their operation should not be used in site huts.
- The heat output of any flueless appliance should not exceed 50 watt/m^3 of the free space of the hut or building.
- There should not be less than 100 mm^2 of fixed ventilation openings divided equally between high and low levels for every 45 watt heat output as well as ventilation required for other purposes.
- All new LPG fired domestic flueless space heaters used in the U.K. will, according to BS 5258, Part 10, be fitted with vitiation detectors which monitor the atmosphere and shut off the gas supply when the oxygen content falls to a dangerous level. Reference 6 recommends that such heaters also be fitted with flame failure devices, and that only heaters fitted with both vitiation detectors and flame failure devices be used in huts and small buildings.
- All LPG burning appliances used in huts and small buildings should operate without producing soot or smell. Where soot or smell are produced the cause should be sought and remedial action taken.
- If a smell of gas is noticed in a site hut, no attempt should be made to light any LPG appliances, electricity should not be used and smoking and other sources of ignition should be prohibited. The gas should be turned off at the cylinder (if valve open) and the hut well ventilated. If the cylinder valve was closed it may have been leaking. The cylinder should then be disconnected and the valve checked.
- Where mobile heaters incorporating an LPG cylinder are used, spare cylinders should on no account be kept in the hut. If the unit is on castors, it should be secured so that it cannot be knocked over or moved inadvertently.

Use of LPG fired cutting and heating torches, roofing irons and other hand tools
- The LPG containers should be upright in use and fixed so that they cannot fall on their sides. For cutting equipment, a purpose-built LPG trolley should be used wherever possible.
- The gas supply should never be cut off by kinking the hose.

- Non-return valves should be used in both oxyen and gas lines to torches; flashback arrestors are also recommended.
- Flames should never be directed onto LPG cylinders.
- Working areas should be kept free of combustible material, and any combustible material that cannot be moved should be screened or covered by temporary insulation.
- Checks should be made for smouldering material immediately after using a hand torch, and again after a suitable period has elapsed. The far side of walls, floors, ceilings and any intervening voids should also be checked.
- A suitable fire extinguisher, ready for use, should always be kept with the equipment.

Use of bitumen boilers and cauldrons
- LPG cylinders should be kept at least 3 m from the burner, and located where they are unlikely to be struck by barrows and other site traffic.
- Flexible hoses between cylinders and burners must be protected from traffic and secured so that they do not constitute a tripping hazard.
- Boilers should not be overfilled or allowed to boil over.
- If a boiler inadvertently boils over, cylinder valves should be promptly closed and cylinders moved to a safe place.
- Any flexible hose damaged by bitumen should be replaced.
- No boiler or cauldron should be left unattended or transported or towed when the burner is alight.
- A suitable fire extinguisher should be available near the equipment (see Chapter 3.3).

Mastic mixers
- Cylinders not in use should be kept at least 6 m from the mixer when working.
- A suitable fire extinguisher should be available near the equipment.

Drying operations
- Great care is needed to avoid local overheating leading to possible fire.
- Heaters in drying rooms should be fitted with protective cages.
- Ventilation must not be restricted by articles being dried.
- LPG cylinders should never be exposed to hot air currents from air heaters.

Floodlighting
- All lighting equipment should be placed on a firm base and the use of flexible hoses should be kept to a minimum.

LPG-fired hot road planing machines
- Operators must be trained in all aspects of the operation, including safety.

3.1.3 Acetylene, hydrogen and oxygen

These are primarily the welding and cutting gases, and are used in construction mainly with hand-held torches—oxy-acetylene for steel and oxy-hydrogen for lead and thin aluminium sheet. Oxy-hydrogen cutting of steel is used by divers under water. The general precautions for the storage and handling of LPG cylinders (e.g. protection of valves, valve caps, leaks, treatment and return of empty cylinders, use of hand trolleys) apply equally to cylinders of acetylene, hydrogen and oxygen. The hazards of welding fumes (cobalt, cadmium, lead, etc.), toxic gases (oxides of nitrogen and ozone) are discussed in Chapter 2.4 and those of UV and IR rays in Chapter 2.5.

Most of the general precautions when using gas torches were discussed in Section 3.1.2. Here we deal primarily with the properties of the gases which are important for their safe storage and handling. Acetylene and hydrogen, being fuels, are considered together first and oxygen is then considered separately.

3.1.3.1 Acetylene and hydrogen

The relevant properties of acetylene and hydrogen are given in *Table 3.1.4*.

Both gases are lighter than air so there is little danger of their accumulating in pits or trenches, although the density of acetylene leaving a cylinder valve, cold and containing solvent vapour, may equal or even exceed that of air. In the case of a large leak of hydrogen in a building, it is likely to form a dangerous concentration under ceilings.

TABLE 3.1.4. Properties of acetylene and hydrogen

Property	Acetylene	Hydrogen
Density of gas relative to air at 15°C	0.90	0.07
Flammability limits in air % V/V Lower	1.5	4.0
Upper	100	74
Heating value, gross at 15°C and 1.0159 bar, kcal/m^3	55545	12088
Ignition temperature °C	305	560
Flame speed (in air) cm/sec	173	320
Minimum ignition energy in air mJ	0.02	0.019
Smell	Ethereal	None

The flammability limits of both gases in air are very wide—much wider than for the LPGs; this increases explosion risks.

The ignition temperature of acetylene is relatively low which means that it is easily ignited by a hot surface.

The flame speeds of both gases are high—several times those of butane and propane. This, again, means that it is easier for an explosion to develop if either gas escapes into the atmosphere. It also means that flames of acetylene and hydrogen are more stable than those of propane and butane. The flames sit more securely on their burners and are less readily blown out.

The ignition energies in air, which measure the ease with which the gases are ignited by electric sparks, are only about a tenth of those of propane and butane. The danger of electrical ignition is thus much greater for acetylene and hydrogen than for the LPGs. One consequence of this is that explosion-proof electrical switchgear, lights and appliances which are safe for use with propane and butane are not safe for acetylene and hydrogen. There is also a greater danger of ignition from static electricity with acetylene and hydrogen compared with the LPGs. Thus the disaster to the airship Hindenberg was attributed to static electricity by the inquiry which followed.

Another dangerous feature of acetylene is that it forms unstable explosive compounds with copper, which are easily detonated by friction or light blows.

Thus the following additional precautions should be taken in the storage and use of acetylene and hydrogen besides those already given for the LPGs.

- Acetylene and hydrogen cylinders should only be stored in open air stores (as discussed for LPG cylinders) and not in buildings.
- No electrical installations of any kind should be permitted in stores where acetylene and hydrogen cylinders are kept.
- Such stores are best lit from elevated directional lamps placed well outside the store area.
- Acetylene and hydrogen cylinders in store should be electrically earthed. This is most readily done by placing them in earthed metal racks with direct electrical contact between the rack and the cylinder.
- If the store is covered with a light roof for weather protection, the roof itself should be well ventilated.
- With these provisos, acetylene and hydrogen cylinders may be kept in separate piles in LPG stores as described in Section 3.1.2.3, provided they are kept at least 3 metres away from LPG cylinders.
- No valves or metal fittings used with acetylene should be made of copper, brass or other copper-containing alloy.

3.1.3.2 Oxygen

Oxygen which constitutes 21% of air by volume is slightly denser than air, so that an escape of oxygen into the air will tend to stratify and remain for some time close to the ground.

The main hazard of oxygen and oxygen enriched air is that all flammable and combustible materials burn far more vigorously, sometimes violently and explosively. Smouldering wood, tobacco, cloth or rope will quickly break into flames in an oxygen enriched atmosphere. Another danger with compressed oxygen in cylinders, valves and pipes is the presence of oil, grease and other organic materials. Only valves and regulators specially designed for use with oxygen should be allowed.

Oxygen cylinders are best stored in the open air in the same way as LPG, but should be separated by at least three metres from cylinders of LPG and other flammable gases.

References

1. ISO, 448-177, *Gas cylinders for industrial use—marking for identification of content*, International Organisation for standardisation, Geneva (available from most national standards organizations). (Closely similar to BS 349).
2. *The Highly Flammable Liquids and Liquefied Petroleum Gases Regulations, 1972*. Statutory Instrument 1972, No. 917 HMSO, London.
3. BS 4250:1975, *Specification for Commercial Butane and Propane*, British Standards Institution, London.
4. LPG INDUSTRY TECHNICAL ASSOCIATION, *An Introduction to Liquefied Petroleum Gases*, William Culross & Son Ltd., Coupar Angus, Perthshire, Scotland.
5. '50 houses evacuated as gas cylinder depot blazes', *Fire Prevention*, July 1971, p.41.
6. HEALTH AND SAFETY EXECUTIVE, *Guidance Note CS 6, The storage and use of LPG on construction sites*, HMSO, London.
7. BS 5482, *Code of Practice for Domestic Butane and Propane Gas Burning Installations*, British Standards Institution, London.
8. LPG INDUSTRY TECHNICAL ASSOCIATIONS, *Code of Practice No. 1. Installation and Maintenance of Bulk LGP Storage at Consumer's Premises* (see ref. 4).
9. *The Times*, 15th January, 1983, p. 3b.
10. HEALTH AND SAFETY EXECUTIVE, *Guidance Note CS 4, The keeping of LPG in cylinders and similar containers*, HMSO, London.
11. HEALTH AND SAFETY EXECUTIVE, *Guidance Note GS 20, Fire Precautions in Pressurised Workings*, HMSO, London.

3.2
Other fire and explosion hazards

Besides LPG and other bottled fuel gases, there are many other possible causes of accidental fires and explosions on construction and demolition sites. These range from the fairly common ones such as fires in timber stores and wooden site huts, to the rarer and more specialized ones such as fires in pressurized workings[1] and risks in the use of cartridge operated fixing tools[2]. We deal here mainly with the commoner causes and mention only the main features of the rarer ones.

Most buildings today, particularly public ones, are subject to strict requirements in their design, construction and use to reduce the risks of fire. Thus the Fire Offices Commitee[3] has set standards of building construction which qualify the owners for low fire insurance ratings. These lay special emphasis on 'fire compartmentation', i.e. the division of hazardous areas and buildings into separate fire compartments, and the use of fire resistant walls, floors, ceilings and doors. Special attention is also given to the provision of adequate means of:

- fire detection and warning
- escape for all persons in the building to a safe place
- emergency fire fighting, including both automatic systems such as sprinklers, deluge systems and inert gas flooding, as well as manual systems—portable extinguishers, hose reels, etc.

During construction, building, maintenance, renovation, extension or demolition, most of these features are absent. Due to incomplete compartmentation (openings in walls, floors, etc.) fire can spread far more rapidly. Built-in protective features of detection, warning and extinguishing are not yet installed or in operation, although every building site should have a system for fire warning. The stairs and floors of escape routes are usually incomplete, and often the only means of escape from the upper parts of buildings under construction are via access scaffolds and their ladders. Against this it must be said that during construction there are usually a number of active workers on hand during working hours who will notice and deal promptly with the initial outbreak of a fire. Outside working hours the site will be deserted, except for a night watchman or security guard. Thus if a fire should break out, whilst there may be property damage, any injuries or loss of life are usually confined to the fire fighters.

The most common accidental fires on construction and demolition sites do not, however, involve the buildings on which work is being done, but site huts, stores of timber and other combustibles, and, of course, rubbish[4].

As this book is primarily concerned with hazards to building workers themselves, with property damage a secondary consideration, we first consider fire itself and its effects on persons exposed to it.

3.2.1 Fire and its effects on personnel

Three conditions are essential components of fires:

- fuel or combustible material
- oxygen (usually from the atmosphere)
- heat to initiate and maintain the fire
  (some external source is needed to initiate the fire).

These three conditions constitute the 'fire triangle'.

More recently a fourth condition, the presence of 'free radicals', has been recognized as essential to flames.

The chemical reactions of flames, known as 'chain reactions', depend on the formation and eventual destruction of 'free radicals', which are incomplete fuel molecules usually with one hydrogen atom missing. With this fourth condition included, the fire triangle now becomes the 'fire pyramid' (*Figure 3.2.1*). All methods of fire prevention and extinction depend on ensuring that one of these conditions is absent or removed.

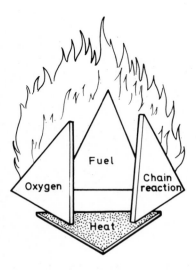

Figure 3.2.1 The Fire Pyramid (National Safety Council, U.S.A.).

The first step is to minimize the amount of combustible material present on the site. Regularly collecting and disposing of all waste materials, offcuts of timber, shavings, packing material, oily rags and surplus solvents and general good housekeeping prevents fires starting or restricts their spread.

When a fire has started air and oxygen may be excluded from the fire by extinguishers which release inert gases (carbon dioxide), or non-inflammable vapours (halons, steam), or by enclosing the fire or covering it with sand. Water is the most effective quenching agent, extinguishing a fire by removing heat and lowering the temperature. Certain types of vapourizing liquid fire extinguishers contain organic bromine compounds which effectively capture and remove free radicals from the flame and thus help to put the fire out.

In planning fire protection measures these four conditions should be borne in mind, since they offer four possible ways of preventing or putting out a fire. If it proves impracticable to employ one method an alternative method based on one of the other conditions may be possible.

Although people may be burnt and injured by fires, the major cause of casualties is not heat but smoke and the products of combustion. Most fires produce water vapour, carbon dioxide and carbon monoxide, whilst many compounds, some highly toxic, which strongly irritate the respiratory system, e.g. aldehydes, phenols and tar distillates, are usually present in the smoke if combustion is not complete.

Complete combustion is needed to avoid production of carbon monoxide and the other organic compounds mentioned.

Other toxic and irritating gases are liberated when certain plastic materials are burnt. Hydrogen chloride and vinyl chloride monomer are given off from PVC, while hydrogen cyanide and other toxic gases may be given off from polyurethanes, especially foams used for heat insulation and furnishings. Fire must thus be considered as much a toxic chemical hazard (Chapter 2.4) as a burn hazard.

3.2.2 Common fire causes on construction sites

The most common fire causes are considered to be:

- rubbish burning out of control
- children playing on site, and arson by older trespassers
- heaters of all kinds, especially clothes drying on heaters
- bitumen boilers
- welding and cutting
- use of blow lamps, especially for removing old paint from woodwork
- electrical faults
- fires starting in chimneys and flues
- fires starting in site huts
- storage and use of flammable liquids

Several of these involve the use of LPG and the welding and cutting gases considered in Chapter 3.1.

3.2.2.1 Rubbish

There are two aspects to the combustible rubbish problem—first its collection and storage prior to disposal and secondly its disposal by burning or other means. All waste, particularly combustible waste, should be cleared up frequently and, subject to limitations of size, placed in metal rubbish bins with close fitting lids until it can be disposed of. This includes floor sweepings, oily rags and waste flammable liquids. If this is not done, combustible waste scattered throughout a building site can form a trail by which fire spreads rapidly.

It may be necessary to segregate collected waste into several categories, each with separate bins identified by suitable colour markings, e.g.

- waste for collection and disposal by local authority,
- site tip, with building rubble,
- toxic or hazardous waste for collection by licensed, waste disposal contractor (see Chapter 2.4),
- waste for burning on site.

All waste awaiting disposal should be kept in an isolated area in the open.

It is strongly recommended that a suitable incinerator be used for the burning of all combustible waste, whose burning is permitted on site, rather than on open bonfire. This reduces the danger of flying sparks, as well as securing more complete combustion and reducing the smoke and fume problem. Burning should take place at least 30 feet (9.1 m) away from huts, stores and equipment, and down wind of where men are working. Rubbish burning should neither be done on windy days nor on very still days, and should be supervised by a responsible man with an adequate supply of water to hand. Fires should not be allowed to burn unattended after the close of work, but should be extinguished and well damped down.

Materials which release toxic fumes on combustion should, as far as possible, be disposed of in other ways.

3.2.2.2 Children playing on site, and arson by other trespassers

These are frequent causes of fires causing property damage as well as injuries. HSE's *Guidance Note GS7*[5] gives practical precautions to prevent accidents to children on construction sites which caused 28 child fatalities during an 18 month period surveyed in 1976. The main measures recommended in this Note apply also to fire prevention, and are primarily security measures. Construction sites should as far as possible be totally enclosed by fences at least 2 m high, not easily climbed, and either boarded or provided with mesh not exceeding 30 mm. Support posts should be securely anchored and access openings provided with gates which should be kept locked whenever the site is unoccupied, and manned when open by security personnel. The fencing should be properly maintained and nothing should be placed near it which would enable it to be easily climbed. Suitable warning notices should be fixed to the fencing.

Where it is not practicable to enclose the entire site, it is recommended that local high fire risk areas, such as hutted compounds and stores for flammable liquids, LPG cylinders, timber and other combustible materials such as expanded polystyrene, be fenced. Plant, if not in a securely fenced compound, should be immobilized. Holes and excavations should be similarly fenced if there is a risk of children falling in.

Security patrols and nightwatchmen should be trained in fire surveillance and emergency fire fighting which form part of their duties.

3.2.2.3 Cigarettes and matches

Smoking should be prohibited in all areas where materials are present which could be ignited by a lighted match or a cigarette end, e.g. carpenters' huts and stores containing flammable and readily combustible materials. 'No Smoking' notices should be prominently displayed in these areas. In places where smoking is allowed, metal ashtrays, partly filled with sand should be provided wherever possible and arrangements made to empty them regularly.

3.2.2.4 Heaters and drying clothes

It is recommended that only fixed convectors or stoves be used in site huts, stores and buildings used during construction, and that the use of portable and radiant type heaters be avoided. All heaters and stoves and their flues (when necessary) should be properly installed in accordance with regulations and codes of practice, e.g. 'Solid Fuel Stoves—Building Regulations, 1965', 'Butane Heaters—British Standard Code of Practice CP 339, Part 2', and 'Electric Heaters, BS 3456 A2'.

Many fires have been caused on building sites by hanging wet clothes too close to heaters. Drying racks and coat stands should be permanently fixed at a safe distance from heaters which should have guards fixed to them to prevent clothes being placed dangerously close.

3.2.2.5 Bitumen and tar boilers

These should always be supervised by an experienced operative and used only in places where material spilled can be controlled. There is a special danger should water enter a hot tar or bitumen boiler. The water vapourizes explosively on contact with hot tar or bitumen, spattering it into the surrounding area, with associated risk of fire or personal injury.

When used for roof work, tar and bitumen boilers should be placed on a thick insulating fireproof base to prevent the roof structure from becoming overheated. A dry powder or foam type fire extinguisher should be at hand, but care must be taken not to direct foam into hot tar or bitumen.

3.2.2.6 Cutting and welding equipment

All combustible material should be removed from the work area or protected by non-combustible mats or blankets or metal screens to prevent sparks and red hot metal fragments impinging on them. Wooden floors may be covered with sand or overlapping metal sheets. Sparks must not be allowed to fall through gaps in floors to the floor below. If welding or cutting has to be done on one side of a wall or partition, all combustible material on the other side which might be ignited must be removed. Bins should be provided for spent electrode stubs. Precautions for gas welding and cutting were given in Chapter 3.1, and precautions for electric welding were discussed in Section 2.5.4.

It is recommended, especially when welding and gas cutting have to be done near combustible materials, e.g. woodwork, that all hot work be terminated one hour before work for the day ceases in the area. A final check for smouldering materials should then be made before leaving. A dry powder or CO_2 extinguisher should be at hand when using welding or cutting gear. Water may be used as a general area extinguisher provided personnel are adequately warned not to direct it onto electric welding equipment, when fatal shocks could be caused.

3.2.2.7 Blow lamps

Blowlamps of the paraffin type should only be filled when cool, and the same applies when changing cylinders or cartridges of the LPG type. After filling and pressurizing a blow lamp it should be checked for leaks (using soapy water if necessary). Blow lamps of the petrol type are considered too hazardous to be used.

All blow lamps should be kept clean and in good condition and air pumps and pressure release valves on the paraffin type should be regularly inspected, maintained and adjusted. Burner jets should be cleaned regularly. Makers' instructions should always be followed, especially as regards lighting and adjustments.

A particular fire hazard arises when using blow lamps for stripping paintwork, as many fires have been started as a result of timber or flakes of old paint being left smouldering after work. As with welding and gas cutting, all work with blow lamps should stop at least one hour before work ceases for the day or before any period when the work place is unattended. A careful check for smouldering material should be made, not only on the side of doors, walls, partitions, etc. where the work has been done, but also behind them or on the other side. Buckets of water or pressurized water type extinguishers should be available.

3.2.2.8 Electrical faults

Electrical faults have been discussed in other chapters. It is advisable that the permanent electrical installation for the building should be installed as soon as possible, in order to minimize the amount of temporary cables and fittings used during construction. All temporary installations should comply with BS CP 1017 'Distribution of Electricity on Construction and Building Sites', and with the Institution of Electrical Engineers' Wiring Regulations which require all temporary wiring to be tested every three months by a competent electrician and a certificate to be issued.

Extension to existing installations must be authorized.

Main switches should, as far as possible, be opened when work ceases, and security floodlighting should be arranged on a separate circuit to enable switches on other installations to be opened at night.

3.2.2.9 Chimneys, flues and plant and vehicle exhausts

It is important to keep combustible materials away from plant and vehicle exhausts and flues of stoves (e.g. used in site huts), heaters, etc. Also, care should be taken to ensure that flue and exhaust gases are not directed on to combustible materials.

As a general rule, fires should not be lit in fireplaces of buildings undergoing renovation or extension. If it is decided to use existing fireplaces they must have been in regular use recently and the chimneys clean, sound and in good condition.

3.2.2.10 Site huts and caravans[6]

As many fires start in temporary site huts used as offices, stores, canteens and sometimes for living quarters, the following special precautions should be followed. These are in addition to those given earlier under rubbish, clothes drying, heaters, flues and smoking:

- huts should be made of non-combustible material
- huts should be placed at least 6 m away from buildings, and spaced as widely apart as possible
- combustible materials should not be stacked close to huts or between them.
- huts should not be placed inside buildings unless there is absolutely no other alternative
- if a hut has to be placed in a building, its base should not be below ground level nor higher than first floor level
- where floors of huts are raised above ground, the space beneath should be enclosed to prevent combustible material accumulating there
- special hot air lockers should be provided for drying clothes.

3.2.2.11 Storage and use of flammable liquids

Flammable liquids are distinguished from 'highly flammable' and 'combustible' liquids by their 'flash points', i.e. the lowest minimum temperature at which sufficient vapour is given off to form a mixture with air which is capable of ignition. There are two methods of determining the flash point, the 'open cup' and the 'closed cup' methods which give results differing by a few degrees; the closed cup method is the one referred to in regulations. Many of the paint and lacquer solvents discussed in Chapter 2.4 are flammable liquids. The classification of the various types according to flash point is given in *Table 3.2.1*.

Highly flammable liquids are defined by U.K. regulations whereas the others are defined by U.S. (National Fire Prevention Association) Codes, hence the overlap. The flash point of a liquid is very important in determining whether in given conditions it is liable to form a vapour which may spread and ignite some distance from its source.

Another important property is ignition temperature, which determines how readily the vapour is liable to be ignited by a hot surface. The flash

TABLE 3.2.1. Classification of combustible liquids by flash point (closed cup)

Classification	Flash point °C Minimum	Maximum
Highly Flammable		23
Flammable Class I		37.8
Combustible Class II	37.8	60
Combustible Class IIIA	60	93.3
Combustible Class IIIB	93.3	

points and ignition temperatures of several organic solvents and other liquids used in construction are given in *Table 3.2.2*.

Many paint and lacquer solvents are in the highly flammable class and their storage, handling and use are subject to the Highly Flammable Liquids and Liquefied Petroleum Gases Regulations, 1972. They present a very high fire hazard since the vapour formed by their evaporation at room temperature (e.g. from an open container) can travel considerable distances over floors until it meets a source of ignition. Many of the vapours are ignited by hot metal surfaces such as a soldering iron or a hot plate at temperatures well below red heat (about 500 °C). The fire then flashes back to the container, igniting its contents and any readily combustible materials in its path. Paints and lacquers containing the solvent are perhaps more dangerous fire hazards than the solvents

TABLE 3.2.2. Flammability classes. Flash points and ignition temperatures of some organic solvents and other liquids used in construction

Class	Liquid	Flash point °C	Ignition temperature °C
Highly flammable	Acetone	−18	535
	Dioxan 1.4	11	366
	Ethanol	12	425
	Hexane	−26	248
	Methyl ethyl ketone	−6	474
	Methanol	11	455
	Octane	13	296
	Hydrocarbon solvent SBP3	−18	
	Toluene	4	552
Flammable but not	Amyl acetate (iso)	35	375
highly flammable	n-butanol	29	340
	Diethyl carbonate	25	
	Nitro ethane	28	
	Hydrocarbon solvent SBP6	28	
	Styrene	30	490
	Turpentine	35	240
	Xylene (mixed)	26	553
Combustible	Cyclohexanol	68	
Class II	Ethoxy ethanol	40	
	Kerosene	38–74	255
	White spirits	40–50	
Combustible	Phenol	75	715
Class IIIA	Diesel oils	88–93	340
	Gas oils	80–92	
	White spirit, high flash	63	

themselves, since they are more exposed when in use and all of the solvent in them evaporates into the air.

Flammable (including highly flammable) liquids should always be kept in closed containers when not actually being used. Special safety cans with a fire arrester in the spout and a spring closing cap are strongly recommended for handling flammable solvents after dispensing from larger containers in which they are bought. Where fair quantities of flammable solvents are used, they will generally be bought in drums. These should be stored upright in a place reserved for that purpose, preferably out of doors in a fenced compound away from buildings. A limited amount of flammable solvents and paints containing them may be stored separate from other combustible materials in a fireproof cabinet in a well-ventilated room with a fire resistance of at least half an hour, and raised door sills. Limits on the quantities of flammable and highly flammable liquids should be decided and laid down by the fire safety officer, and prominently displayed in notices in the store. Drums of solvents for use should be fitted with taps and placed on a purpose-designed rack, slightly inclined to the horizontal (with the tap at the lowest point), and a drip tray underneath. They are best located in an open sided shelter on a hard base with a bund or low retaining wall all round to contain massive spills. The shelter should be provided with a sloping corrugated roof and be situated well away from other buildings or materials.

Flammable solvents from drums are first filled via their taps into safety cans after loosening the vent plug, which must be on the upper side of the drum. The vent plug is tightened again after filling the can. The solvent is then transported in the safety can to the point where it is used (e.g. in a paint shop) which should always be a well-ventilated place, at least 20 feet (6.1 m) from any naked flame. Lighting in solvent stores and places where flammable solvents are used should be flameproof and certified for use in hazardous zones in accordance with BS 5345, 'Code of Practice for the selection, installation and maintenance of electrical apparatus for use in potentially explosive atmospheres'.

The importance of ventilation in all areas where solvents and paints, lacquers, adhesives and other materials containing them are used was stressed in Chapter 2.4 from the point of view of health hazards. It is equally important in preventing fire risks.

3.2.3 Fire organization and first aid fire fighting[7]

3.2.3.1 Organization

A senior person with experience and training in fire prevention should have responsibility for fire precautions at site level. On large sites he will be designated 'Fire Precautions Manager'. He should be able to advise the company on legal requirements and practical aspects of fire safety, and liaise with the local fire brigade, security officials and insurance companies.

His responsibilities should include:

- making all necessary arrangements to inform and train all site workers on the actions to be taken in the event of fire, especially those detailed for fire fighting and rescue;

- the preparation of budgets for all fire protection work for which the contractor is responsible, and looking after the ordering of extinguishers, warning equipment, rescue apparatus and any special items needed;
- liaising with engineers and architects to ensure that supplies of fire water from the permanent supply (mains and risers) are available as soon as possible during construction, and that there is unimpeded access for brigade fire appliances right onto the site;
- carrying out regular inspections to check that the various precautions discussed in Section 3.2.2 are being carried out, with special attention to rubbish collection and disposal and site security.

In the event of fire breaking out, the action needed falls under four main headings, for which adequate arrangements should have been made in advance:

- fire detection and warning
- escape of personnel
- rescue
- emergency fire fighting

It is usually advisable for the Fire Precautions Manager to form a small Committee of site personnel to assist him in making practical arrangements on these points, and to facilitate full personnel involvement and cooperation.

3.2.3.2 Fire detection and warning

Whilst a number of different types of automatic fire detectors are available for permanent installations—factories, commercial premises, public buildings, etc.—these are generally unsuitable for a rapidly changing construction site. One is therefore primarily dependent on the human element for detection of fires. For sites which are completely unmanned at night time some form of portable fire and intruder alarm, operated by light beams or infra-red light and linked to the nearet police or fire station, may be considered.

Once a fire has been detected by personnel on site there should be some agreed warning system which they can use and some central fire control and reporting station on site. On most construction sites a hand operated gong or series of gongs appropriately spaced is probably a more suitable alarm than an electrical system.

The central reporting and control station should be near to a telephone which can be kept free or rapidly cleared in the event of fire. A proper system of warning all persons on site, as well as the local fire brigade and police station of the outbreak, should be worked out by the Fire Precautions Manager, and all personnel trained in its use and implementation.

3.2.3.3 Escape

All personnel on a construction site should be able to escape safely on their own feet in the event of a fire, without recourse to electric hoists, etc.

which are liable to be put out of action by a fire and may merely trap the occupants. The general principle is that two alternative means of escape should be available at opposite ends of the work place, so that if one is blocked by the fire there is a good chance that the other will be clear. Whatever the means of escape, they should be sufficient for all workers who have to use them to escape quickly, i.e. within two and a half minutes. This is not usually a serious problem on low buildings of two or three floors, but it is often difficult on tall buildings under construction or repair. If the Fire Precautions Manager is unable to find a satisfactory solution alone he should discuss the problem with the local fire authority and his own engineers.

It is not enough in a sound fire plan merely to ensure that everyone on site has two safe fire escape routes, knows where they are and how to use them. It is equally as important to know whether everyone on site has escaped from a serious fire situation, and whether a search and rescue operation has to be mounted, in which other lives may be risked.

Hence all escape routes should terminate at a safe assembly place where those escaping can report, a head count be taken and any missing personnel quickly identified. Suitable people, such as time clerks, who are not likely to have been exposed to the fire and who have lists of all persons present on the site, should be trained to act as 'checkers' in the event of a fire alarm, and should proceed at once with their lists to the agreed assembly place to check all persons escaping. Similarly all persons present on site should be trained to report to the checker after escaping from a fire and before dispersing.

3.2.3.4 Rescue

Rescue of trapped persons from a fire is a highly skilled task generally requiring special clothing, breathing apparatus and other equipment. Rather than risking other workers' lives, it is generally best left to the professionals (fire brigade). There are some circumstances, however, where the local fire brigade personnel cannot be expected to effect a rescue. Work in a tunnel in a compressed air atmosphere is an example. For such work the contractor usually has to organize and train both rescue and fire fighting teams from his own staff—men in particular who are used to working in a compressed air atmosphere and familiar with all its problems.

3.2.3.5 Emergency fire fighting and appliances

Portable extinguishers and buckets of water should be available at points clearly marked 'Fire Point' on a construction site for dealing with smaller fires. Since water in buckets and water-type extinguishers may be frozen in cold weather, non-freezing solutions (as recommended by the manufacturers of the extinguishers) should be used in winter.

The principal types of extinguisher, with materials on which they should be used and notes on their use are given in *Table 3.2.3*. The Fire Precautions Manager should discuss the number and types of extinguisher

TABLE 3.2.3. Types of portable fire extinguishers (BS 5306, Part 3:1980 and BS 5423:1980)

Type (main)		Water	Foam	Dry powder (general purpose)	Halon (generally 1211)	Carbon dioxide gas
Pressure Source	Gas cartridge	√	√ Mechanical foam	√		√
	Stored pressure	√	√ Mechanical foam	√	√	
(Alternate)	Chemical reaction	Soda-acid. Now less common	√ Chemical foam			
Body colour signal red with following coding →		Signal red throughout	Pale cream	French blue	Emerald green	Black
Recommended wording (in addition to other information stated in BS 5423: 1980)		'For wood, paper, textile and similar class A fires'	'For liquid fires'	'For all fires except metal fires and except gaseous fires having an explosion risk'	'For liquid, electrical equipment fires and if no explosion risk, gaseous fires'	'For liquid and electrical fires and, if no explosion risk, gaseous fires'
Warning markings (except for extinguishers specially developed and tested against these risks)		'Do not use on live electrical equipment'. 'Do not use on burning liquids'	'Do not use on live electrical equipment'		'The fumes given off are dangerous, especially in a confined space'	
Other risks and special notes		Soda-acid type especially damages paper, textiles and other *delicate* materials. Gas cartridges contain either carbon dioxide or nitrogen or a mixture of both. The body of the extinguisher is only pressurised when the cartridge is released. In stored pressurised types, the body of the extinguisher is pressurised with compressed air or other gas, usually to 8.3 bar.	Generally incompatible with dry powders. Special foams needed for liquids *which mix* with water. Chemical foams are formed by mixture of two liquids which generate gas inside extinguishers. Mechanical foams (gas cartridge and stored pressure types) rely on special air injector nozzle. Light water foam is generally most effective.	Not effective against fires where part of burning surface is shielded against powder discharge. Reignition possible *when discharge ceases*. Special powder extinguishers are available for metal fires (e.g. magnesium)	Reignition possible when *discharge* ceases. Permitted halons are: Bromochloromethane 1011; Bromochlorodifluoromethane 1211; Bromotrifluoromethane 1301; Dibromotetrafluoroethane 2402. Also following mixtures: Halon 1011 with 75% volume of 1.1.1 Trichloroethane; Halon 1211 with maximum of 20% mole of dichlodifluoromethane as expellant.	Reignition possible when discharge ceases. Danger of asphyxiation and carbon monoxide poisoning when used in confined spaces. Static electricity (gas ignition hazard) when liquid carbon dioxide is discharged (i.e. when extinguisher is used upside down).

Notes: (1) Classes of Fire (BS 4547). *Class A.* Fires involving solid materials, usually of an organic nature, in which combustion normally takes place with the formation of glowing embers. *Class B.* Fires involving liquids or liquefiable solids. *Class C.* Fires involving gases. *Class D.* Fires involving metals. *Avoid multiplicity of types of extinguishers* and standardize on as few types as possible. Train operators to use extinguishers. Ensure that the extinguishers are safely located near to fire risks to be protected and not subject to extremes of temperature.

(2) Various sizes from about 1 kg to a maximum weight of 23 kg. Obsolete types, such as invertable water and foam extinguishers, and vapourising liquid extinguishers with carbon tetrachloride methyl

to be used with the local fire brigade before ordering, and should arrange with the Chief Fire Officer to instruct his personnel in their proper use.

Most construction sites have water points and small hoses for mixing concrete and mortar and hosing down. These hoses are useful for emergency fire fighting, but on large sites proper 1 inch (25.4 mm) hose reels permanently connected to a water supply should be provided.

3.2.4 Special fire and explosion problems

Special fire and explosion problems arise in construction from the use of explosives in demolition, excavating and tunnelling, from the use of cartridge operated fixing tools, from work in compressed air atmospheres, particularly when open flame torches have to be used, and from the use of reactive or explosive chemicals such as sodium chlorate and some peroxides. These problems are treated only briefly below since they affect only small numbers of construction workers and because adequate guidance and training is available from other sources.

3.2.4.1 Use of explosives in demolition, excavating and tunnelling[8,9]

There are few activities so subject to control by regulations as the transport, storage and use of explosives. They cannot be bought without a police certificate. Storage and even unpacking of explosives are strictly controlled, with limits on the amount of any type that may be stored.

The use of explosives in construction is subject to section 19 of the Construction (General Provisions) Regulations, 1961 which states, "Explosives shall not be handled or used except by or under the immediate control of a competent person with adequate knowledge of the dangers connected with their use, and steps shall be taken to see that when a charge is fired, persons shall be employed in positions in which, so far as can reasonably be anticipated, they are not exposed to risk of injury from the explosion or from flying material".

The Factories Act, 1961, Section 146, empowers inspectors to enter and examine factories where it is believed that explosives are stored and used. In fact all stores, stocks and records may be inspected at any time by H.M. Inspectors of Explosives, the Local Authority explosives inspector, a police inspector or inspector of mines and quarries. Under the Health and Safety at Work, etc. Act, 1974, Section 33, sub-section 4(c), it is an offence to acquire, possess or use an explosive article or substance in contravention of any of the relevant statutory provisions.

Other legislation which affects the use of explosives include:

- The Explosives Acts 1875 and 1923, and Orders in Council and Orders of Secretary of State made under them (Guide to the Explosives Act available from HMSO);
- The Construction (Health and Welfare) Regulations, 1966;
- Mines and Quarries (Explosives) Regulations, 1959;
- Conveyance of Explosives Bye-Laws, SI No. 230;
- The Noise Abatement Act, 1960;
- The Clean Air Act, 1960;

- The Public Health (Recurring Nuisances) Act, 1969;
- The Control of Pollution Act, 1974.

Detailed guidance is given in BS 5607:1978[8]. This is essential reading for managers, engineers and supervisors involved in the use of explosives in construction. After two general sections, it contains sections dealing with the use of explosives in:

- tunnelling and shaft sinking
- demolition
- underwater blasting
- land clearance and excavation

A shorter guide is given in Section 8 of 'Construction Safety'[4].

Few serious accidents involving the use of explosives have appeared in the safety statistics of the U.K. construction industry. The HSE's biennial report, 'Construction Health and Safety, 1982' recommends, however, that 'closer control over the competency of men responsible for the storage and use of explosives in demolition work should be exercised by means of certificates of competence associated with defined standards of training'.

At the same time, the use of explosives in demolition is increasing, partly at least because their use is often safer and has fewer health problems than other demolition methods. Some of the main precautions needed when using explosives in demolition were given in Section 1.5.4.9. It is, however, quite impossible to cover here all the detailed technical and legal requirements which are essential for the safe use of explosives. To give only a selection here could lead to the danger of the reader thinking that 'that is all he needs to know'. Pope's lines apply:

"A little learning is a dangerous thing;
Drink deep, or taste not the Pierian spring.
There shallow draughts intoxicate the brain
And drinking largely sobers us again."

To conclude this section, some general considerations are given which apply to the safety of all personnel on a site where explosives are used.

Rules have to be formulated in advance by the site agent in conjunction with the safety manager, explosives engineer and shot firer, which take into account the special conditions of the site and spell out the duties of all who are involved with the use of explosives. These rules should cover the following:

- The operation of a warning system to alert all personnel when explosives are about to be used;
- The installation of an adequate set of warning notices which will be seen and read by anyone approaching the danger zone;
- The sounding of distinctive warning signals before and after each blasting operation;
- The posting of clearly visible and recognizable sentries at the boundary of the danger zone who should signal to the shot firer that the area is clear before firing commences. Re-entry of personnel should not be allowed until the 'All Clear' has been sounded;
- Notification of shot firing times to be posted on notice boards.

Shot firers must be at least 21 years old and properly trained and experienced in the particular type of work to be done. They should only be appointed in writing by the site agent, and a set of rules governing the work to be done should be prepared at the same time.

3.2.4.2 Use of cartridge-operated fixing tools[2,10]

Fixing tools operated by explosive cartridges are widely used in construction, particularly for fixing cladding to structures by means of metal pins. Their use is subject to most of the regulations mentioned in the previous section. Since their energy is about the same as that of a bullet from a .22 rifle, accidents arising from their use are usually serious. They should only be used by trained operators, preferably those who have been trained and tested by the manufacturer of the tool and who have passed a competence test in its use. Their hazards are explained in detail in an HSE Guidance Note[2] and BS 4078.[10]

There are two types of tool:

- Direct acting tools, similar to a pistol, in which the cartridge is fired by a trigger, and in which the pin is propelled directly by the gas formed in the explosion;
- Indirect acting tools, in which a piston is interposed between the cartridge and the pin. Some of these are spring and trigger operated, while others are operated by hitting the end of the tool with a heavy hammer.

Some of the main hazards of these tools are:

- tool held at angle to working surface, causing pin to ricochet. Although most tools are designed so that they can only be fired when the tool is pressed hard against the working surface, and its axis perpendicular to the surface, trouble can still arise if the surface is grooved, corrugated or irregular in shape.
- pin hits a hard object beneath the surface (e.g. a hard pebble in a concrete block), ricochets internally, causing splinters or the pin to fly out. To prevent accidents from this cause a splinter guard should be fitted to the tool.
- too powerful a cartridge is used, causing the pin to pass through the wall, panel, etc. and perhaps hitting someone on the other side. Whilst this is generally avoided by following the marker's recommendations, trial fixes should be done for any new work, using the weakest cartridge first.
- tool used on hard or brittle material such as tile, marble or cast iron, which shatters into flying fragments.
- use of unsuitable pin, which may itself shatter. Makers can supply pins of different shapes and materials with recommendations as to their use, but tests should always be carried out before a pin is selected for a particular job.
- operator uses tool whilst on a ladder or other precarious working place, and falls when the tool is fired due to the reaction of the

cartridge exploding. Cartridge operated tools should only be used from a firm and stable position.
- injury caused through faulty maintenance of the tools or in extracting an unfired cartridge from the tool without proper training or the correct withdrawal tool. Only mechanics with special training on cartridge operated fixing tools should be allowed to maintain them.

Other important precautions when using cartridge operated fixing tools are:

- approved eye protectors (complying with BS 2092:1967) should be used by all persons handling cartridge operated fixing tools, including loading, firing, unloading and maintenance;
- the area where cartridge operated fixing tools are being used should be protected. No bystanders should be allowed in the vicinity, and the sides of the working area may have to be screened off;
- tools should only be loaded as the last operation prior to firing, and should never be left unattended when loaded;
- tools should not be used in areas where there is flammable vapour or the risk of a dust explosion
- cartridge operated fixing tools and their cartridges and pins should always be stored in their proper containers under lock and key, with proper security arrangements.

3.2.4.3 *Fire and explosion hazards in pressurized workings*[1]

Like the use of explosives in construction, this is a specialized subject which only affects a small percentage of construction workers (special tunnel and caisson workers), whose hazards were considered in Section 1.5.6. The hazards are exceptionally severe and the precautionary measures include so many points of detail that all involved with the subject need a complete guide such as the recent Guidance Note issued by the HSE. Any short resumé which could be given here would necessarily be incomplete and could lead to a false sense of security. The main factors which magnify the fire and explosion hazards of pressurized workings are:

- combustible materials ignite more easily and burn more vigorously in compressed air than at atmospheric pressure;
- the necessity for air locks and decompression of personnel leaving compressed air workings severely hinders evacuation, rescue and fire fighting;
- the time for which self-contained breathing apparatus can be used in compressed air workings is progressively reduced as the air pressure in the working chambers increases;
- the problems of smoke and toxic combustion products are enhanced by the confined nature of the workings;
- in many tunnels there is the additional risk of ingress of flammable gas (especially methane) from the surrounding strata.

The main focus of fire prevention measures in pressurized workings lies in carefully minimizing and controlling all possible sources of fuel and heat within the workings. Special considerations also apply to:

- emergency planning, organization and training
- fire warning and communications
- evacuation and rescue
- fire fighting
- electrical supply and emergency lighting

3.2.4.4 Reactive and explosive chemicals

A few of the chemicals used in construction pose special fire and explosion risks, in addition to the fuel gases discussed in Chapter 3.1 and the solvents discussed in Section 3.2.1.

- *Sodium Chlorate*, used as a weed killer, is a strong oxidizing agent which can form explosive mixtures with organic materials such as sawdust, resin powders, and coal dust. These can even be detonated by friction or shock waves.

  Clothing and paper which has been impregnated with sodium chlorate solution becomes highly dangerous when dry, since it is very easily ignited, burns almost explosively and is likely to produce fatal injuries to the wearer. Paper which has been in contact with sodium chlorate during storage should be burnt immediately and not allowed to accumulate.

  Persons handling or using chlorate solutions should wear impervious clothing and rubber or PVC footwear, which should be thoroughly washed immediately after work to remove spilled material. Any personal clothing contaminated by spillage of chlorate or chlorate solutions should be removed and rinsed immediately.

  Special care is needed when storing sodium chlorate to keep it separate from other chemicals, particularly mineral acids, and in sound airtight containers, where it is not exposed to rain, heat or damp conditions.
- *Organic peroxides*, such as benzoyl peroxide, which is used as a catalyst in some resin formulations, are highly flammable and in some cases explosive. The manufacturers' precautions should be followed carefully and the same general measures taken as for sodium chlorate.
- *Styrene monomer*, used with unsaturated polyesters in making glass reinforced plastic objects, can polymerize spontaneously in storage. This reaction is accompanied by the evolution of considerable amounts of heat and vapour which can not only burst containers but is highly flammable. This is unlikely to happen unless the material is stored in bulk in a warm place and at the same time exposed to air.
- *Aluminium powder* in contact with iron oxide, as may happen when rusty steelwork has been painted with aluminium paint, can react strongly so that the whole mass becomes red hot. This reaction can be initiated by heat or a strong mechanical blow. Special care is thus

needed when using aluminium paint on steel to remove all rust carefully before applying the paint.

Of the paint and varnish solvents discussed earlier, the *nitroparaffins* (nitromethane, nitroethane and nitropropane) can burn very violently and the first of these can actually explode. Special care is thus needed in their storage and handling.

- *Hydrogen peroxide*, discussed in Chapter 2.4, as a chemical health hazard, can decompose vigorously in contact with some finely divided solids, such as manganese compounds. This reaction leads to the evolution of large quantities of oxygen with the consequent risk of breakage of bottles containing the peroxide. It must be stored in a cool place in dark bottles and not allowed to become contaminated.
- *Calcium carbide*, used in acetylene generators for welding purposes, reacts spontaneously with water or moisture to produce the inflammable and explosive gas acetylene. Containers of calcium carbide should be kept sealed until required, in sound condition in a cool, dry place. If a container is opened and only part of its contents are needed, great care must be taken to keep the remaining contents dry, and the lid replaced with an airtight seal.

References

1. HEALTH AND SAFETY EXECUTIVE Guidance Note GS 20, *Fire Precautions in pressurised workings*, HMSO, London.
2. HEALTH AND SAFETY EXECUTIVE, Guidance Note PM 14, *Safety in the use of cartridge operated fixing tools*, HMSO, London.
3. FOC, *Rules for the Construction of Buildings*, available from the Fire Offices Committee, Aldermary House, Queen Street, London, EC4.
4. Section 5, 'Fire Prevention' of *Construction Safety*, National Federation of Building Trade Employers, London.
5. HEALTH AND SAFETY EXECUTIVE Guidance Note GS 7, *Accidents to children on construction sites*, HMSO, London.
6. FOC, 'Recommendations for Builders or Contractors Operations' (see ref. 3).
7. UNDERDOWN, G.W., *Practical Fire Precautions*, 2nd Ed, Gower Press, Farnborough, Hants, England (1979).
8. BS 5607, *Safe use of explosives in the construction industry*, British Standards Institution, London.
9. Section 8, 'Explosives' of *Construction Safety* (see Ref. 4).
10. BS 4078:1966, 'Cartridge Operated Fixing Tools', British Standards Institution, London.

PART 4
SPECIAL PROBLEMS OF CONSTRUCTION AND THE FUTURE

Introduction

The poor health and safety record of construction as compared with manufacturing industry is clear from previous chapters, particularly in the statistics given in Chapter 1.3. To reinforce this point some comparable figures for Japan are given in the Table below.

Total accidental and fatal accident incidence rates (per million man hours worked) in construction and other industries in Japan, 1980 and 1981. (FAIR figures in brackets)

	1980	1981
All industries	3.59 (0.02)	3.23 (0.03)
Coal mining	35.29 (0.28)	29.54 (1.64)
Construction (general)	6.67 (0.08)	4.55 (0.09)
Tunnel building	15.90 (0.12)	10.75 (0.16)
Building works	5.87 (0.07)	3.90 (0.09)

In both Japan and the U.K. fatal accidental injury rates are about four times higher in construction than in manufacturing industry, although not as high as in coal mining. These and the previous statistics given in Chapter 1.3 have a financial significance which tends to be obscured.

Unfortunately few figures are available on the costs of occupational accidents, particularly in construction. However, from Ontario it was reported[2] that in 1967 and 1969 the *direct* cost of personal injuries in construction as paid by the Workman's Compensation Board was approximately equal to the gross profit of the industry. The total costs of these personal injuries were estimated to be about five times this figure. The high costs of personal injuries in construction probably only persist because they are borne by society as a whole rather than by the organization which employs the injured workers.

In this concluding part of the book, we try first to identify some of the features of the construction industry which create special problems and then consider possible remedies for the future.

4.1
Special safety and health problems of the construction industry

The construction industry has many special features which adversely influence the safety and health of those who work in it as compared with workers in manufacturing industries. Whilst these influences are felt in all countries, they are specially serious in developing countries, which are considered in the latter part of this Chapter. These special features must not be ignored by those inclined to castigate the industry for its poor safety and health record, nor used as an excuse by those prepared to accept things as they are. They need to be studied objectively if successful counter-measures are to be taken.

Questions of safety and health, working conditions and the working environment in the construction industry were reviewed in a recent report to the Building, Civil Engineering and Public Works Committee of the ILO presented at its tenth session in Geneva, 1983[1]. This chapter draws extensively on this material.

The special problems of construction arise from the following causes:

- temporary duration of work sites
- seasonal employment
- extensive use of migrant labour
- small size of most construction firms
- extensive use of sub-contractors
- effects of weather, including spells of long working hours to compensate for bad weather
- high labour turnover
- welfare problems of construction workers
- competitive tendering

4.1.1 Temporary duration of work sites

The temporary duration of work on individual sites, together with the rapidly changing character of the work and the hazards associated with it are in complete contrast to regular factory production work and form the most serious hazard to safety. They lead to the following problems:

1. The time available for site visits and inspections by safety and health inspectors to see particular operations in progress is very limited. On top of this, hazards are continuously changing as excavations go deeper or walls grow higher, so that what was safe yesterday may no longer be safe today. Safety precautions thus need to be continuously controlled by someone who stays on the site. Managements cannot rely (as they often do in repetitive factory processes) on periodic visits by inspectors to ensure that all safety regulations and necessary precautions are observed.
2. Safety measures must be organized 'from scratch' on every new site. This makes the development of a methodical approach to safety a constant uphill struggle.
3. Most workers are new to the site and the teams they work with. They frequently have to adjust themselves to new surroundings, new bosses, new workmates and new working conditions. Much of the work has to be done before they have time to get to know and understand each other, their skills, limitations and habits.
4. Many workers are inadequately trained, especially in safety and health matters, on arrival and however good their on-site training, the time they spend on site is often too short for the training to have much impact on their performance.
5. The cost and time involved in acquiring, installing and utilizing simple but necessary safety equipment are high when compared to the value of and time needed for the job for which the safety equipment was required.

4.1.2 Seasonal employment

Seasonal employment (and unemployment) in construction usually results from climatic reasons, although many workers from rural backgrounds in developing countries habitually return home for work at harvest time. A common result of seasonal unemployment is that, on returning to work, many workers are in poor physical condition, and have also forgotten many of the hazards of the job and safe working methods they had learnt earlier. Where building work is seasonal because of weather conditions there are also likely to be periods of good weather, where excessive hours are worked, leading to tiredness, loss of attention and higher accident rates.

4.1.3 Extensive use of migrant labour

The employment of migrant labour has always been a characteristic of large construction sites. Migrant workers may come from considerable distances, often from other countries, and may be unfamiliar with the local language, dialect and customs. And, although skilled and well paid construction engineers are in the same boat, most migrant construction workers come from the poorer sections of their communities. Examples of migration among construction workers are: Irish workers in England; Turks and Yugoslavs in the Federal Republic of Germany; Algerian,

Moroccan and Spanish workers in France; Italian and Spanish workers in Switzerland; Indians, Pakistanis and others in the Gulf States; Rajastanis in Delhi and workers from the Sertao in the large Brazilian cities of Rio de Janeiro, Sao Paulo and Brasilia.

The scale of such migrations should not be underestimated. In Switzerland, 60% of the constructional labour force in 1978 were foreigners[2]. In France, 24% of all construction workers (about 1.5 million) in March 1980 were foreign, and in Germany, 20% out of a total of about 1.25 million in 1981[3]. Such extensive use of migrant labour in construction, many of whom have only the most rudimentary knowledge of the language of the employing country, is a serious obstacle to communications and safety training and can only have a serious adverse effect on safety and health statistics.

4.1.4 Small size of construction firms

The large number of small firms engaged in construction was clearly shown in the U.K. in the statistics given in Chapter 1.3. It is a general feature of all countries, and applies most to building maintenance, and least to civil engineering, where the scope and nature of the work has led to the development of some large international firms. Often the firms are family concerns where most workers come from the same family. In Switzerland[4], for example, it is reported that nearly 80% of all construction enterprises have fewer than 20 employees each.

The high proportion of small undertakings is a handicap to the spread and adoption of safe working practices.

First, they cannot afford the services of safety specialists or instructors, so that there is little opportunity for organized safety instruction, whether on or off the job. (The development of 'Group Safety Schemes' to cope with this problem is discussed in Appendix A.) Secondly, it is much more difficult for official safety and health inspectors, who are generally employed on a national basis, to inspect the work and practices of a large number of small undertakings than a smaller number of medium-sized and large ones. Thirdly, many small businesses are short of capital and under greater pressure than large ones to cut costs at the expense of safety.

The situation was put in a nutshell by J.D.G. Hammer, H.M. Chief Inspector of Factories in the U.K., in an introduction to Construction—Health and Safety 1979–80. "The scene is dominated by a few large companies with good records in health and safety, supported by a sizeable group of middle-sized concerns who, within their resources, take health and safety seriously but contrasted with a much larger number of medium and small employers who are far less committed to the achievement of healthy and safe working."

4.1.5 Extensive use of sub-contractors

The specialization of activities on building sites has been a main factor leading to the extensive employment of sub-contractors, even by large civil engineering companies. This causes many problems of coordination, planning, allocation of safety responsibilities and communication. The main responsibility for planning and safety rests with the main contractor, who should insist that all necessary safety requirements are written into the sub-contract and the tender document. But unless proper provision is made for their cost and these costs are allowed for, it may be doubted whether even inclusion of the safety requirements in the sub-contract will be taken seriously.

4.1.6 Effects of weather

The problems for human beings working in an unprotected natural environment as compared to a controlled factory environment were discussed in Chapter 2.6. Apart from direct safety and health hazards arising from extremes of temperature, precipitation and wind, the same factors often adversely affect the worker's state of mind and general attention, making him more accident-prone. Yet weather also has a third important effect on safety and health in construction. This arises from the tendency in most construction firms to work unusually long hours during spells of good weather to make up for time lost in bad weather. A correlation between long working hours and high accident rates was well established among munitions workers in the U.K. during World War I, and H.M. Vernon[5] showed that they had two and half times as many accidents in a twelve hour day as in a ten hour day.

Several countries have legal limits to the number of hours of overtime that may be worked. Thus in Spain, the amount of overtime may not exceed 2 hours a day, 15 hours a month or 100 hours a year. In Poland and Venezuela, the legal maximum is 120 hours a year, in France in the public works sector, 145 hours. In the Netherlands, workers over 55 years old are not allowed to work overtime[1].

It is, however, doubtful how far such restrictions are adhered to in practice, since most workers are happy to work overtime which carries bonuses of 50 to 100% over their normal hourly rates. It is thus fair to say that on most construction sites a considerable amount of overtime is worked during months of good weather, while activity is reduced or even suspended completely for several months during winter. Another reason for occasional long working hours is the need to finish certain operations which, once started, must be carried out and completed without interruptions.

4.1.7 Clandestine work

'Moonlighting' or 'double jobbing', generally illegal, appears particularly widespread in construction, judging from frequent press comments, although for obvious reasons few statistics are available about its true scope. Sometimes the work is undertaken by workers from other sectors, who, having successfully carried out work of decoration, repair or minor alterations in their own homes, find similar services appreciated by relatives and friends until in a short time they have a flourishing tax-free spare-time business. At other times the work is undertaken by construction workers themselves, in the evenings, over the weekend or during holidays. Undoubtedly one main reason for this is the ready market for such services, which can usually be offered at lower cost than a firm would charge. Some indication of its extent was given in a recent study in France where a number of building workers were interviewed in confidence[6]. The average worker interviewed was found to be doing no less than 12 hours a week of building work outside his normal employment.

Clandestine work has two adverse effects on safety and health in construction:

1. By lengthening the daily hours worked, the care and attention of the worker are bound to suffer, rendering him more accident prone, as the wartime study on munitions factories in the U.K. clearly showed.
2. Clandestine workers often lack essential safety equipment for their work, and indulge in dangerous improvisation.

4.1.8 High labour turnover

Construction has a particularly high labour turnover when compared with other industries. This is due partly to the mobility required of construction workers, who may be engaged on several widely separated sites in any one year, and partly to the short duration of most jobs on particular sites. Thus many construction workers have frequently to move from site to site in search of fresh work and are virtually nomads. This is no new situation, and was described thus by Karl Marx[7]:

> "Nomad labour is used for various operations of building and draining, brick-making, lime-burning, railway-making, etc. A flying column of pestilence, it carries into the places in whose neighbourhood it pitches its camp, smallpox, typhus, cholera, scarlet fever, etc."

Whilst the situation today has improved in most industrialized countries, that in many developing countries which are struggling to industrialize is probably still as bad as Marx described, although the pestilences have changed. Put simply, a high labour turnover in any job is not conducive to a good safety and health record. It leads to workers leaving a job before they are properly conversant with most of its details, including safety, whilst the employer is constantly faced with the need to train new workers. Human relations in the work force also suffer, and opportunities for workers' safety committees to be established and function usefully are much reduced.

4.1.9 Welfare problems

Work on temporary construction sites creates special problems in the supply of food, housing, transport and other amenities, all of which are easier to provide for those in regular employment in factories and offices.

These welfare problems are compounded by the other special factors discussed in the previous sections, particularly the widespread use of migrant labour in construction.

The welfare problems of construction workers have long been recognized by workers and employers organizations alike. Thus the International Federation of Building and Public Works unanimously adopted a resolution in 1950[9] declaring itself fully in favour of the promotion of the welfare of workers in the construction industries and public works sector to the maximum extent economically possible and practicable, as a contribution to industrial peace and social justice. Whilst it is generally agreed that a wide variety of measures must be taken to provide satisfactory welfare facilities for construction workers, the vast differences in working and environmental conditions on construction sites makes it difficult to generalize. The welfare problems mentioned at the beginning of this section are discussed in turn, in the light of practices and regulations in different countries.

4.1.9.1 Food

For construction work near towns and industrial centres contractors need to ensure that workers are able to have hot meals, either by making available premises where they can prepare, heat and eat their meals or by arranging for them to eat in a canteen, restaurant or even at home.

For sites further from towns, mobile or temporary field kitchens and canteens, with adequate shelter, wash basins, showers and toilets should be provided.

On remote sites where workers have to be housed in camps, it is essential that stores be provided where workers can buy food, beverages and other basic commodities at fair or reduced prices, and that kitchens, canteens, washing and toilet facilities are also available.

Employers are usually obliged by law to provide and cover the cost of site catering and associated facilities, depending on the location of the site and the number of workers employed. The extent of such obligations varies from country to country. In Italy[1], building contractors are required (under collective agreements), if requested by at least 20 workers, to supply on remote sites a properly equipped and manned kitchen and canteen of which the meals and their ingredients are monitored by a workers' committee. A cook must be provided for every 50 workers. In India and Pakistan on the other hand, enterprises are obliged to set up a canteen only when more than 250 workers are employed.

It is important that workers be involved in the organization of catering and other services to ensure understanding and mutual trust. Whilst there is an onus on employers to provide the services and even subsidize the cost of food, workers must expect to pay a fair price for meals and similar facilities. Where no meals are provided on remote sites, or where workers prefer to bring their own food, there is usually an obligation for employers to provide food allowances as part of their employees' pay. In some countries (Libya and Saudi Arabia) employers are prohibited from substituting cash allowances for the meals they are required to provide[1].

4.1.9.2 Housing and shelter

Construction sites are often so far from workers' homes that they have to be accommodated near their work. While housing and welfare facilities on construction sites are sometimes satisfactory, they frequently fall far short of those most other workers are used to. The contractor often experiences real difficulties in providing adequate temporary accommodation as a result of lack of space, lack of piped water, sewers and electricity, or difficulties in making connections to these services. Discontinuity of the work and variations in the size and composition of the work force compound these problems. Due to the fact that construction work is mostly outdoors, and workers are exposed to all sorts of bad weather, their need for warm and adequate housing is greater than most other workers, and they also often need shelter from bad weather during working hours.

Most countries have regulations which require construction employers to provide shelters, as well as accommodation on remote construction sites. In Yugoslavia[1], for example, the Federal Committee of the Building Workers' Union drew up a framework agreement in 1978 under which the living quarters of single workers must comply with the following minimum standards:

- four beds at most per room, each with a floor area of 10 to 15 square metres
- at least one bath room for every 12 persons
- a common kitchen of at least 0.2 square metres per person
- central heating

In practice conditions vary widely. Caravans, trailers and similar mobile homes, or easily dismantled huts are frequently used in industrialized countries. In many poor developing countries, however, building workers are simply left to fend for themselves on remote sites.

Migrant and seasonal workers especially suffer most from inadequate housing and other welfare facilities on construction sites.

4.1.9.3 Transport to and from work

Transport for construction workers may involve short distances when workers travel daily between their homes and the site, and long distances when workers live in temporary accommodation near the site, but visit their homes and families at weekends or at holiday times.

Many construction workers, particularly in industrialized countries, own their own means of transport—cars, motor bicycles, bicycles, etc. Public transport is seldom adequate for daily journeys between home and construction site. In many cases transport is, or has to be, provided by the employers, often free of charge, either by using the company's own vehicles (cars, minibuses, vans or lorries which may also be used for conveying materials), or by hiring the services of a carrier. Transport in lorries is usually uncomfortable and of a much lower standard than most workers would expect.

In many countries construction workers are paid a transport allowance, or in cases where public transport is used, their fares may be refunded partly or wholly. In the U.K. the daily transport allowance, payable up to 50 km, is proportionate to the distance which the workers must travel[1]. When long distances have to be travelled for periodic family visits by workers on remote sites, there is often a collective agreement or legislation which requires the enterprise to assist their employees in making such journeys, financially or otherwise.

Special means of transport have to be used for construction work in the mountains or offshore, e.g. cable cars, small boats and helicopters, and special means of transport has to be provided when weather conditions (snow, ice, etc.) render normal routes impassable. Thus, all in all, workers' transport raises more problems in construction than in most other industries.

4.1.9.4   Other welfare amenities and services

The need for other welfare amenities and services is particularly evident on isolated sites where considerable numbers of workers have to remain for several weeks or months at a stretch. In most industrialized countries, legislation or collective agreements oblige contractors to provide, in addition to accommodation, canteens and sanitary facilities, recreational, cultural and medical amenities such as libraries, cinemas, television rooms, sauna baths, chapels or mosques and hospitals or sanitoria. On sites where workers' families are living, creches, day nurseries and schools may have to be provided for the children. Even in developing countries these facilities may be well or even lavishly provided by foreign contractors for their expatriate staff and their families. An example visited by one of the authors was a community of European engineers engaged in the construction of a dam and tunnel at an altitude of nearly 4000 metres in the Peruvian Andes to suppy a large irrigation project. The main contractor was Swedish. The staff lived with their wives and families in comfortable pre-fabricated bungalows, although pregnant mothers were sent to lower altitudes for the last months of pregnancy and the birth of their children. The community was provided by the contractor with a canteen, auditorium for stage and cinema, a medical centre and a small school with two full-time teachers for children up to 12. Most of the expatriate staff were well satisfied with their conditions.

However, the living conditions of manual workers on remote construction sites in many developing countries show great shortcomings. The situation of female and child labour on construction sites in India is particularly bad, as Section 4.2 shows.

4.1.10   Competitive tendering[10]

The problem of ensuring that contractors are not financially penalized for making proper provision for safety costs when tendering for jobs is repeatedly encountered. Even in the petrochemical industry, where both the clients and the contractors are usually large and well known companies, the problem persists, but is at least recognized. In the case of hard-pressed local authorities, struggling with government imposed spending cuts, and smaller industrial firms on the brink of survival in a business recession, the problem is usually more acute, and seldom comes to the surface.

Some indication of the problem and how it is dealt with in petrochemical and process plant construction was given by P.W.P. Anderson, Group Safety Adviser to Matthew Hall & Co in an article in *Protection* in 1980. This attempted to provide objective answers to several relevant questions.

The first of these was the extent to which the inclusion of safety costs in a tender and their acceptance by clients actually reduce accidents during major construction projects. Based on statistics available, it was found that lost time accident frequency rates which ranged from 2.5 to 6.0 per 100 000 man hours worked on contracts where no provision for safety costs had been made in tenders (and hence little safety planning had been done), could be reduced to an altogether lower range of 0.2 to 1.0 per 100 000 man hours worked on projects where proper safety planning and costing had been done and the costs accepted by the client.

The second was the extent and manner in which safety measures and their costs are included in the wording of contracts. Such contracts fall largely into two types:

- contracts between client and main contractor
- contracts between main contractor and his subcontractors

In the first type, a wide range of formats were found, which depended largely on the attitude of the client. Some took a responsible line, accepting their safety obligations squarely, whereas others clearly wanted to economize at the expense of safety.

In the second type, greater uniformity existed as most main contractors issued similar 'Health and Safety Rules and Conditions' to their subcontractors.

The next questions concerned the definition of safety costs and their identification at the tender stage. It is often difficult to distinguish between true safety costs and ones which would have to be met in any case to comply with regulations.

A brief summary of safety costs, which were broken down into more detail by Anderson, grouped them under the following headings:

1. Wages and salaries of safety officer, safety representatives and of the construction manager, superintendents, state registered nurses and work force for their time spent on safety and health related activities.
2. Safety training costs
3. Costs of safety incentive schemes
4. Costs of protective equipment provided

For a project with high safety standards and low lost-time accident frequency rates, the identifiable safety cost amounted to between 5 and 6% of the total labour cost.

Finally came the sixty-four thousand dollar question of how to prevent a contractor who makes a responsible provision for safety costs in his tender from being placed at a disadvantage by competitors who make no such provision. This was the most important question of all, and no generally accepted solution existed. Anderson suggested three possible remedies:

(a) Seek industry agreement to use standard wording in contracts to draw out detailed safety duties and safety cost responsibilities;
(b) Amend HASAWA to enforce (a)
(c) Seek industry agreement that safety costs will not be included at all in tenders, but be identified and presented separately for payment by the client on a Cost Reimbursable Basis.

Competitive tendering clearly poses one of the most important and thorny issues for the health and safety of construction workers. It is not just one for the petrochemical and process industries and their large process plant contractors. It faces the construction industry as a whole, both nationally and internationally. Some suggestions for action at all levels to improve the situation are given in Chapter 4.3, 'The Future'.

4.1.11 Special problems in developing countries

The very high toll of accidental injuries and disease among construction workers in many developing countries has been commented on in earlier chapters. Few actual statistics are available, but where they are, as in *Table 1.3.1*, they show average fatal accident frequency rates three times higher in developing countries than in industrialized ones. The most general conclusion reached is that the safety performance in construction everywhere correlates roughly with the living standards of the poorer sections of the communities from which construction workers are recruited. Many of the codes of practice and standards of developing countries, which in some cases are quite extensive, have been adopted 'carte blanche' from various industrialized countries, with little critical evaluation or participation in the process of standardization by those who will use them. The countries have only a fraction of the legal apparatus needed to control modern construction methods, such as exists in the countries where the methods originated. The safety and health inspectors for the industry are generally inadequate in numbers and powers. There is also a tendency in some countries to set up several inspection authorities to cover the same ground under different ministries, thus dissipating limited funds and resources and leading to confusion.

The following deficiencies (mentioned also in other chapters) are common in developing countries:

1. Bamboo or wooden scaffolding is employed on tall buildings without proper guard rails or toeboards.
2. Deep trenches and excavations, usually hand dug, have near-vertical sides, regardless of soil conditions. Shoring is provided only when needed to protect an adjacent building from subsidence, and not for protection of the workers. The soil removed from the excavation is piled close beside it, thereby adding to the risks of a collapse.
3. There is a general dearth of protective equipment of all kinds, and only supervisory workers are provided with hard hats.

Such problems in developing countries appear to be worst when modern building designs originating in industrialized countries are used. Construction of buildings traditional to the country, and seldom more than two stories high, appears much safer, as does that of modern buildings whose design has been specially developed for the local conditions and climate.

The various welfare facilities needed for construction workers, particularly migrant workers, are often quite inadequate. Poverty, unemployment and absence of social welfare facilities in poorer developing countries are powerful incentives to migration in the first place. Premises, where provided, are over-crowded and unhealthy, poorly ventilated, lighted and heated, with inadequate supplies of wholesome water, drainage and sewers. The situation in India seems particularly bad according to recent trade union reports quoted in Reference 1, although equally bad situations undoubtedly exist in other countries but do not come to our attention because of language difficulties. Thus in India 'no housing, medical services, cultural and recreational amenities or stores where workers can buy foodstuffs and other products exist in 90% of the places where public and private construction work is carried on'.[8]

'The situation is especially bad for female labour since the women employed in the large public works sponsored by the State, where they account for between 40 and 50 per cent of the workforce, are deprived of housing, medical care and sanitary installations.'[11] The writer can confirm the truth of this from four months spent in New Delhi in early 1982. Many workers of both sexes, as well as young children, were living on site in the most makeshift accommodation, sometimes in the incompleted structures they were helping to build, sometimes inside large concrete drainage pipes. Women and children were working alongside men doing unskilled work, such as carrying loads of sand, gravel or cement, or breaking up bricks. Often they were exposed night and day to the hazards of the site—dust, noise and the risk of being struck by vehicles or falling objects. To one side would usually be a group of workers with various injuries who could only support themselves by begging from passers-by.

References

1. Report II to the 10th session of the Building, Civil Engineering and Public Works Committee of the International Labour Organisation, 'The Improvement of Working Conditions and of the Working Environment in the Construction Industry', Geneva, 1983.
2. ANON, 'Problems of Foreign Construction Workers employed in European Countries', Paper prepared for ILO meeting of Experts, Geneva, 1979, quoted in Ref. 1.
3. Documents provided by the French and West German Governments to the Committee on Housing, Building and Planning of the United Nations Economic Commission for Europe, quoted in Ref. 1.
4. Information supplied to the ILO by the Swiss Federal Office of Industry, Arts and Crafts and Labour, quoted in Ref. 1.
5. VERNON, H.M., 'An Experience of Munition Factories during the Great War', *Occupational Psychology*, **14**, 1–14 (1940).
6. LE BARS, J.J. and OTHERS, 'Essai d'analyse des causes sodio-economiques du travail au noir dans les metiers du batiment, 'Societe d'Etudes pour le Development Economique et Social, Paris, July, 1980.
7. MARX, K., *Capital*, p. 681, English translation by Moore and Aveling, reprinted by Allen and Unwin (1946).
8. General report of the Co-ordination Committee of Workmen of Construction and Erection Engineering Industry, Eastern Region (Calcutta, 1980).
9. See ILO 'Welfare in the construction industry', Report II, Building, Civil Engineering and Public Works Committee, Third Session, International Labour Office, Geneva, 1951, p. 1.
10. ANDERSON, P.W.P., 'The effect of safety costs on the competitiveness of tenders', in *Protection* (London), March, 1980, p. 21.
11. Information supplied by the Indian Self-Employed Workers' Association to the International Labour Office, Geneva.

4.2 Training

4.2.1 General

Having discussed the special health and safety (HS) problems of the construction industry, it is natural to ask, 'How far can they be mitigated by special training?' If we consider other occupations which pose special hazards such as the fire services, mining, army parachutists and nursing lepers, the importance given to training is quickly apparent. So too are the resulting improvements in HS, although these tend to be regarded as the norm and taken for granted.

For workers in the armed forces and various emergency services who spend much of their time on 'stand by' it is much easier to find and pay for the time spent in training than it is in a competitive activity like construction. None-the-less, we can fairly assume that if steel erectors, roof workers and scaffolders were to spend as much of their working time in training as firemen, their accident records would improve. The better safety records of medium and large construction enterprises compared with small ones is partly due to better training. The same applies when we compare safety records of construction workers in India and Sweden. But it is not just the training of the workers that matters. The training and attitudes of managers, engineers and architects and supervisors are also crucial.

HS training is still too often thought of as an appendix to be tagged on to the end of the individual's trade or professional training which provides him with the basic skills and expertise needed for his job, or to develop further specializations. A variety of programmes cover needs ranging from the practical and mainly manual skills of general building operatives to the properties of materials and building science needed by engineers and architects. Much of the training is carried out in technical colleges by teaching staff whose experience of the health and safety hazards to which their students will be exposed is limited. If their teaching is devoid of safety content and the hazards of construction are ignored or treated as unimportant, there is a danger of creating wrong attitudes in the minds of the students from the start, and making the task of those giving the HS training later more difficult.

All technical training in construction is involved with the ultimate safety of the buildings or other works of construction and their users, but unless a special effort is made it tends to be less concerned with the safety of the workers. The reasons are mainly historical. To the great engineers of the early nineteenth century, the lives of construction workers counted for little. Today many of those most concerned with training in the construction industry strongly deplore this separation of HS training from general occupational training. They are insisting, for instance, that designers of buildings be trained to pay far more attention to the questions of how they are to be erected and maintained safely. Designers can no longer ignore such matters or assume that no matter how their structures are designed they will somehow get erected.

Employers today are obliged under the Health and Safety at Work Act, 1974, to provide such information, instruction, training and supervision as is necessary to ensure so far as is reasonably practicable the health and safety of their employees. All except the smallest employers are also required to prepare written HS policies which among other things identify the special needs for safety training in their organizations and show how they are satisfied.

The needs for HS training in construction are far from uniform. To ensure that the training is appropriate, one has to examine carefully the jobs carried out by different groups of personnel to identify the various skills needed, including their health and safety components. With each group one needs to consider the extent to which they are responsible for the creation and control of hazards which others may be exposed to, and the extent to which they themselves are exposed to hazards created by others. Directors, managers, planners and designers are potential creators of hazards which may affect others, but to which they themselves are scarcely exposed. Drivers of cranes, trucks and dumpers, plant operators, scaffolders and electricians are potential creators of hazards to which both they and others are subjected. The general labourer in construction may create some hazards to which others are then exposed, but generally he is on the receiving end of hazards created by others.

The classification given in this book of hazards to which workers in different trades and occupations are exposed helps to identify safe working practices and training objectives and to formulate detailed syllabi. It must, however, be again emphasized that such HS training should be fully integrated into the structure of the vocational courses. As an example of this lack of integration, many text books still in use contain chapters giving preparative details for mixtures of cements, mortars and protective and decorative coatings which say nothing about their hazards or how to handle them safely. For this one needs to consult a separate chapter on HS, or often a separate publication entirely.

A further problem is the number of trades (and skills) which are practised without any formalized training or qualifications. Demolition is a prime example. There is still a high proportion of construction workers who have received no formal training whatever, but have simply picked up their knowledge and work habits from other untrained and unqualified workers. There is evidence that this proportion of untrained workers is actually growing at the present time. The existence of such a large body of untrained workers has created a special need for HS instruction both at the workplace and in classes where films and other audio-visual aids can be used.

Although employers, as already mentioned, are under a legal obligation to provide such HS training, trade unions too are playing an important part, especially in the training of workers' safety representatives. More details are given later in this chapter.

4.2.2 Training for different vocational groups in construction

The following vocational groups were listed in a recent CITB publication[1] which gave the main points to be covered in the HS training of each group:

Group 1   Directors and Principals
Group 2   Site Managers, including Safety Supervisors, Foremen and Gangers
Group 3   Operatives and new entrants
Group 4   Planners and Designers
Group 5   Safety Officers and Advisers
Group 6   Safety Representatives

The general training needs of each group are discussed below. Unfortunately the subject is too complex for these to be discussed in detail.

4.2.2.1 *Directors and principals*

HS attitudes within a company are largely determined by those of its directors. Since the Robens report the need has generally been recognized for making HS matters within the company the personal responsibility of one nominated director, with a clear and unbroken chain of responsibility and communication on HS matters to him through the whole management/supervisory structure, and including the most junior employees.

The main headings under which the HS education of directors should be organized include:

Company Involvement
The Law
Accident Prevention
Occupational Health
Involvement of Operatives
Personnel Training in Accident Prevention

Those directors of construction companies who have special responsibility for HS matters can generally be expected to have acquired a basic knowledge of the subjects in earlier stages of their careers, e.g. as site managers. Much of their additional training will come from personal reading of books such as this one, and from attendance at seminars and conferences organized by the trade and professional organizations to which they belong. For those requiring intensive study of particular subjects, some of the safety management training courses (one week, residential) run by the Brooklands School of Management in Weybridge (UK) are suggested. Current course subjects include:

Law/Techniques of Safety Management
Behavioural Science
Occupational Health and Hygiene
General Science

4.2.2.2 Site managers and supervisors

This is a broad group covering anyone from the contracts manager to the trades foreman. Thus no universal training scheme can be recommended. Trade foremen who give direct instructions to operatives are in a key position to reduce accidents. Among training courses for site managers is the CITB Site Management Safety Course (one week) which is intended for site agents, safety supervisors, general foremen and others in a supervisory or controlling role. The course contents are:

Health and Safety at Work, etc. Act and allied legislation.
Accident prevention, investigation and reporting procedures.
Site documentation, registers, inspections
Occupational Health and Hygiene—personal protective equipment
General hazard identification
Safe working practices in relation to:
  Scaffolding and means of access
  Working in excavations
  Lifting equipment—cranes and forklifts
  Small plant and portable tools
  Abrasive wheels
  Woodworking machines
  Cartridge operated tools

Similar courses are run by BAS (Building Advisory Service) sponsored by the Building Employers' Federation.

Other safety training courses for managers and supervisors are organized by the Brooklands School of Management, Weybridge, in the following fields:

- Structural steelwork
- Oil and chemical plant construction (primarily for member companies of the Oil and Chemical Plant Constructors Association)
- Engineering construction (primarily for member companies of the Engineering Employers Federation)

As well as employing direct 'in line' supervisors, contractors employing more than 20 persons (on one or several sites) have, since 1961, been obliged by law to appoint a suitably qualified and experienced safety supervisor to advise on the observance of all applicable regulations and to supervise the safe conduct of the work. They should, as far as possible, be qualified members of the Institution of Occupational Safety and Health and have passed its qualifying examinations. They should be specially well informed on legal requirements and their training should include the following legal subjects:

Statute and Common Law
Health and Safety at Work, etc. Act, 1974
Factories Act, 1961 (as applicable to construction)
Offices, Shops and Railway Premises Act, 1963
Control of Pollution Act, 1974
Construction Regulations (4 parts)
Notification of Accidents and Dangerous Occurrences Regulations, 1980, and subsequent amendments.

Their general safety training should provide them with a thorough working knowledge of:

Company health and safety policy
Principles of accident causation and prevention
Accident reporting and investigation
Control at the place of work
Site documentation, registers, inspections
Health hazards, especially those affecting the eyes, skin, lungs and hearing.

They must have a thorough knowledge of all potentially hazardous activities (e.g. timbering excavations) and equipment (e.g. cartridge operated tools) with which workers under their supervision may become involved.

4.2.2.3 Operatives and new entrants

New employees in a firm and new entrants into the industry are particularly vulnerable to its hazards, especially if they have had no previous occupational training at a recognized training establishment. Such training should include special attention to the HS hazards and instruction in safe practices in their particular trades and occupations.

A comprehensive range of practical courses is available at the CITB's national training centres (particularly Bircham Newton). The general subjects covered include:

Earth moving
Crane operation (including courses for slinger/banksmen)
Fork lift truck operators
Roadworking machines
Bar bending and steel fixing
Scaffolding
General construction operations
Painting
Gas distribution

Many others are carried out at various technical colleges, throughout the country. In many cases the courses include practical and, as necessary, theoretical examinations, and the issue of appropriate certificates of competence.

Managers and directors of construction companies should in liaison with their safety supervisors and advisers lay down minimum training standards and qualifications for the operators of all equipment which, if improperly used, can create special hazards to other workers, as well as for workers in occupations where risks are specially high (e.g. demolition and removal of old asbestos from buildings).

To supplement such special training and provide a rapid and effective appreciation of hazards in particular jobs, a number of useful films and other audio-visual aids are available for hire or purchase through the CITB. Current film titles include:

Safety in Construction—The Cost of Chaos
Falsework
Heads you Lose
No Questions Asked
Eyes Down
The Drummond Report (builders hoists)
Check Mate (Scaffolding)
Working over Water
Road Works Ahead
Safe Handling
Powered Hand Tools
Welding Safety
Where's Danny? (roof painting)
You can't be too careful (renovation of old buildings)

Slide/tape packages are available with the following titles:

Ladder Safety
Short Fall
Mobile Towers
Roof-Top Roulette
Just One Slip

Such films and audio-visual aids provide one of the most effective and economic means of safety training. They should be followed by discussions led by safety advisers/supervisors and workers' safety representatives.

Mobile training units equipped with teaching aids including safety films and slide/tape programmes are available from BAS Management Services. Larger construction companies have established training centres, while some have their own mobile training units.

Induction training in accident prevention is especially necessary for new entrants, and proper provision for it must be made. The responsibility for its organization rests largely in the hands of the director responsible for HS matters. It should be carried out in a formalized manner and as far as possible by the company's own safety supervisor or adviser in close collaboration with company training personnel and site managers. Suitable headings for an induction training course for new entrants include:

Hazards on site
Precautions against such hazards

Protective clothing
Health
The Company's duties and responsibilities
The employee's duties

4.2.2.4 Planners and designers

Many of the architects, estimators and planning engineers have done their basic training in a university or technical college. Their responsibility for the functional safety of the building and plant which they design has long been recognized, but many of the courses in operation before 1974 paid insufficient attention to the need to design buildings and plant so that they could be erected, maintained and ultimately dismantled safely, or to provide for access of workers and heavy plant. Directors responsible for HS matters should stimulate discussion on these matters among their senior architects and engineers and be on the alert for opportunities of sending them to useful seminars held within the industry on these topics.

4.2.2.5 Safety officers and advisers

Training programmes for full time safety officers and advisers in the construction industry have been developed by the Building Employers' Federation, the Federation of Civil Engineering Contractors, the Construction Industry Training Board and the Institution of Occupational Safety and Health in cooperation with the Health and Safety Executive. The main centre of instruction in the UK is the Brooklands School of Management, Weybridge (referred to earlier in Section 4.2.2.2).

4.2.2.6 Safety representatives

Under the Safety Representatives and Safety Committees Regulations, 1977, employers are obliged to give employees appointed as safety representatives by recognized trade unions sufficient time-off with pay to train for their work as safety representatives. The trade unions are playing an important role in providing such training in the form of short courses of a week or a weekend at their existing education centres. Thus UCATT (Union of Construction Allied Trades and Technicians) is currently running three courses:

Health and Safety Stage I (2–3 days)
Health and Safety Stage II (2–3 days)
Advanced Health and Safety (5–6 days)

The first two are provided on a regional basis this year (1984–85) at the following locations:

Edinburgh University
Bristol University
St Helens College, Merseyside

The last is provided on a national basis at Ruskin College, Oxford.

Other trade unions involved in construction also run courses for safety representatives, e.g. the EETPU (Electrical, Electronic, Telecommunication and Plumbing Union) provides training at Esher College, Thames Ditton, Surrey, UK.

The TUC itself runs two courses, Health and Safety Stage 1 and Health and Safety Stage 2 for safety representatives.

Each course is of 10 days, usually arranged as one day per week for 10 weeks. These are run at local colleges and WEA centres. The aims are to "help participants to develop the confidence and skills they need to be effective on health and safety issues. These include:

- involving members and getting their support
- identifying key problems of health and safety, including hidden hazards
- finding and using information about hazards and safety standards
- negotiating with management and getting things done."

The UCATT courses have come into being because both employers and trade unions in the construction industry felt the need for more specialist training for its safety representatives than can be provided by the general TUC course.

As well as trade union courses, the CITB offers the Construction Site Safety Course for safety representatives with a very similar content to a course offered to safety supervisors.

4.2.3 How effective is safety training in construction?

From the training available for the construction industry as described in the last section, one might be tempted to think that all is well with HS training in construction. The acid test, of course, is to look at recent trends in accident statistics in the industry. They tell a different story—orders and employment down, accidents up. The apparent reasons for this are not hard to discover. Most of the training discussed in the foregoing applies mainly to medium-sized and large firms. It appears to have had minimal impact on small construction firms, family and one-man businesses. The unfortunate truth is that in a harsh recessionary climate less responsible firms using cheaper labour with little or no safety training can undercut firms who insist on employing only workers who are properly trained and qualified.

One answer to this problem is to license all construction firms and impose quite strict standards of worker training before a licence is granted, and prohibit all unlicensed construction. Needless to say, with so much 'do it yourself' amateur construction work going on, that is quite impossible as a general solution today. In certain areas, however, e.g. demolition and in the removal of asbestos from buildings, this idea is gaining more acceptance. We as a nation prefer to work by consensus and to employ voluntary solutions to problems even if it takes a little longer to reach them than more draconian measures would achieve. One principal reason against the early introduction of a widespread licensing system is the immense proliferation of bureaucracy that would follow. A gradual extension of licensing, both of the construction worker and of the construction firm, appears, however, to be necessary and inevitable if the present deterioration in the safety record of the industry is to be reversed.

References

1. CONSTRUCTION INDUSTRY TRAINING BOARD, 'Patterns of Training', Chapter 2.2 in *Construction Safety*, December, 1982.

4.3
The future

This chapter was drafted by one of the authors in 1982 for an ILO Conference at Geneva and forms part of the report referred to as Reference 6. It is reproduced by kind permission of the International Labour Office, Geneva.

4.3.1 General remarks

It would be easy to make optimistic forecasts of improved world OSH standards in the construction industry, but the matter has to be looked at realistically. There is a very wide gap at present between the standards of the most advanced industrialized countries, such as Sweden, and of the poorest and least industrialized, for which few reports are made or statistics kept. There seems to be a rough correlation between accident frequency rates among construction workers and the general standard of living or the life expectancies of their communities. This is not to say that occupational injuries in construction work are a dominant factor which limit the average life expectancies in a community; it is simply a fact that poor workers living in an environment where there are many risks of early mortality are prepared to take greater risks to earn a living.

The world population can be expected to continue to grow for the next 40 years at least. This, together with the current stagnation in living standards, generally suggests that there are no real grounds for optimism about any immediate improvement in OSH in construction. Such improvements as we may see through better safety organization and training, particularly in those activities with greatest risks to safety and health, and through technological changes which improve safety and health—in short, the doing of all those things which are needed to improve the OSH record in construction on a world scale—may, at most, merely enable us to prevent any further worsening of the situation. This rather gloomy prospect should spur us to increase our efforts to meet the challenges.

4.3.2 Action at the undertaking level

One of the most disturbing tendencies today is the award of construction contracts and subcontracts to firms able to quote lower prices than more reputable competitors simply by failing to include health and safety features which would slow down their work or increase their costs. Often they employ labourers who are here today and gone tomorrow and pay them piece rates or bonuses for dangerous work, instead of taking the necessary safety and health precautions. Such firms usually spend such a short time on site and the government safety and health inspectorates are so thinly stretched that official safety regulations are ignored and broken with impunity.

Managements of more safety-conscious firms and organizations with better records have a common interest with workers and workers' organizations in exposing such firms and their safety records, especially those that manage to escape the vigilance of the OSH inspectorates. Managements can probably do this most effectively through their employers' federations.

Another step which these firms should press for is to have OSH specifications and budgets written into the contract documents and tenders. This would undoubtedly command a wide measure of support if the issues were properly explained. If the bare cost of the work to be done can be separated from the cost of the OSH measures needed to ensure that it can be done safely, it becomes much easier to arrive at a true and fair comparison of tenders[11]. Moreover, since a large part of the cost of occupational injuries and diseases is eventually borne by governments, there is every justification for government aid in meeting this part of the cost. Furthermore, if governments were to cover a good share of the cost of the necessary OSH measures shown separately in the contracts, they would have a strong interest in seeing that the money so earmarked is spent, and spent widely and properly. In this way, we might hope to see a more critical evaluation of the cost effectiveness of OSH measures by all concerned. It is, of course, necessary to have the support of the owners and backers of the construction projects, since it is in their names that the tender documents are issued in the first place.

Measures such as these seem absolutely essential if managers generally are to play a fully effective role in promoting OSH in construction activities. They should have no qualms or hesitation in spending, say, 10 per cent of their time on OSH and controlling losses due to accidents or in appointing competent safety managers, engineers and specialists to devote their full time to these activities. In many cases OSH work and loss-control activities pay for themselves so that safety managers may justifiably say that 'safety pays'. Accident-related property loss and injuries nearly always go together.

Unfortunately, however, profits and safety do not always coincide. Sometimes a very serious hazard, such as asbestos, lies hidden for years while doing its deadly damage, largely because any suitable alternative is costlier to those immediately involved in its manufacture and use, although the cost to the community as a whole may be much higher.

Managers as a whole can do a great deal to improve OSH in construction by better planning of work and anticipating hazards before they arrive, as they inevitably will if the work is organized in an ad hoc and haphazard manner. If the work is properly planned and the hazards are analyzed, it will often be found that the first plan selected is not the best or safest and that several plans have to be examined on paper to arrive at a really safe and economical one. But once such a plan has been worked out on paper, tried out and found to work as expected, it is generally possible to make use of it again on similar contracts in the future. Architects and designers, as well as the client, should pay more attention to ensuring that their buildings, etc. can be constructed and maintained safely. To do this architects and designers should have practical experience and understanding of the construction methods to be used. If a structure is designed which needs complicated falsework or scaffolding, the designer should be required to advise the client in the selection of tenders to ensure that the contractor has the ability to provide the temporary works needed.

Action to promote safety and health in the firms engaged in construction should be seen not as a legal embarrassment to plague already overburdened managers, but as a normal and necessary part of the firm's activities in which workers and management have a common interest. The responsibilities of managers and safety supervisors and the role of safety representatives and joint safety committees have already been discussed at some length.

Firms should provide new employees and young persons with proper information and training in connection with their work and the hazards to which they are liable to be exposed. For key occupations with a high hazard potential both to the worker and to others, only workers who have undergone an approved course of training and passed a proper test of competence in the job should be employed.

In many countries construction firms are legally obliged to provide their employees with work clothing meeting specified conditions for protection against bad weather and against contamination by toxic or dirty or greasy materials which they encounter during the course of their work, as well as to launder and maintain the work-clothing. It is not always easy to implement such a policy, particularly in developing countries where labour-intensive methods are used, labour output is low and there is a high turnover of the workforce. All permanent employees should, however, be issued with appropriate work clothing, taking care that it is acceptable to them and that they will have no problems (including ones of religion and local tradition) in wearing and using it.

Employers' organizations should co-operate in all these matters with recognized workers' organizations, with governments and with the ILO and other international bodies, such as the WHO, which may be concerned with the OSH problems of their industry.

4.3.3 Action at the national level

Governments, employers and workers' organizations in every country should be involved in the promotion of occupational safety and health in the construction industry.

4.3.3.1 Government action

Governments need to set standards that are acceptable to the community as a whole and legislate to ensure that the most important points in the standards can be legally enforced with penalties for non-compliance. They also need to set up one or more inspectorates with unquestionable authority to visit construction sites and all people and organizations concerned with the safety and health of construction workers in order to satisfy themselves that the legislation is being complied with and to take legal action against any organization infringing it. The inspectorates should consult and be consulted by employers and workers and should take a broader view of their responsibilities than a strictly legal one. The official safety and health organization which employs the inspectorate should also collect and publish statistics and regular reports on safety and health in the construction industry, investigate accidents, highlight areas where problems exist and, in consultation with workers and employers, prepare guidelines, codes of practice and, where needed, initiate legislation to cover particular activities, hazards and problems. It should assist in education and research in safety and health matters and, where needed or appropriate, undertake these activities itself. Governments need to provide for the treatment, rehabilitation and compensation of construction workers who have suffered injury or ill health in the course of their work. They also need to compensate the worker's survivors in the event of his death, or ensure that these benefits are properly provided for (e.g. by the employers, through a reputable insurance company).

Governments, through their appropriate safety and health organizations, need to exchange views and information with similar bodies in other countries as well as regional and international organizations, such as the ILO, giving them the benefit of their experience. They should follow the highest ethical standards in the receipt and transfer of new technologies to ensure that the protection provided to workers in the technology-receiving country is as good as that provided to those in the country exporting the technology and that all the necessary legislation, codes of practice, instruction and protective measures are applicable or are available in the country receiving the technology[1].

4.3.3.2 Action by employers' organizations

Employers' organizations should provide a forum for the discussion of safety and health problems within their industry and include a proper safety and health content in the training courses which they run. They should aim at setting a high standard of occupational safety and health among their member firms and exercise such control as may be possible and necessary over any of their members who deliberately flout these standards for financial advantage. Their action should include the employment of safety officers, industrial hygienists, nurses, first-aid personnel, etc. where appropriate[2].

They should co-operate with governments and official OSH organizations in the setting-up of standards and codes of practice. They should play an active part in drawing up model contracts for construction work to be carried out, containing not only proper provision for OSH measures needed, but also provision that these be clearly budgeted and paid for. They should also institute a system for checking and auditing the implementation of the OSH measures.

4.3.3.3 Action by workers' organizations

Workers' organizations are primarily concerned with the safety, health and welfare of their members and their dependants. They are not responsible for the hazards faced by their members at work or for the protection provided by their employers, but they should be in a position to appoint workers' safety representatives from among their own members in the firms employing them to act as watch-dogs on matters of safety and health on their behalf. The legal scope and responsibilities for safety representatives should be clearly defined. Workers' organizations in countries which do not yet have such a system should strive to obtain the legal right to appoint workers' representatives. Legalized safety councils that include workers' safety representatives are often the result of the appointment of workers' safety representatives[3].

The main needs at the national level are:

1. Strengthening of national OSH agencies for enforcement, inspection and reporting.
2. Strengthening and widening of OSH legislation as needed to allow national OSH enforcement agencies to work effectively.
3. Improvements and modernization of existing codes of practice and standards on critical subjects, including scaffolding, asbestos and employment of women and children on construction sites.
4. Government support for the compulsory inclusion of OSH items in construction contracts.
5. Better information for and education of the public and construction workers, in particular on OSH matters.
6. Better liaison on OSH matter with other countries, particularly the more industrialized ones from which much of the construction technology is acquired, as well as with the ILO and other international agencies. This should include better reporting of OSH statistics to the ILO and the adoption of a uniform international system of reporting and definitions of terms used.

4.3.4 Action at the international level

In considering action at the international level, one tends to think automatically and perhaps exclusively of the International Labour Office. This is natural since the ILO is the only international tripartite organization representing the interests of workers, governments and employers alike.

Many large civil engineering organizations, as well as insurance companies, work on an international or semi-international basis, and workers and employers in the construction industry are represented on special international bodies. International financial organizations, which supply credit for construction activities in developing countries, such as the World Bank, have, or should have, an interest in the safety of the construction workers on projects which they are helping to fund. In addition to the international organizations, there are regional organizations such as the EEC, which are interested in OSH in the construction industries of their member countries and in improving communications in order to solve particular problems. While it is clearly difficult for any one international organization to coordinate the roles of others, it is clear that co-operation is needed to avoid the dangers of duplication and indeed of making opposing or incompatible recommendations to member countries and organizations. This point was brought out at a recent ILO symposium where several participants felt that there was a need for the various international agencies, particularly those belonging to the United Nations system, to act in a more coordinated manner on matters of occupational safety and health and working conditions.

The interest of the International Labour Office in OSH in the construction industry has as its focal point the Building, Civil Engineering and Public Works Committee of the ILO. Although OSH problems in the construction industry are very serious, at its Ninth Session in 1977 the Building, Civil Engineering and public Works Committee showed even greater concern[2] about the cyclical and seasonal nature of employment in the industry—problems which have now become even more pressing than they were in 1977. Special concern was expressed about two problems of the construction industry, asbestos and the premature ageing of construction workers. At its 67th Session (June, 1981), the International Labour Conference adopted the Occupational Safety and Health Convention (No. 155) and its Recommendation (No. 164). While applicable to all industries, this Convention and Recommendation are of special importance to the construction industry in view of its poor OSH record.

At the national level, both instruments urge governments to adopt coherent national policies on OSH to minimize hazards inherent in the working environment. Both stress the need for adequate regulations, adequate inspectorates and enforcement officers, and adequate penalties for violations of laws and regulations. They call for the establishment of procedures for the notification of occupational accidents and diseases and the production of annual statistics on them. Article 13 of the Convention provides that "a worker who has removed himself from a work situation which he has reasonable justification to believe presents a serious danger to his life or health shall be protected from undue consequences."

At the undertaking level, both instruments endeavour to increase the involvement of employers in the OSH conditions of their workers, as well as providing for the appointment of workers' representatives. Article 19 (f) of the Convention contains an important provision which, if implemented in practice on a wide scale, could result in a rapid raising of OSH standards throughout industry: "A worker reports forthwith to his immediate supervisor any situation which he has reasonable justification to believe presents an imminent and serious danger to his life or health; until the employer has taken remedial action, if necessary, the employer cannot require workers to return to a work situation where there is continuing and imminent danger to life or health".

The ILO has produced only one Convention[3] dealing specifically with

OSH in construction and it is now 45 years old. It has, however, produced a code of practice[4] and two guides on OSH in construction[5].

There are several areas where the ILO might extend its activities:

1. *Statistics.* There is a general need to work out a common basis for the reporting of occupational injuries and diseases by different member countries, as well as a need to break these statistics down by occupation, so that the risks faced, for example by painters and roof workers in Switzerland and Thailand, could be compared.
2. *OSH clauses and budgets in contracts.* It needs to be recognized that a major stumbling block to improved OSH in construction is the general lack of provision for OSH measures in construction contracts. To be effective such provision should include not merely what is to be done but also what it will cost. The same applies to subcontracts.
3. *Teaching on hazard identification and analysis.* Since OSH in construction tends to be an emotive subject, much of our thinking and writing on it remains in an emotive stage. More objective and quantitative thinking is needed, which could perhaps be stimulated by the provision of courses or workshops run by the ILO on hazard identification and analysis and the development of safe systems of work. These might be given at the national or regional level.
4. *Training.* The industry, as it becomes more mechanized and expands, will feel a growing need for a trained workforce. This training, while basically of a vocational nature, will also need to reflect occupational safety and health needs. It would appear that training in this field geared to some internationally acceptable standard would be desirable.
5. *Codes of practice and standards.* Most member countries particularly those using imported technologies, need additional OSH-related standards and need to improve and modernize existing ones. It would obviously be advantageous if standards could be harmonized and unified. The ILO has published its own code of practice for the construction industry, as well as compendia of occupational safety practices in building work and civil engineering to serve as guides. There seems, however, to be a need for more detailed guidance on a number of special items. Examples are the use of wooden scaffolding above a certain height and more detailed specifications for tubular metal scaffolds.
6. *Regional work on OSH activities in construction.* There seems to be a need for more work at the regional level, including research, for obtaining a clearer picture of the position in various countries in the regions and efforts to pool information and experience within the region with a view to improving standards, legislation and inspection and to bringing these into line as far as possible, not only on a regional basis but also with unified international goals.

References

1. See ILO, 'Occupational safety, health and working conditions and the transfer of technology', Proceedings of the Inter-regional Tripartite Symposium on Occupational Safety, Health and Working Conditions Specifications in Relation to the Transfer of Technology to the Developing Countries, Geneva, 23–27, November, 1981 (Geneva, 1982).
2. ILO, 'General report', Report I, Building, Civil Engineering and Public Works Committee, Ninth Session, Geneva, 1977; and 'Official Bulletin', (Geneva, 1977) Series A, No. 4, pp. 259–76.
3. The Safety Provisions (Building) Convention, 1937 (No. 62).
4. ILO, *Safety and health in building and engineering work*, International Labour Office, Geneva (1972).
5. ILO, 'Building work—A compendium of occupational safety and health practice', OHS series 42 and Civil Engineering work, ditto, OHS series 45, International Labour Office, Geneva (1981).

Appendix A
Sources of help

Several sources of information, training and specialist advice have been referred to in various chapters of this book. Here we try to provide brief details on those international and national UK organizations which are specially concerned with safety and health aspects of the construction industry. Following the section on UK organizations, we give a few details on some other national and regional organizations which have specially come to our attention, although no attempt is made to cover organizations in countries other than the UK.

A.1 International organizations

A.1.1 ILO

The International Labour Office (ILO), a special United Nations Agency with headquarters in Geneva, is a 'tripartite organization' representing governments, workers and employers in all member countries. Its main concern is with the improvement of working conditions everywhere, for which purpose it maintains regional and national offices in a number of countries. Among other activities it carries out special surveys and issues reports, assists in training programmes, promotes research, holds conferences and publishes standards and codes of practice, including the three referred to in the introduction to this book[1,2,3]. The ILO also passes resolutions and conventions on important issues which affect the safety and health of workers, and member countries are asked to ratify and abide by these. In addition to a permanent staff, it recruits experts for short and medium term assignments to assist governments of member countries who request help in matters within its competence.

Its activities in the field of construction come under the control of its 'Building, Civil Engineering and Public Works Committee' which, in turn, is guided by the Governing Body of the ILO. This committee has had a number of major international meetings since its inception, the most recent being its tenth session in 1983 which was devoted to 'The Improvement of Working Conditions and of the Working Environment in the Construction

Industry'[4]. This was closely linked to the wider ILO International Programme for the Improvement of Working Conditions and Environment, known as 'PIACT'.

Another branch of the ILO is the International Occupational Safety and Health Information Centre (CIS). The Centre keeps an up-to-date computerized data bank and publishes regular abstracts in English and other languages of publications on matters of occupational safety and health (OSH), covering the whole gamut of working activities. From these it prepares on request computerized bibliographies on OSH matters on any one of a large number of subjects.

A further service introduced by the ILO in 1981 was the International Occupational Safety and Health Hazard Alert System, for which it received a grant of $250 000 from the US Government. Governments participate in this network system by designating a competent authority to issue and receive information regarding newly recognized occupational health and safety hazards. Many of these alerts have related to the use of particular chemicals, of which the construction industry uses its fair share.

A.1.2 WHO

The World Health Organization, another United Nations Agency, which is also based in Geneva, is involved in many health problems throughout the world, including occupational ones. In these it liaises closely with the International Labour Office.

A.1.3 ISO and IEC

The International Organization for Standardization (the ISO) and the International Electrotechnical Commission (the IEC), both based in Geneva, are concerned with creating and issuing international standards which affect the whole range of industrial activities and electrical technology. Many of these standards originate in advanced industrialized countries and, subject to international agreement, are adopted, often with modifications, as international standards. These are often adopted by developing countries who need standards in connection with new technologies which are transferred to them. Nearly all standards have some safety significance, although this may be only incidental, and only a few are primarily concerned with safety and health.

ISO safety standards cover such subjects as safety colours, safety signs, powered industrial trucks and earth moving machinery, but as yet their coverage is far less extensive than those of national standards organizations, such as the British Standards Institution. The difficulty in securing international agreement between countries which have evolved different national standards is an obvious obstacle to progress.

A.1.4 Le Comité International de Prevention des Risques Professionnels du Batiment et des Travaux Publics

Le Comite International de Prevention des Risques Professionels du Batiment et des Travaux Publics is based in Paris and arranges exhibitions and conferences on the safety of machinery and equipment used in construction. Whilst international in character, it is mainly active in France and markets served by French equipment.

A.2 UK organizations

UK organizations involved in matters of health and safety in construction can be classified broadly under the following headings:

1. Government and official but independent national organizations
2. Employers' organizations
3. Workers' organizations
4. Joint bodies involving two or more of (1), (2) and (3) above
5. Universities and technical colleges
6. Voluntary organizations
7. Others

A.2.1 Government and official but independent National Organizations

The principal organizations involved may be considered to be:

The Law itself
The Health and Safety Commission and Executive
The British Standards Institution is an independent body with national status which receives some financial support from the government.

Many other national and local government departments and offices are variously involved, sometimes as direct employers of construction workers, sometimes as owners of property and clients for whom contractors work. On the other hand, they may act as regulatory authorities concerned with the standards of buildings and public works as regards function, appearance, public safety and hygiene, etc.

A2.1.1 The Law

There are several Acts of Parliament dealing with the health and safety of workers and a number of Regulations made under their authority which deal specifically with Construction. The Criminal Courts are mainly concerned with prosecutions brought by the Health and Safety Executive against employers (and occasionally employees) for infringements of statutory requirements, so that the man in the street seldom comes face to face with 'The Law' except when a prosecution is in progress. He is, however, expected to know the law so far as it concerns him, and he can usually get correct advice freely from members of the Health and Safety Executive, or if he wishes by paying for the services of a lawyer.

Many of the regulations which affect construction are discussed in considerable detail, and in some cases reproduced, verbatim, together with an interpretation for the benefit of readers, in works such as *Construction Safety*.

Construction impinges on so many walks of life and activities that it is hardly surprising that it may become involved in a vast range of Acts, Regulations, Certificates and official forms. The principal regulations which affect construction are given in Table A.1. Other Rules and Orders affecting construction are given in Table A.2.

Lists of prescribed Registers, Certificates and Forms which apply to construction, and of other Acts of Parliament and Regulations which sometimes apply, are also given in Reference 5.

Regulation 5 of the Construction (General Provisions) Regulations is specially important in requiring every contractor who employs more than 20 men to appoint one or more Safety Supervisors who are familiar with the legislation, supervise its observance and promote safe conduct at work.

The change in the nature of Regulations which came about after the passing of the Health and Safety at Work etc. Act, 1974 and the creation of the Health and Safety Commission and Executive should, however, be

TABLE A.1. Principal regulations which apply to safety and health in construction

Acts and Regulations	SI	No.
Factories Act 1961, as amended by the factories Act 1961 etc		
(Metrification) Regulations 1983	1983	978
Abrasive Wheels Regulations 1970	1970	535
Asbestos Regulations 1969	1969	690
Construction (General Provisions) Regulations 1961	1961	1580
Construction (Lifting Operations) Regulations 1961	1961	1581
Construction (Health & Welfare) Regulations 1966	1966	93
Amendment	1980	1248
Amendment	1981	917
Construction (Working Places) Regulations 1966	1966	94
Highly Flammable Liquids and Liquefied Petroleum Gases		
Regulations 1972	1972	917
Protection of Eyes Regulations 1974	1974	168
Amendment	1975	303
Woodworking Machines Regulations 1974	1974	903
as amended by the Factories (Standards of Lighting)		
(Revocation) Regulations 1978	1978	1126
Health and safety at Work, etc. Act 1974		
Safety Representatives and Safety Committees Regulations 1977	1977	500
Control of Pollution Act 1974		
Mines and Quarries Act 1954		
Offices, Shops and Railway Premises Act 1963		
Control of Pollution (Special Waste) Regulations 1980	1980	1709
Ionising Radiations (Sealed Sources) Regulations 1969	1969	808
Food Hygiene (General) Regulations 1970	1970	1172
Control of Lead at Work Regulations 1980	1980	1248
Diving Operations at Work Regulations 1981	1981	399
Work in Compressed Air Special Regulations 1958	1958	61

noted. The HSE are themselves empowered to issue Regulations, Approved Codes of Practice and Guidance Notes which provide guidance on objectives and how they are to be achieved without being as specific over detail as many of the Regulations made under the Factories Act, 1961. One problem often found with these older Regulations was that they could not anticipate changes brought about by the use of new materials and technology.

The HSE Approved Codes of Practice so far cover the following subjects:

Safety Representatives

TABLE A.2. Other Rules and Orders which apply to safety and health in construction

Rules and Orders	SR	No.
Factories (Electrical Energy) Regulations 1908	1908	1312
Electricity (Factories Act) Special Regulations 1944	1944	739
Engineering Construction (Extension of Definition) Regulations 1960	1960	421
Work in Compressed Air (Amendment) Regulations 1960	1960	1307
Breathing Apparatus, Etc. (Report on Examination) Order 1961	1961	1345
Ionising Radiations (Sealed Sources) (Radiation Dosemeter and Dose Rate Meter) Order 1961	1961	1710
Ionising Radiations (Sealed Sources) (Radiation Dose Record) Order 1961	1961	1713
Ionising Radiations (Sealed Sources) (Health Register) Order 1961	1961	1714
Ionising Radiations (Sealed Sources) (Transfer Record) Order 1961	1961	1715
Construction (General Provisions) Reports Order 1962	1962	224
Construction :Lifting Operations) Reports Order 1962	1962	225
Construction (Lifting Operations) Prescribed Particulars Order 1962	1962	226
Construction (Lifting Operations) Certificates Order 1962	1962	227
Building (Inspection of Scaffolds) Reports Order 1962	1962	237
Construction (Lifting Operations) Prescribed Particulars (Amendment) Order 1962	1962	1747
Construction (Lifting Operations) Certificates (Amendments) Order 1964	1964	531
Construction (Notice of Operations and Works) Order 1965	1965	221
Young Persons (Certificates of Fitness) Amendment Rules 1965	1965	868
Work in Compressed Air (Prescribed Leaflet) Order 1967	1967	112
Electricity Regulations 1908 (Competent Persons Exemption) Order 1968	1968	1454
Engineering Construction (Extension of Definition) (No. 2) Regulations 1968	1968	1530
Electricity Regulations 1908 (Portable Apparatus Exemption) Order 1968	1968	1575
Certificates of Appointment of Factory Inspectors Order 1971	1971	1680
Diving Operations (Diver's Fitness Register) Order 1972	1972	1942
Work in Compressed Air (Health Register) Order 1973	1973	5
Industrial Tribunals (Improvement and Prohibition Notices Appeals) Regulations 1974	1974	1925
Fire Certificates (Special Premises) Regulations 1976	1976	2003
Fire Precautions Act 1971 (Modifications) Regulations 1976	1976	2007
Fire Precautions (Application for Certificate) Regulations 1976	1976	2008
Fire Precautions (Factories, Offices, Shops and Railway Premises) Order 1976	1976	2009
Fire Precautions (Non-Certificated Factory, Office, Shop and Railway Premises) Regulations 1976	1976	2010
Health and Safety (Enforcing Authority) Regulations 1977	1977	746
Health and Safety at Work, etc. Act 1974 (Application outside Great Britain) Order 1977	1977	1232
Pneumoconiosis, etc. (Workers Compensation) (Determination of Claims) Regulations 1979	1979	727
Social Security (Industrial Injuries) (Prescribed Diseases) Regulations 1980	1980	377
Notification of Accidents and Dangerous Occurrences Regulations 1980	1980	804
Safety Signs Regulations 1980	1980	1471
Health and Safety (First Aid) Regulations 1981	1981	917
Pneumoconiosis, etc. (Workers Compensation) (Payment of Claims) (Amendment) Regulations 1982	1982	1867
Health and Safety (Fees for Medical Examinations) Regulations 1983	1983	714
Asbestos (Licensing) Regulations 1983	1983	1649
Pneumoconiosis, etc. (Workers Compensation) (Payment of Claims) (Amendment) Regulations 1983	1983	1861

Control of Lead at Work
Work with asbestos insulation and asbestos coating
First Aid
Classification of dangerous substances for conveyance in road tankers and tank containers
Petroleum Spirit (Plastic Containers)

Non-compliance with the principles set out by the HSE in these Approved Codes of Practice can lead to the imposition of 'improvement notices', 'prohibition notices' and legal sanctions just as would a breach of one of the older style Regulations.

A.2.1.2 The Health and Safety Commission and Executive

Most UK readers are familiar with the Health and Safety Commission and Executive which were created in 1974 as a consequence of the Robens Report (1972) and HASAWA (1974). The HSC was set up to take over the responsibilities for health and safety at work from the fragmented official authorities which preceded them. It consists of a Chairman and six to nine members appointed by the Secretary of State for Employment. The Executive is the operational arm of the Commission. It controls a unified inspectorate which was previously split under factories, mines and quarries, nuclear installations, alkali and clean air, explosives and agriculture. Besides the work of inspection, it has major responsibility for research, information, education and advice in health and safety matters. It controls the Employment Medical Advisory Service. Both the Commission and the Executive can bring proceedings in Courts.

The HSE has inspectors for the construction industry attached to its 21 area offices throughout the UK and to coordinate their work it has established a Construction National Industry Group in the London Area South office, located at 1 Long Lane, London SE1 4PG. The addresses of these area offices, as well as a list of HSE publications relevant to construction are available from the HSE. These fall under the categories of:

Guidance Notes
Advice to Employees
Approved Codes of Practice
Health and Safety at Work Series Booklets
Health and Safety (new series) Booklets
HSE Special Reports Relevant to Construction
Health and Safety Commission/Construction Industry Advisory Committee—Guidelines on the implementation of safety policies

Many of these publications have been referred to in this book, and the recommendations of three JAC sub-committees on Scaffolding, Demolition and Steel Erection are reproduced here as Appendix B.

A.2.1.3 The British Standards Institution

The British Standards Institution is responsible for national standards covering the entire field of industrial activity. The main functions of standards are to promote

454 Sources of help

- Rationalization of product sizes, designations and test methods
- Quality, to ensure that products and services are fit for their purpose
- Safety, especially applied to standards on protective equipment such as respirators and goggles, but nearly all standards have safety implications, and the use of sub-standard building materials, components and methods is generally unsafe.

British standards are of four main types:

- Symbols and glossaries, as an aid to communication and understanding. Drawing symbols for ventilating installations are an example.
- Methods, to specify the way activities should be performed, such as sampling, measurement, testing and calculation.
- Specifications give the characteristics of products in terms of size, shape, materials and function, as well as procedures for checking compliance with these requirements.
- Codes of practice recommend accepted good practice as followed by competent practitioners. They are written for guidance only and are not intended to provide exclusive solutions to particular problems. Many codes of practice published by the BSI make a contribution to safety. Although they do not have the same legal authority as Approved Codes of Practice published by the HSC they are often quoted in evidence in prosecutions for unsafe practices.

Several of the very large number of British Standards have been quoted or referred to in different chapters of this book.

The British Standards Institution has issued a special booklet, 'British Standards and the Construction Industry', as well as a 'Sectional List of British Standards' (SL 16) which apply to Building.

A.2.2 Employer's organizations

The two main employers' organizations in the construction industry are:

The Building Employers' Federation
The Federation of Civil Engineering Contractors

In addition, the following employer's organizations cover specialized fields of construction:

The Oil and Chemical Plant Constructors' Association
The British Constructional Steelwork Association
The Engineering Employers' Federation

The Building Employers' Federation sponsors the comprehensive loose-leaf two volume publication *Construction Safety*[5]. This contains useful and informative sections dealing with HASAWA, 1974, Safety Representatives and Safety Committees, Policy Organization, Administration and Training.

All employers' organizations cooperate in training matters with:

The Construction Industry Training Board,
The Engineering Industry Training Board, and the
Institution of Occupational Safety & Health

Residential safety training courses for the construction industry are organized at a variety of centres, particularly at Brooklands School of Management which forms part of Brooklands Technical College, Weybridge, and at Bircham Newton Training Centre, Norfolk. The first of these caters specially for the training of Safety Officers and Managers, whilst the second concentrates more on practical courses for operatives whose work is critically important for the safety of others, e.g. scaffolding, timbering, falsework and cranes.

All of the employers' organizations have one or more officials with full-time responsibility for safety and training and some of the employers' organizations run special safety courses for their own managers, supervisors and safety officers at one of the training centres mentioned.

The employers' organizations cooperate further in promoting and supporting CIRIA, the Construction Industry Research and Information Association, which is supported partly by government funds, partly by members' subscriptions, partly by contracts undertaken and by the sale of publications. Its researches range from literature studies to field and laboratory trials. A proportion of its work is on health and safety topics. Its 'Guide to the Safe Use of Chemicals in Construction' is included in *Construction Safety* and referred to in Chapter 2.4 of this book.

A.2.3 Workers' organizations

The main workers' organizations in the construction industry are those unions which are party to the National Joint Council for the Building Industry and to the Civil Engineering Construction Conciliation Board. These are:

UCATT	Union of Construction and Allied Trade Technicians
TGWU	Transport and General Workers Union
GMWU	General and Municipal Workers Union
FTAT	Furniture, Timber and Allied Trades Union

Workers' organizations are involved through safety representatives and safety committees, especially following the Safety Representatives and Safety Committees Regulations, 1977. Under this a recognized trade union may appoint safety representatives from among the employees where one or more are employed by an employer by whom the union is recognized.

The main functions of safety representatives are:

1. to investigate potential hazards and dangerous occurrences at the work place and to examine the causes of accidents at the work place.
2. to investigate complaints by any employee he represents relating to that employee's health, safety or welfare at work.

In order to carry out these functions, safety representatives need training, for which purpose they are entitled to have paid time off.

The unions themselves organize training courses for their members who undertake the duties of safety representatives, generally at their permanent training centres, under tutors specially appointed for this purpose.

Training at a national level is provided by the Trades Union Congress, who also publish a *Health and Safety Short Course*.

The TUC also supports an extensive research and information service and library through the TUC Centenary Institute of Occupational Health, which is part of London University.

A.2.4 Joint bodies

Four joint bodies have already been mentioned:

- The Construction Industry Training Board
- The Engineering Industry Training Board
- The National Joint Council for the Building Industry
- The Civil Engineering Construction Conciliation Board.

The last two have concluded a number of agreements, known as 'Working Rule Agreements' between employers and employees concerning wages and working conditions and other matters of mutual concern. Agreements on Safety Representatives and Safety Committees for instance are covered by the NJCBI National Working Rule 7A and by the CECCB Working Rule 18A.

Safety Groups of several kinds exist in all parts of the UK.

First there are groups of contracting firms, manufacturers of construction equipment, safety officers, trade unionists and representatives of educational bodies which meet periodically to exchange information.

Secondly there are similar groups which in addition provide safety training.

Thirdly there are Group Safety Schemes, generally consisting of smaller firms employing between 20 and 100 operatives, which are too small to employ their own full time safety officers. Such groups will form a management committee, with the assistance of the BEF, to employ a Safety Officer on behalf of the group to advise each member on safety matters, make regular inspections of each member's sites, investigate accidents and liaise with external bodies.

The Joint Advisory Committee on safety and health in the construction industries was set up with government support prior to the creation of the HSE, with representatives of employers and trade unions to study and make recommendation on 'safety blackspots' in the industry. It usually appointed sub-committees as working parties for this work. This committee has now been reconstituted as the Construction Industry Advisory Committee of the Health and Safety Commission.

A.2.5 Universities and technical colleges

Most universities and technical colleges offer training courses in civil and other branches of engineering, and a number of technical colleges offer courses in various branches of building. All have, or should have, some safety content.

The special safety courses at Brooklands School of Management and at the Bircham Newton Training Centre for the Construction Industry have already been mentioned.

Degree courses in occupational hygiene are offered at the London School of Hygiene and Tropical Medicine by the TUC Centenary Institute of Occupational Health, whilst degree courses in safety and hygiene are offered at the University of Aston in Birmingham.

A.2.6 Professional and voluntary organizations

The Institution of Occupational Safety and Health is a professional association to which most UK safety managers and officers belong. It has various grades of membership, all of which, except 'student' grade, have qualifying requirements by examination and/or relevant experience. Special courses (mostly part time) for membership examination are run by a number of technical colleges. Its official journal is *The Safety Practitioner*, published monthly by Victor Green Publications Ltd.

The main voluntary UK organizations are the Royal Society for the Prevention of Accidents (RoSPA), and the British Safety Council, both of which run training schemes and publish books, journals, posters, etc. on safety-related topics. Both rely for the main part on members' subscriptions for finance, and both have been rather eclipsed in recent years by the HSE which generally provides more authoritative information at lower cost.

Training in first aid has traditionally been provided by the St John Ambulance Association and Brigade, the St Andrews Ambulance Association and the British Red Cross Society. Since the Health and Safety (First Aid) Regulations, 1981, came into force, a number of other organizations have also been approved to carry out First Aid training.

The British Society for Social Responsibility in Science publishes a periodical and various booklets on particular hazards including a useful one on vibration which was referred to in Chapter 2.5.

The Industrial Safety (Protective Equipment) Manufacturers Association publishes biennially a reference book of Protective Equipment, as well as organizing exhibitions and conferences.

Other journals and regular publications include Health and Safety at Work (Maclarens), the Industrial Safety Data File (United Trade Press) and Health and Safety Information Bulletin (Industrial Relations Services).

A.2.7 Other UK organizations

A number of specialized organizations which are not directly construction oriented but which are primarily concerned with particular aspects of health and safety, exist in the UK. The names and addresses of several of these which are particularly related to health and the working environment have been given in Part 2 of this book. Others which should be mentioned here are:

The Fire Prevention Association
The Contractors Plant Hire Association
The LPG Industry Technical Association

A.3 Other national organizations

Since most industrialized countries have a range of organizations similar to those in the UK which are concerned in one way or another with health and safety in construction, it is obvious that the total number must be very large indeed. The names and, in some cases, addresses of many organizations whose published contributions have been abstracted by CIS (see Chapter 2.1) are published with the abstracts. Alternatively, the International Labour Office can supply the names and addresses of many government, employers' and workers' organizations concerned with occupational safety and health throughout the globe. Only a few organizations which have come specially to the writers' attention are listed here.

A.3.1 USA

The National Institute for Occupational Safety and Health (NIOSH), Cincinatti, Ohio, plays a very similar role to that of the HSE in the UK, and publishes a number of reports, guides and standards in addition to its monitoring activities.

The National Safety Council, 444 North Michigan Avenue, Chicago, Illinois 60611, occupies a position somewhat similar to that of RoSPA and the British Safety Council in the UK. Of its numerous publications, its *Accident Prevention Manual for Industrial Operations*, of which a new edition is published every two or three years, is a first class reference book on all aspects of industrial safety, construction included.

A.3.2 Canada

The Construction Safety Association of Ontario came to our attention as a lively safety ginger group in the Canadian Construction Industry. Its informative annual report gives a good introduction to the safety scene in Canadian construction.

A.3.3 Brazil

The Ministry of Labour, Sao Paulo, publishes a regular bulletin of accident statistics with a report on similar lines to the annual reports of our HSE. The position in Brazil and in most South American countries is clouded by the high proportion of 'unofficial' and 'shanty town' building activities, mostly carried out by untrained labour with substandard materials, and in fact mainly illegal.

A.3.4 Japan

The Ministry of Labour, Tokyo, publishes regular statistics and reports which in detail appear to match those to which we are accustomed in the UK. Article 72 of a collective agreement in force in the Japanese construction industry in 1982 states 'The Company shall appoint a general safety and health supervisor, over-all safety and health controller, fire prevention supervisor at the establishment where such personnel are needed and grant them necessary power to perform their duties'.

A.3.5 India

A very critical report highlighting the bad working conditions in the Indian Construction Industry from the Centre of Indian Trade Unions, New Delhi, has reached one of the authors through the ILO. Perhaps the only encouraging feature is that such reports actually exist and come to light, and may eventually spur public conscience in the country to remedial measures. The conditions are largely a reflection on the widespread poverty in the country which is common throughout much of Asia, Africa and South America.

A.3.6 Kenya

An interesting agreement between the Kenya Association of Building and Civil Engineering Contractors and the Kenya Building, Construction, Civil Engineering and Allied Trade Workers Union has reached one of the authors through the ILO. The areas covered are much the same as those with which UK readers would be familiar, and it is clear that awareness of safety and working conditions in construction certainly exists in Kenya, even though a great deal of progress has still to be made.

A.3.7 France and EEC

The 'Institut National de Recherche et de Securite pour la prevention des accidents du travail et des maladies professionnelles', 30 Rue Olivier-Noyer, 75680 Paris, publishes a number of reports on accident causes and their prevention, including one seen by one of the authors on falls from height in the building industry and public works industry in the European Economic Community. Its conclusions are very similar to those reached in this country and point to the need for better enforcement of regulations on the use of proper scaffolding and fall protection devices, and on phasing out the use of portable ladders as work places and their replacement by mobile tower scaffolds.

References

1. ILO, *Safety and health in building and civil engineering work*, International Labour Office, Geneva (1972).
2. ILO, *Occupational Safety and Health Series 42, Building Work*, International Labour Office, Geneva (1979).
3. ILO, *Occupational Safety and Health Series 45, Civil Engineering Work*, International Labour Office, Geneva (1981).
4. ILO, *The Improvement of Working Conditions and of the Working Environment in the Construction Industry*, International Labour Office, Geneva (1983).
5. CONSTRUCTION INDUSTRY TRAINING BOARD, *Construction Safety*, (Two volume looseleaf manual, updated annually by issue of new and revised pages), BAS Management Services.

Appendix B
Recommendation of advisory committee on construction safety

Action precipitated at industry level is extremely important in promoting improvements in health and safety standards. While this was formally acknowledged by the Committee on Safety and Health at Work chaired by Lord Robens, 1970–1972, the then Factory Inspectorate with the sponsorship of the Department of Employment had established a number of advisory bodies in this vein. Under the chairmanship of a senior Factory Inspector nominated representatives from the employers and employees sides of an industry with the assistance of expert advisors undertook general and specific reviews of their respective fields.

One such body was the Joint Advisory Committee (J.A.C.) on Safety and Health in the Construction Industries. This has now been superseded by CONIAC, the Construction Industry Advisory Committee.

The procedure of the J.A.C. was to work through subcommittees. These were provided with a particular brief on which a report usually with recommendations would be made. The Reports, so prepared, were particularly relevant as representative of the Industry's view.

Reference has been made in this book to three Reports by the J.A.C. relating to Safety in Steel Erection (Chapter 1.5.1); Scaffolding (Chapter 1.5.2.5); Demolition (Chapter 1.5.4). The main recommendations of these Reports follow. After each set of recommendations notes are given indicating where positive action has been taken. No comment does not mean a total lack of activity but that any action taken forms part of some general requirements not limited to the particular subject matter under consideration.

B.1. Safety in steel erection (1979)
The main reommendations were:

1. A campaign should be mounted jointly by the Health and Safety Executive, employers' organizations and unions to influence attitudes and improve safety practice in steel erection work by publicizing the high accident levels in the industry.
2. The steel erection contractor should provide, for site use, a clearly expressed and comprehensive description of his proposed erection procedure together with details of temporary works and methods of ensuring stability at all times.

3. Contractors should appoint suitably qualified and experienced persons to act as co-ordinators of all steel erection projects.
4. Bonus schemes should be made dependent on the performance of planned safe working methods.
5. Additional legal measures should be introduced to cover the special problems created by the stacking and storing of materials on construction sites.
6. Consideration should be given by the Health and Safety Executive to the setting up of a mandatory registration scheme by which the competency of crane drivers could be established and appropriate certificates issued.
7. Jacking systems and pulling devices should be covered in the same manner as lifting appliances by an addition to the Construction (Lifting Operations) Regulations 1961.
8. A medical examination for rigger/erectors (including trainees) should be mandatory to demonstrate their fitness to work at heights and in arduous conditions. These examinations should be carried out before commencement of work in that capacity and thereafter at regular intervals. Failure to meet the standard should prohibit the employee from working at heights.
9. The Health and Safety Executive should give consideration to publishing a Guidance Note, covering safety practice in steel erection, based on the BS Code of Practice Safety in Erecting Structural Frames (BS 5531:1978).
10. An improvement is desirable in Regulation 38(3) of the Construction (Working Places) Regulations 1966, which should be clarified to secure, where appropriate, the supply and use of safety belts and to remedy the present anomaly.
11. The steel erection industry, in co-operation with the footwear industry, should initiate research leading to production of suitable boots for the use of steel erectors.
12. Safety helmets should be worn by everyone on a steel erection site.
13. Safety officers who at any time are concerned with steel erection should be given specialized training to deal with the particular problems of this work.
14. Consideration should be given by the Health and Safety Executive to employing more inspectors with responsibilities relating to the construction industry.
15. A safety training film on steel erection should be produced by the Engineering Industry Training Board to remedy the absence of this valuable form of training aid.

Notes

Recommendation 9—a series of four Guidance Notes are in the course of preparation by the Health and Safety Executive. These will deal with initial planning and design, site management and procedures, working places and access, training and legal requirements. The series will therefore cover material outlined in other recommendations.

Recommendation 12—is being actively pursued as part of a wider campaign promoting the wearing of safety helmets.

462 Recommendation of advisory committee on construction safety

Strengthening of controls in other areas covered by the Report is being achieved through the implementation of the general responsibilities and enforcement powers contained in the Health and Safety at Work etc. Act 1974.

B.2. Safety in Scaffolding (1974)
The main recommendations were:

1. There should ultimately be a compulsory certificate of competence for all who erect, substantially alter or dismantle scaffolds, whether they be full-time or part-time. Scaffolders who erect, substantially alter or dismantle scaffolds less than five metres high should be exempted. Existing training centres at Bircham Newton and Mitcham should be examples for further training centres and standards of competency of each grade of scaffolder should be agreed. Testing centres should be established to maintain common standards. In the meantime, scaffolders should carry a card to show the record of training and experience which they have acquired and this should be counter-signed by employers at each successive stage. People who carry out thorough examinations should be suitably qualified and trained to standards required by the complexity of the scaffolds they examine. They should attend approved courses at training centres and reach approved standards.
2. Hirers and owners of scaffolding should be urged to use the model forms of contract documents among other things to alert both of the parties as to their responsibilities for safety of the scaffold. The contract documents should give ample description of the purpose and detail of the scaffold including a section on safety. Notices giving the limitations with respect to, for example, height, loadings or cladding, should be prominently displayed on all but the very smallest scaffolds so that those who erect, alter, use and dismantle the scaffold are made aware of those limitations.
2. It is recommended that there should be a new legal requirement in Regulation 22 of the Construction (Working Places) Regulations 1966 that before use after erection or substantial alteration, and every seven days thereafter there should be a thorough examination of scaffolds, the results of which should be recurded in F91 Part 1 Section A as at present. A checklist should be included in any future Schedule to Regulation 22 of the Construction (Working Places) Regulations 1966 but in the meantime a checklist should be included in the notes in F91 Part 1 Section A to give guidance on important points in the procedure for thorough examination. It was agreed that it was desirable to inspect scaffolds daily, but because of the considerable administrative difficulties it is recommended that this should only be included in a code of practice and not be made a legal requirement.
4. Designers of new buildings and structures should be encouraged to incorporate suitable means of erecting and tying scaffolding which will be necessary for maintenance work during the life of the building or structure.
5. Research, in addition to that which is carried out within companies for their own information, should be encouraged. The current research at

Oxford commissioned by the Science Research Council and by the Construction Industry Research and Information Association should yield valuable information both for the designer and the user of scaffolds.

6. Generally, advisory literature on scaffolding was sufficient for the needs of the industry although guidance should be produced on methods of dismantling scaffolds. However, those who produce literature are strongly urged to ensure that it reaches the audience at which it is aimed.
7. Wider use should be made of the mobile training units provided by the National Federation of Building Trades Employers and the Federation of Civil Engineering Contractors with the co-operation of the Construction Industry Training Board and Unions. The example of in-company safety training schemes operated by some member companies of the National Association of Scaffolding Contractors should be copied by other firms. Where it is not possible for employers to carry out a safety training scheme within their own organization, they should make greater use of the training courses at scaffolding safety training centres which are at present available.
8. Current statistics are not sufficiently meaningful to identify or solve the real problems of providing safety for scaffolders. We recommend that a comprehensive survey of all accidents to scaffolders should be carried out over a period of one year by the Factory Inspectorate and that a short code of practice on the erection, alteration and dismantling of scaffolding for the benefit of scaffolders should be devised.
9. The existing provisions of Section 81 of the Factories Act 1961 with respect to compulsory reporting of Dangerous Occurrences should be extended to include scaffold collapses whether injury is incurred or not.

Notes

Recommendation 1—On the 1st January 1979 the Construction Industry Scaffolders Record Scheme came into effect. This forms part of the Working Rules for the Construction Industry as promulgated by the National Joint Council for the Building Industry and the Civil Engineering Construction Conciliation Board and is administered by the Construction Industry Training Board. The scheme categorizes scaffolders as trainee, basic or advanced dependent upon time service and attendance on approved training courses. Individual record cards are issued signifying the holder's category.

The Scheme has no formal legal status and this has no doubt contributed to the Scheme not being uniformly adopted or enforced throughout the Industry.

Recommendation 3—An amended F91 Part, Section A, Record of Inspections of Scaffolding has been introduced. This embodies a checklist of points to be covered when inspecting a scaffold.

Recommendation 9—has been incorporated in the Notification of Accident and Dangerous Occurrences Regulations 1980. The exact requirement relates only to scaffolds more than 12 metres high which results in a substantial part falling or overturning.

These Regulations are due to be superseded owing to the effect of changes in the wider sphere of accident reporting for Social Security purposes. It is not envisaged that this will alter the requirement for the reporting of scaffold collapses.

Control in other areas covered by the Report is being achieved through the general requirements of the Health and Safety at Work etc. Act 1974. Also the National Association of Scaffolding Contractors has produced model forms of contract documentation and a form of handing over certificate. Unfortunately the latter, which contains information relating to imposed loadings on a scaffold and other safety related material is not widely used.

B.3. Safety in Demolition (1979

The main recommendations were:
1. Information on accidents in demolition work should be improved.
2. There is a need to present the law relating to demolition, and guidance on its meaning, in a single publication.
3. A survey of the physical characteristics of the buildings and plant to be demolished should be mandatory and a written record of the survey should be kept available for inspection.
4. A duty should be laid on the client when inviting tenders for a contract to supply the contractor with all the information available to the client about potential hazards involving the property to be demolished or dismantled.
5. A contractor should be under a duty not to start work until he has a certificate from the gas and electricity authorities that all known supplies have been cut off from the site except those required for the execution of the work.
6. The Department of the Environment should issue advice to local authorities having control over space around demolition sites, urging them to separate the public from the demolition work to the maximum possible extent.
7. The present statutory duty to notify the commencement of demolition work should be reviewed with the object of reducing the present six-week limit to three weeks. Extra staff to inspect a proportion of these smaller sites should be made available by the Factory Inspectorate.
8. A formal training scheme on a regional basis should be established for the demolition industry. Priority should be given to training those who are in immediate control of demolition work, and there should be, amongst other things, a standardized training programme and a record of training.
9. The requirements relating to supervision and competency in Regulations 39 and 41 of the Construction (General Provisions) Regulations 1961 should be strengthened by requiring a higher standard of supervision of the demolition of structures, particularly where there is a risk of a collapse that might endanger a person; and by requiring that those exercising supervision shall have received training in the correct techniques of demolition, the prevention of accidents and the precautions against risks to health.

10. Greater thought should be given to the planning of each phase of demolition work, and employers when choosing a method of work at the planning stage should keep to the order of preference listed in the report.
11. Safety helmets should be worn by everyone on a demolition site.
12. The person responsible for carrying out the statutory weekly inspection of a lifting appliance used for balling should be trained to identify defects resulting from balling.
13. The attention of the Health and Safety Commission (HSC) is drawn to the committee's views on the licensing of plant drivers and operators. It is suggested that the Commission might wish to consider initiating a study of this problem.
14. When 'hot work' is to be undertaken in the demolition or dismantling of plant containing explosive or flammable substances the demolition contractor should invariably institute a written permit-to-work system. Consideration should be given to whether the law governing this type of process should be strengthened.
15. Closer control over the competency of men responsible for the storage and use of explosives in demolition work should be exercised by means of certificates of competence associated with defined standards of training.
16. A suitable procedure should be instituted so that new buildings which are either:
    (a) post-tensioned structures, or
    (b) unbonded stressed structures, or
    (c) structures which are progressively stressed as construction proceeds and further dead load added,
    could be readily identified. The plans deposited with local authorities of such buildings should be used to provide information in connection with the appraisal of the building's structural characteristics prior to demolition.
17. In any revision of the law relating to the inspection of scaffolding used in demolition work suitably stringent inspection requirements should be applied.
18. In dealing with the risks arising from the presence of asbestos dust on demolition sites the committee recommends that the Advisory Committee on Asbestos should consider the following proposals:
    (a) that before work starts on a site the sample analysis of 'lagging' asbestos should be obligatory;
    (b) that stronger measures should be taken to ensure that there is strict control of the removal of asbestos in demolition; one possibility which should be considered is that of changing the present requirements in Regulation 6 of the Asbestos Regulations 1969, about notification of processes involving crocidolite, by extending their application to the removal of lagging containing any form of asbestos;
    (c) that the possibility of some form of licensing of contractors who undertake work involving the extensive removal of asbestos should be considered;
    (d) that research should be initiated into the risks produced by, and the control of, asbestos dust;

(e) that a long-term medical survey of sufficient numbers of demolition workers to establish the incidence of asbestosis or other medical conditions caused by asbestos should be undertaken.
19. A 'check-list' should be prepared for supervisors giving brief guidance on the actions and precautions to be taken when asbestos is found to be present on a site.
20. Men whose work exposes them to lead fumes should be medically supervised and a man offering himself for employment should only be accepted after undergoing a pre-employment medical assessment in relation to lead.
21. An advisory publication for the demolition industry on health hazards in demolition work should be prepared.
22. Private clients should follow the lead given by the Government by letting contracts only to firms who are members of the Demolition and Dismantling Industry Register.

Several other recommendations are made in the report concerning, for example,
(a) the factors to be taken into account in the preparation of legislation for demolition work.
(b) the use of 'aerial' platforms.
(c) anchorage points for safety belts and harnesses.
(d) lifting appliances used for balling operations.

Notes

Recommendation 2—a series of four Guidance Notes are being prepared by the Health and Safety Executive covering planning, the provision of information, legislation, technical aspects and health hazards of demolition work. In addition BS 6187:1982, Code of Practice for Demolition provides much useful and practical information. The Code also includes a synopsis of legislation appertaining to this type of work.

Recommendation 11—is being actively pursued in the wider context of the wearing of safety helmets on construction sites.

Recommendation 18—in line with the general strengthening of the legislative controls dealing with asbestos demolition processes are covered by the following:

Approved Code of Practice and Guidance Note—Work with asbestos insulation and asbestos coating.
Asbestos (Licensing) Regulations 1983

Recommendation 20—is covered by the Control of Head at Work Regulations 1980 and Approved Code of Practice.

Recommendation 21—is to be incorporated in the series of Guidance Notes mentioned in Recommendation 2.

Recommendation 22—Government has since the Report withdrawn its support for using only Contractors on the Demolition and Dismantling Industry Register.

Index

Access platforms, mobile, 73–74
Accident,
  ILO classification, 31
  meaning, 5, 30
  prevention in tunnelling, 106–108
  reports, 172
  UK classification, 32
Accommodation,
  for clothing, 303–304
  temporary, 305, 426
  use of LPG in, 394
Acetic acid, 225
Acetone, 182, 186, 208
Acetylene, 199, 383, 396–397
Acids, 211–214
Acoustic intensity, 239
Acoustic trauma, 248
Acrylic polymers, 223–224
Adeno-carcinoma, 160
Adhesives, 54, 187, 206
Adobe, 16
Aerial platforms, 466
Agriculture (Poisonous Substances) Act, 1952, 188
Air Compressors, 236
Air sampling and analysis, 312–321
  asbestos, 317
  chemical hazards, airborne, 312–321
    diffusion method, 318
    official methods, 313–314
    physical methods, 318
  flammable gases, 318
  lead, 314–317
Alcoholism, 178, 284, 309
Alcohols, 202, 208
Alkalis, 211, 214
Allergens, 173, 183–184
Allergy, 176, 183–184
Aluminium, 216
  Powder, fire hazard, 415–416

Alveoli, 177
American Conference of Governmental Industrial Hygienists, 168, 188
American Industrial Hygiene Association, 168
Ammonia, 182, 199
Ammononiacal copper carbonate, 216
Amyl acetate, 182
Amyl nitrite, 201
Analysis,
  Air, 187
  Blood, 187
  Urine, 187
Anchorage points for safety belts and harnesses, 367, 466
Angiosarcoma, 181
Animal hazards, 284
Ankylostomiasis, 160, 285–286
Anthrax, 287
Architects, 11, 25, 60, 76, 134
Arsenic, 179
Arson, 402
Asbestos, 177, 228–229, 441
  air monitoring, 317–318
  bulk sampling and analysis, 317–318
  diseases,
    asbestosis, 40, 136, 179
    mesothelioma, 40
    mortality from, 40
    environmental sampling, 318
  dust, protection from, 347, 352
Asbestos (Licensing) Regulations, 1983, 466
Asbestos Regulations, 1969, 465
Asphalters, work hazards, 184
Asphyxiants, 149, 198, 394
Asphyxiation, 103, 172, 387
Asthma, 149, 183, 217, 227
Atmospheric monotoring, 107
  physical methods, 318
  reagent tubes, 320

468　Index

Audiograms, 246, 259
Audiometry, 246, 323, 326
Audio visual aids, 433, 436

Babel, tower of, 12
Bacteria, 287
Bar benders, 26
Barium peroxide, 225
Barrier creams, 233, 303, 358
　ultraviolet radiation, 180
Basements, 92
Beat conditions, 299–300
　beat elbow, 39, 40, 300
　beat hand, 39, 40, 267, 299
　beat knee, 39, 40, 299
Beech, 227, 289
Bel, 239
Benzene, 177, 181–182, 206, 208, 210
Benzoyl peroxide, 223
Beryllium, 188, 216, 217, 233
Biological hazards, 284, 327
Biphenyls, chlorinated, 220
Bircham Newton Training Centre for the Construction Industry, 435, 454, 456
Bisphenol A, 222
Bitumen,
　boilers and cauldrons, 395
　fire risks, 403
Blood, 180, 181
Blowlamps, 386, 404
Body protection, 359–364
Bone marrow, 182, 208
Bosun's chairs, 117
Brazil, 14, 16, 48, 458
Brazing, 187, 216, 386
Breathing apparatus, 346, 348–351
　compressed air line, 349
　fresh air hose, 348
　self-contained, 351
Bricklayers, 43, 133, 167
　employment statistics, 22, 23
　hazards of work, 43, 44, 136, 137, 161, 295
British Constructional Steelwork Association, 454
British Safety Council, 457
British Standards Institution, 453–454
Bronchitis, 5, 103, 183
Brooklands School of Management, 434, 437, 455, 456
Brucellosis, 287
Building, 10
　high rise, 11, 12
　houses,
　　industrialised countries, 13
　　primitive rural, 16
　　shanty town, 13
　maintenance, 13, 14
　meaning, 10, 11
　prefabricated, 12
Building Advisory Service (BAS), 434, 436

Building, Civil Engineering and Public Works Committee of the ILO, 445, 448
Building Employers Federation (BEF), 434, 437, 456
Building Research Establishment (BRE), 254
Burba, 299
Buisitis, 103, 142, 299
Bursters, chemical, 93
Butane, 198, 201, 383–385
Butanol, 208
Butanone, 209
Byssinosis, 39, 149, 161–162

Cadmium, 217, 233
Caisson, 93
　disease, 101
　work, 101
Calcium,
　carbide, 416
　chloride, 215
　formate, 220
　hydroxide, 214
Candidiasis, 288
Canteens, 302, 425
Capsulitis, 295
Carbon dioxide, 198, 199, 212
Carbon monoxide, 50, 181, 199–200
　monitoring, 318, 394
Carbon tetrachloride, 209, 210
Cardio-vascular system, 136–137, 181, 182–183
Careers officers, 166
Carpenters, 43, 133
　employment statistics, 22, 23
　hazards of work, 44, 45, 184, 295
Cartridge-operated fixing tools, 54, 236, 399, 413–414
Casella London Ltd, 225, 313
Catalysts, 223
Cellulitis, 295, 299
Cement, 183, 187
　dust, 183
　workers, 135
Census, 1971, 20, 128
Chadwick, Edwin, 101–102
Chemicals, 186
　disposal, 230–231
　explosive, 415–416
　protective clothing, 231–232, 363–364
　reactive, 415–416
　storage, 229–230
　working with, 230
Chemical Works Regulations (1922), 313
Children, 402–403
Chimneys, 91, 117, 405
Chlorine, 179, 182, 200
Chlorobenzene, 210
Chloroform, 210
Chromium, 135, 149, 233
　compounds, 53, 179, 188
　oxide, 217

Index 469

Circular saws, 42, 236
Civil engineering, 17
Civil Engineering Construction Conciliation Board, 463
Cladders, 43
  hazards of work, 54
Clandestine work, 423
Clerk of Works, 25
Clothes drying, 304, 403
Coal tar pitch, 182–183 220–221, 235
Cobalt, 135, 233
Codes of Practice, 446, 451
Collagen, 227
Comitè International de Prèvention des Risques Professionals du Batiment et des Travaux Publics, 449–450
Committee on Safety and Health at Work (Lord Robens), 460
Compressed air working, 108–110
  air locks, 108–109
  decompression sickness, 101, 137
  fire and explosion hazards, 414–415
  medical examination, 108
  medical locks, 109–110
  rules for decompression, 109
  welfare facilities, 110
  workers health, 108
Concreters, 25
  hazards of work, 53, 135, 136, 184
Concrete additives, 187, 218
Concrete breakers, 236, 256
Concreters,
  hazards of work, 53, 135, 136, 184
Confined spaces, 49, 92
Construction,
  building, 10
  civil engineering, 17
  companies, 422
  labour turnover, 424
  meaning, 9
  migrant labour, 421–422
  offshore installations, 18, 110
  output, 14
  power plant, 17
  process plant, 17
  seasonal employment, 421
  Standard Industrial Classification, 10
  subcontracting, 422–423
  temporary nature of work, 420–421
Construction camps, 301–302
Construction (General Provisions) Regulations, 1961, 451, 464
Construction (Health and Welfare) Regulations, 1966, 411
Construction Industry Advisory Committee (CCNIAC), 460
Construction Industry Research and Information Association (CIRIA), 187, 229, 252, 455, 463
Construction Industry Scaffolders Record Scheme, 463

Construction Industry Training Board (CITB), 28, 66, 433, 437, 454, 456
Construction (Lifting Operations) Regulations, 1961, 461
Construction Safety Association of Ontario, 458
Construction Site Safety Course, 438
Construction (Working Places) Regulations, 1966, 117, 461, 462
Control Limits (CL), 189
Control of Lead at Work Regulations, 1980, 466
Control of Pollution Act, 1974, 87, 236, 252, 412, 435
Control of Pollution (Special Wastes) Regulations, 1980, 87
Copper, 186
Corneal distrophy, 182
Cranes, 52, 100
Crane drivers, operators, 22, 134, 167, 306, 432
Crawling boards, 76
Creosols, 221, 269
Creosote, 184
Critical wind speed alarm system, 330
Cross sensitisation, 184
Cumene peroxide, 223
Cyclohexane, 208
Cyclohexanol, 182, 203, 209

dB(A) scale, 244
DDT, 220
Deafness, *see also* Noise,
  age induced, 247
  conductive, 247
  cortical, 247
  nerve, 247
  occupational, 142, 162, 236
Decibels, 239
Decompression sickness, 101, 103
Decorators, 43, 133, 167
  hazards of work, 49
Demolition, 18, 19, 83–93, 432, 464–466
  ball, 88–89
  chemical works, 92
  current trends, 92–93
  enclosed spaces, 92
  explosives, 90–91, 411–413
  fatal accident-analysis, 85
    incidence rates, 84
  gas works, 92
  hazards, 83–84, 184
  Joint Advisory Council on, 84, 464–466
  knocking down, 89–90
  oil refineries, 92
  pre-contract survey, 86
  pre-tensioned structures, 92
  post-tensioned structures, 92
  pulling down, 90
  pushing down, 90
  safe system, 85–93

Demolition (*cont.*)
  steel structures, 91
  storage tanks, 92
  supervisors,
    competence, 88
    inspections, 88
  workers, 167, 184
    competence, 88
Demolition and Dismantling Industry
  Register, 466
Department of Health and Social Security,
  145, 148, 162
Dermal route, 179–180
Dermatitis, 38, 40, 135, 173, 179, 183, 208, 223
Dermatologist, 165, 173–174
1.4 Dichloro-benzene, 220
1.2 Dichloro-ethane, 209, 210
1.1 Dichloro-ethylene, 210
Dieldrin, 220
Diet, 137, 302
Diethylene triamine, 222
Diglycidyl ether, 222
1.4 Dioxan, 203, 209, 210
Diphenyl methane di-isocyanate, 225
Disablement resettlement officers, 166
Diseases,
  notifiable, 150–159
  occupational, 39–40, 146–162, 299–300
  prescribed, 39, 40, 150–159, 209, 287,
    299–300
Disinfectants, 184
Dismantling, 18
Divers, 113–115
  certificate of fitness, 113, 115
  log book, 113
  medical examinations, 115
  qualifications, 115
  training, 113
Diving, 110–117
  accident, fatal, 116
  bell, 113
  contractors, 112
  equipment, 115–116
  health hazards, 137
  rules, 114
  supervisors, 113
  team, 113–114
Diving Operations at Work Regulations,
  1981, 111–112, 167
Donkey jacket, 361
Drains, 87, 93
Drivers, 306, 432
Drug addiction, 309
Dusts, 216–218, 225–229
  cement, 183
  metal pigment, 187, 216–218
  mineral, 228
  nuisance, 227
  respirable, 225
  semi-nuisance, 228
  wood, 45, 184

Dyestuffs, 184

Ears,
  hazards to, 137, 183
  muffs, 258, 355
  noise helmets, 355
  plugs, 258, 354–355
  protection, 354–355
  structure of, 242–244
Ecyema, 179, 183
Electric(al),
  accidents, 280–281
  burns, 50, 280
  cables,
    locating devices, 96, 277
    overhead, 83, 377
    underground, 95–97
  faults, 404
  fires, 404
  shock, 50, 277–278
    protection from, 281
    treatment for, 279
  tools, portable, 281
  welding, 281–282
Electrical, Electronic, Telecommunication
  and Plumbing Union (EETPU), 438
Electricians, 432
  employment statistics, 20, 22, 23
  hazards of work, 51–52
  standard mortality ratios, 132
Electricity, 276–284
  effect of voltage, 278
Electromagnetic radiation, 236, 267–276
  infra-red, 269–270
  ionising, 274–276
  lasers, 270–274
  ultra-violet, 268–269
  visible light, 269
Empire State Building, 11, 12
Employment Injury Benefits Convention,
  1964, 144
Employment Medical Advisory Service
  (EMAS), 160, 169, 453
Employment, seasonal, 4, 420–421
Enclosed spaces, entry, 92
Engineering Employers Federation, 434, 454
Engineering Industry Training Board, 454, 450
Epichlorlydin, 222
Epidemiologist, 128, 165, 173
Epoxy resins, 222
Ergonomics, 295
Ergonomist, 165, 174
Escape (fire), 408
Esters, 202, 209
Ethanol, 208, 209
Ethanolamine, 210, 219
Ethers, 202, 209
Ethyl,
  acetate, 182
  butyrate, 209
  silicate, 219

Ethylene glycol, 209
Ethylene glycol mono-ethyl ether, 203
European Economic Community (EEC), 188, 445, 459
Excavations, 93–101
  access to, 99
  biological hazards, 100
  contaminated ground, 100
  explosives, use in, 91, 411–413
  fencing, 98
  flammable atmosphere, 99
  flooding, 97
  hazards of work in, 94, 100, 429
  inspections, 100–101
  side protection and support, 97–98, 429
  site investigation, 95
  spoil, 98–99
  toxic atmosphere in, 99
  underground services, 95, 96
  unidentified objects, 99
Excavators, 52, 100
Explosion (hazards), 186
Explosives, 90–91, 411–413
  engineers, 26, 90
Eyes, 138, 172, 180, 182, 209, 271
  protection of, 233–234, 341–346, 414
Eye shields, 274

Face protection, 352–353
Face screens (shields), 341, 346
Factories Act, 1961, 172, 188, 233, 411, 435, 463
Fall protection equipment, 364–367
  anchorages, 367
  arrest devices, 367,
  safety belts and harnesses, 365–366
  safety lanyards, 365–366
  safety nets, 366
Falsework, 62, 74–75
  Bragg report, 46, 74
  meaning, 62
Farmers Lung, 164, 287
Favelas, 14–16
Federation of Civil Engineering Contractors, 437, 463
Films (safety), 433, 436
Fire, 381–416
  causes (hazards), 186–401
  detection, 408
  escape, 408–409
  extinguishers, 393–395–409
  -fighting, 409–411
  LPG hazards, 387
  organisation, 407–408
  rescue, 409
  triangle, 400
  warning, 408
Fire Offices Committee, 399
First aid, 171, 172
  boxes, 172
  personnel, 171–172
  room, 167

Flammable gases, 99
Flammable gas monitors, 318
Flammable liquids, fire risk, 405–407
Flammability limits, 385, 396, 405
Flame speed, 396
Flash point, 406
Fleas, 179, 285, 286
Fluorine, 149
Fluxes, fluoride containing, 210
Food, 425
Foot protection, 358–359
Fordham Cooper, W, 278, 279
Formaldehyde, 179, 182, 200
Farm workers, 43
  hazards of work, 44, 184
Foundations, 90, 93
Frequency rate, 32
  fatal,
    by occupation, 38
    comparisons, 34–36
  meaning, 32–33
Frost, 294
Fumes metal, 187, 216–218
Fungi, 287
Fungicides, 187
Furnace bricklayers, 44, 142
Furniture, Timber and Allied Trades Union (FTAT), 455

Gamma rays, 275
Gas cutters, 43, 133, 187, 345, 386
  hazards of work, 50
Gas torches, 394–396
Gases, 189–202
  asphyxiant, 198–199
  flammable, 99
  toxic, 99
General and Municipal Workers Union (GMWU), 455
Genitals, 183
Glanders, 287
Glaziers, 23, 43
  hazards of work, 54
Gloves, 211, 356–358
Glues, 187
Gonioma Kamossi (African boxwood), 289
Goggles, 274, 345–346
Graunt, 2, 3
Ground,
  types, 95
  workers and work hazards, 93
Grouts and grouting compounds, 187, 206
Guard rails, 66, 76, 79

Haemoglobin, 181
Hail, 294
Hammers, 236
Hand,
  inflammation of tendons, 39, 40
  protection, 355–358
    barrier creams, 358
    gloves, 356–358

Hard hats; *see* Safety helmets, 335, 353
Harnesses, safety, 366
  types, 366
Hay fever, 183
Hazard, meaning, 5, 31
Hazards, health, 121–367
  biological, 284–289
  chemical, 186–235
  physical, 236–281
  respiratory, 176–179
  skin, 179–180
  weather, 289–294
Head,
  protection, 353–354
  vibration, 263
Health and Safety at Work Act, 1974, 171, 188, 337, 411, 432, 435, 462, 464
Health and Safety Commission, 453
Health and Safety Executive, 437, 450, 451, 453, 460, 461
  Construction, Health and Safety, 123–124
Health and Safety Exposure Limit, 264
Health and Safety First Aid Regulations, 1981, 172
Hearing Loss; *see* Noise, 247
Heat stroke, 5, 290
Heaters, 403
Heating, ventilating and air conditioning engineers, hazards of work, 49, 50
Helmets, 354
  acoustic, 354
  airstream, 354
  shot blast, 354
  welders, 354
Hepatitis, viral, 177
Herbicides, 177
Hexamethylene di-isocyanate, 224
Hexane, 203, 208
High Risk Occupations, 57–119
Highly Flammable Liquids and Liquefied Petroleum Regulations, 383
Hod carriers, 300
Hookworm, 285
Housing, 426
Humidifier fever, 288
Huts, 394, 401, 426
Hydrocarbons, 207–208,
  chlorinated, 181, 202, 207, 209–210
Hydrochloric acid, 212
Hydrogen, 212, 383, 396–398
  chloride, 177, 209, 211,
  cyanide, 200–201, 212, 401
  flouride, 200, 211, 212
  peroxide, 213–214, 416
  sulphide, 103, 105, 179, 201, 212
  measurement in air, 318
Hypothermia, 294

Ignition,
  energy, 397
  temperature, 396, 405

Incidence rate, 32
  UK statistics, 37
Incinerators, 402
India, 63, 66, 74, 430, 431, 459
Industrial hygiene, 168
Industrial Safety (Protective Garment) Manufacturers Association, 341, 363, 457
Inhal route, 176–179, 187
Inorganic compounds (health hazards), 211–216
Insecticides, 179
Inspection, excavations, 100–101
Inspectors,
  building control, 25
  Health and Safety (Factory), 25, 166
Institute of Acoustics, 323
Institution of Civil Engineers, 17, 87
Institution of Occupational Safety and Health, 434, 437
Insulating foams, 187, 222
International Occupational Safety and Health,
  Hazard Alert System, 449
  Information Centre (CIS), 454
International Organisation for Standardisation (ISO), 341, 449
Ionising radiations, 268, 274
Ionising Radiations (Sealed Sources) Regulations, 1969, 167, 275
Iron, corrugated, 17
Iron oxide, 216
Irritants, 180, 187
Iso-amyl acetate, 209
Iso-butanol, 208
Iso-cyanates, 179, 221
Iso-propanol, 203, 208

Jiggers, 286
Joiners, 43, 133
  hazards of work, 44, 184, 295
Joint Advisory Committee on Safety and Health in the Construction Industry, 456, 460

Kata thermometer, 331
Ketones, 180, 181, 209, 219, 220
Kidney, 180, 181, 209, 219, 220
Klety on life expectancy, 41, 42
Knee pads, 300

Labour,
  migrant, 4, 420–422
  self employed, 4, 27
  turnover, 424
Labourers,
  employment statistics, 22
  hazards of work, 55, 137, 184, 295
Ladders, 59, 369–377
  erection, 374–375
  extension, 374–375

Ladders (*cont.*)
  inspection, 376
  long, plain rung, 374–375
  purchasing, 372
  records, 373
  roof, 76, 82, 375
  short plain rung, 374
  statutory requirements, 370–372
  step ladders, 375
  trestles, 375
Lances, thermol (oxygen), 93, 111
Laquers, 206, 210
Lasers, 268, 270–274
  gallium arsenide, 271
  hazards, 138
  helium-neon, 271
  recommendations, 271, 274
Latin America, 3, 13–16
Law Reform (Personal Injuries Act), 1968, 148
Lead, 178, 181, 182, 188, 217
  carbonate, 216
  control for outside workers, 316
  medical examination, 167
  monitoring in air, 314–317
  paint, 217
  tetraethyl, 216
Lestospirosis, 287
Lethal dose, 187
Leucaemia, 208
Lice, 179, 285
Lift engineers, 43
  hazards of work, 55
Lifting,
  appliances, 82
  manual, 297–298
Light, visible, 269
Lighting, 269
  artificial, 269
  levels, 269
Lime, 214
Liquefied Petroleum Gases (LPGs), 48, 201, 382–395
  burning equipment, 392–394
  cylinders, 388
  hazards, 387
  hoses, 392
  regulators, 392
  storage, 388–391
  torches, 394
  transport, 388
  unloading, 388
  use, 385–387, 391–395
Liver, 180, 181, 208, 209, 219, 220, 223, 288
Local authorities, 252, 258
London School of Hygiene and Tropical Medicine, 457

Magnesium fluorosilicate, 215
Manganese, 178, 182, 186
  oxide, 216–217

Mastic mixers, 395
Mattock men, 19
Medical examinations, 167, 168
  divers, 115
  lead, 167
  pre-employment, 167
  routine, 168
  workers in compressed air, 108
Medical Research Council, 225, 249
Mental ability, 307
Mercury, 178, 182, 218
Mesothelioma, 40
Metal (gas) cutting, 187, 216, 345, 386
Methane, 103
Methanol, 208
Methyl,
  ethyl ketone, 182, 203, 209
  methacrylate, 224
  styrene, 223
Methylene chloride, 202, 209
Micro organisms, 287
Migraine, 183
Migrant-labour, 421
Mineral oil, 184
  mist, 184, 220
Mines and Quarries Act, 1954, 216, 345, 386
Ming tombs, 2
Mobile work platform, 73, 92
Monday morning feeling, 186
Mortality, occupational, 126
Mosquitos, 16, 287
Motivation, employee, 309
Motor ability, 307
Mucous membrane, 177, 209

Raphthalene, chlorinated, 220
Raphthalene di-isocyanate, 224
Narcotic, 208
National Association of Scaffolding Contractors, 463, 464
National Federation of Demolition Contractors, 87
National Federation of Master Steeplejacks and Lightning Conductor Engineers, 117
National Institute for Occupational Safety and Health (NIOSH) (USA), 458
National Joint Council for the Building Industry, 455, 465
National Joint Utilities Group, 96
National Safety Council (USA), 458
Nervous system, 137, 180, 182, 209, 210, 223
Nettlerash, 183
Nickel, 135, 217
Night work, 295
Nitric,
  acid, 211, 213
  oxide, 201–202, 213
Nitrocellulose, 206, 208

Nitro
 ethane, 210
 glycerine, 149
 methane, 210
 paraffins, 202, 210, 416
 propane, 202, 210
Nitrogen, 108, 199
Noise; *see also* Sound, 236–259
 broad band, 245
 Code of Practice, 248–252, 323
 exposure limits, 248–252
 helmets, 355
 induced hearing loss, 160, 247–248,
 level monitoring, 257–258, 322–323
 levels on construction piles, 252–253
 meaning, 236
 permanent threshold shift, 249
 piling, 255
 power tools, 256–257
 reduced site vehicles, 52, 254–255
 stationary plant, 257
 survey, 244
 temporary threshold shift, 248
 titinitis, 247
 Woodworking Machines Regulations, 1974, 253
Nonanes, 208
Nose, 177, 209
Notification of Accidents and Dangerous Occurrences Regulations (1980), 463
Nurses, 171
Nystagmus, 182

Oak, 227, 289
Occupational diseases,
 byssinosis, 39, 161–162
 ILO list, 146–147
 notifiable, 150–159
 pneumoconiosis, 39, 161–162
 prescribed, 39, 40, 150–159, 209, 287, 299–300
Occupational Exposure Limits, 169–171, 188–197, 204–205, 311
Occupational health, 123, 124
 meaning, 127
 monitoring, 311–331
 sources of information, 175–176
Occupational health risks, 163–309
 animal, 284–285
 chemical, 186–229
 fungi, 288
 micro-organisms, 287–289
 parasitic, 285–287
 physical, 236–282
 poisonous plants, 289
 viruses, 288
 weather, 289–295
Occupational health physician, 165–168
Occupational hygienist, 165, 168–171
Occupational health nurses, 165, 172

Occupational Safety and Health Convention (noise) 1981, 455
Occupations, construction, numbers employed, 20–28
Octane, 208
Oil and Chemical Plant Contractors Association, 455
Optic neuritis, 182
Oral route, 180
Organ of balance, 263
Organic compounds, 202–211, 218–221
Organo tin compounds, 219
Ornithosis, 288
Osteo-arthritis, 263
Outriggers (for suspended scaffolds), 73
Overalls, 234
Oxygen, 186, 383, 397
 measurement in air, 318
Ozone, 179, 182, 201–202, 216

Paint, 187, 206, 210
 spraying, 187, 216, 232
Painters, 43, 133, 167, 184
 colic, 48, 217
 employment statistics, 22, 23
 hazards of work, 47–49, 135, 137, 139, 184, 210
Paracelsus, 186
Paraquat, 221
Parasitic hazards, 285
Pentachlorophenol, 219
Peroxides, 186, 209, 415
Personal dose meters, 275, 311, 320, 322, 325
Personal hearing protectors, 257, 354–355
Phenol, 218–219
 chlorinated, 219
Phosgene, 202
Phosphone, 199
Phosphoric acid, 211, 213
Photosensitisation, 184, 269
Physiotherapist, 299
Pile drivers, 236, 255
Pile driving operatives, 43
 hazards of work, 52, 255
Piling, 255
Pitch pine, 45, 289
Pits, 92, 164
Plants,
 mobile, 52, 100
 operators, work hazards, 43, 52–53
Plasterers, 25, 167
 hazards of work, 53, 135, 142
Plumbers, 43
 employment statistics, 22, 23
 hazards of work, 49, 138–139
Pneumoconiosis, 39, 44, 149, 161–162
Pneumonia, 103
Poisonous plants, 289
Pole belts, 365
Polyesters, 222–224
Polyurethanes, 206, 221, 224–225, 401

Power tools, 256
Presbycusis, 207
Pressurised workings, 414
Principal regulations applying to safety and health in construction, 451
Propane, 198, 201, 383–385
Propanol, 208
Protective clothing and equipment, 231, 333–367, 429
  body, 359–364
  chemical, 212, 213, 215, 219, 225, 230–231, 363–364
  ears, 258–259, 354–355
  eyes, 233–234, 341–346
  face, 352–353
  fall arrest devices, 364–367
  footwear, 234, 358–359
  foul weather, 361–362
  goggles, 345–346
  hand, 233, 355–358
  head, 353–354
  high visibility, 362–363
  legal requirements, 335–341
  overalls, 234
  philosophy, 333–335
  respiratory, 232–233, 346–352
  spectacles, 345
Protection of Eyes Regulations (1974), 233, 337
Psychologist, industrial, 165, 174–175
Psychosomatic syndrome, 137
Pyrethsums, 220
Pyridine, 219

Radioactive materials, 188
Rain, 294
Raynauds phenomenon, 259, 263
Reduced comfort boundary, 264
Relative age gradient (RAG), 128, 132
Relief valves, 385, 387, 388
Rescue (fire), 409
Resins, 187, 221, 225
  epoxy, 222
  styrene-polyester, 222
Respiratory protection, 232–233, 346–352
  breathing apparatus, 348–351
    airline, 211, 232, 348–351
    self-contained, 351
    respirators, 351–353
Respiratory system and hazards to, 136, 176–179
Rheumatism, 5
Riggers, 43, 61, 132
Roof,
  edge protection, 78–81
  ladders, 375
  maintenance, 75–78
  types, 75
Roof falls,
  edges, from, 76
  fragile, through, 76, 78

Roof falls (*cont.*)
  internal structures, from, 77
  lifting appliances, near, 82
  sloping, from, 82
  special shapes, from, 82
Roof work, 22, 75–83, 184
  access for, 83
  adverse weather, in, 83
  overhead cables, 83
Rotherol and Mitchell Ltd, 225, 313
Royal Society for the Prevention of Accidents (ROSPA), 457
Rubbers, 213, 225, 233
Rubbish, fire hazard, 402
Russell, Bertrand, 126, 132
Rust inhibitors, 188

Safe, meaning, 5
Safety,
  belts, 365–366
  boots (and shoes), 358–359
  cans, 407
  groups, 456
  harnesses, 365–366
  helmets, 61, 334–335, 353–354
  lanyards, 365–366
  nets, 365
  specialists (engineers, managers, officers), 15, 165, 437
Safety Representatives and Safety Committees Regulations, 1977, 437, 455
Salts, 214–216
Sand flies, 287
Sanitary conveniences, 303
Scables, 179, 285
Scaffolders, 66, 167, 300, 432
  hazards of work, 66–67
  CITB registration, 66
Scaffolding, 60–75, 462–464
  causes of failure, 70, 73
  ILO recommendations, 63–66
  Joint Advisory Council Report, 74, 462–464
  ladder, 63
  meaning, 62
  proprietary systems, 73
  terms, 69
  trestle, 62
  tubular metal, 66–72
  types, 68–69
  wood and bamboo, 12, 62, 429
Seasonal employment, 421
Sealants, 206
Sensitisers, 180
Sensitisation (cross and photo), 184
Shanty Towns, 3, 13
Shipbuilding and Ship Repair Regulations, 1960, 313
Shock waves, 242
Shopfitters, hazards of work, 46
Shot (and sand) blasting, 232

Index

Site agents, 25
Skin, 43
  care of, 172
  creams, 358
  hazards to, 135–136, 142, 179–180, 212
Slipped discs, 5, 296–299
Small size of construction firms, 24, 422
Smell, senses of, 201, 387
Smoking, 403
Snakes, 285
Snow, 294
Social Security, 144
Sodium,
  chlorate, 211, 215, 415
  hydroxide, 211, 214
  hypochlorite, 214–215
  phenate, 218, 219
  silicate, 215
Solvents, 48, 180, 182, 187, 202–211
Sound; *see also* Noise,
  decibel scale, 244–245
  filters, 244
  frequency, 238–239
  intensity, 239
  level meters, 244–245, 323–325
  personal dose meters, 325–326
  power level, 239, 244
  pressure, 239
  velocity, 237–238
Space heaters, 385
Spacial perception, 307
Spectacles, 274, 345
Spiders, 286, 287
Spine, 138, 263, 296–299
Standard mortality ratio (SMR), 128–134
Stationary plant, 257
Statistics, 30–42
  accident causation, 37
  accident frequency rate, 32, 33
    fatal, comparison by country, 34–36
    fatal, by occupation, 38
  accident incidence rate, 32
    UK totals, 37
  employment, 20–29
  occupational disease, 127–134
  relative age gradient (RAG), 128, 132
  standard mortality rate, 128–134
Stearic acid, 221
Steel erection, 59–61, 460–462
  falls of materials, 60
  falls of persons, 59
  fatal incidence rate, 37, 58
  instability of structure, 60
  Joint Advisory Council Report, 61, 460–462
Steel erectors, 167
  employment statistics, 22
  standard mortality ratios, 132
Steeplejacks, 26, 117–118, 167
  scaffolds, 75, 117
Stenching agents, 384

Step ladder, 375
Stomach,
  dietary habits, 137
Stone cleaning fluids, 187
Stonemasons, 43, 133, 161
  banker mason, 44
  fixer mason, 44
  hazards of work, 44, 136
Storekeepers, work hazards, 43, 55
Stretchers, special, 172
Stress,
  job classification, 306–307
  mental, 305–309
  selection tests, 307–308
  signs of, 306
  testing, 308
  training, 308
Stroboscopic effects, 269
Styrene, 182, 223
  monomer,
    fire risk, 415
  polyester resin, 222–224
Subcontractors, 4, 422–423
Sulphur dioxide, 202
Sulphamic acid, 213
Sulphuric acid, 213
Sunlight, 180
Suspension, 263
Sweden, 30, 34, 35, 353–354, 431, 440

Tanks, 92
Target organs, 176
Temperature, 291–292
  black bulb, 291
  dry bulb, 291
  wet bulb, 291
Temporary duration of work sites, 420–421
Tendering, contract, competitive, 4, 420, 427–429, 442
Tenosynoritis, 295, 300
Tetanus, 100, 288
Tetrachloro ethane, 209
Tetraethyl lead, 179, 216
Tetrahydrofuran, 209
Thatch, 16
Third world, 12, 13–18
Threshold Limit Values, 169, 188, 311
Threshold shift, 248
  permanent, 248
  temporary, 248
Threshold shock voltages, 279
Ticks, 286
Ties, window for scaffolds, 70, 72
Tilers,
  employment statistics, 22
  floor, 43
  hazards of work, 54, 137, 142, 184, 210
  wall, 43
Timbermen, 100
Time weighted average, 189
Tinnitus, 247

Toilet facilities, 303
Toluene, 179, 208
Toluene di-isocyanate, 183, 224
Top men (demolition), 19
Toxins, 180
Trade Unions, 166
Training, 431–439
  directors and principals, 433–434
  divers, 113
  operatives, 435–437
  planners and designers, 437
  safety officers and advisers, 427
  safety representatives, 427–428
  site managers and supervisors, 434–435
Transport (for workers), 426
Transport and General Workers Union (TGWU), 455
Trestles, 62, 375, 376
Tri-butyl tin compounds, 219
1.1.2-Trichloroethylene, 179, 202, 209, 210
Tricresyl phosphate, 182
Triethylamine tetramine, 222
Tse-tse fly, 287
Tuberculosis, 42, 287,
Tumours, 268
Tunnelworkers, hazards of work, 7, 103, 136, 164, 295
Tunnelling, 101–110
  accident prevention, 106–108
  explosives, 411–413
  sub-aqueous, 105–106
  types of, 102
Tunnels, 93, 103–105
Turpentine, 182, 208

Ultra-violet radiation, 180, 184, 268–269
Union of Construction Allied Trades and Technicians (UCATT), 437, 455
University of Aston in Birmingham, 457
Urea-formaldehyde foamed resins, 225

Vanadium, 217, 233
Varnishes, 187, 206
Ventilation, 107
Ventricular fibrilation, 278
Vibration, 83, 103, 149, 236, 259–267
  arm, 263–264, 327
  body, 139, 262–263, 327
  exposure limit, 264
  fatique decreased efficiency boundary, 264
  forced, 260
  free, 260
  frequency, natural, 260
  hand and arm, 263–264, 327
  monitoring, 326–327
  physical characteristics, 260

Vibration (*cont.*)
  reduced comfort boundary, 264
  resonance, 262
  standards, 264
  white finger, 259, 263
  whole body, 139, 262–263, 298, 327
Viruses, 287
Vinyl chloride (monomer), 181, 202

Washing facilities, 303
Water, drinking, 302–303
Waterproof clothing, 361
Waterproofing compounds, 187
Weather, 61, 83, 284, 289–295, 423
  forecasts, 294
  hazards, 328
  monitoring, 328–331
  protection, 295
Weedkillers, 187
Weils disease, 100, 164, 180, 287
Welders, work hazards, 43, 50–51, 133, 137, 142, 216, 345
Welding,
  hazards of, 50, 187, 216
  fire risks, 403
  rods, 212
  steel, 216
Welfare facilities, 300–305, 427
  accommodation, 305
  canteens, 302, 425
  clothing, 303–304
  construction camps, 301–302, 426
  drinking water, 302–303
  sanitary conveniences, 303
Welfare problems, 424–427
Wet bulb globe temperature (WBGT), 292, 331
Wind chill index, 292–293, 331
Wind measurement, 329
Window cleaning, 370
Windspeed, 293–294
Woodworking Machines Regulations, 1974, 46
Woodworking machinists, work hazards, 43, 46, 133
Work in Compressed Air Special Regulations, 1958, 167
World Health Organisation, 144, 442, 449
Writer's cramp, 149, 267

X-rays, 167, 275, 328
Xylene, 182, 203, 208

Zinc, 218
Zinc fluorosilicates, 215

COSHH SUPPLEMENT

The Control of Substances Hazardous to Health (COSHH) Regulations 1988 (S.I. 1988, No. 1657)

These regulations are issued with two Approved Codes of Practice (ACOPs), "Control of Substances Hazardous to Health' and "Control of Carcinogenic Substances'. Two further ACOPs are issued separately, one for the Control of Vinyl Chloride Monomer (VCM) and one on Fumigation. The regulations replace most previous UK legislation on personal protection against harmful substances which may be encountered at work.

The scope of COSHH includes prohibitions, identification of substances with health hazards, assessment of risk, control of exposure, monitoring exposure at the workplace, health surveillance, informing, instructing and training personnel. Construction processes especially affected by COSHH include cement mixing and handling, applying and removing materials containing asbestos and other mineral fibres, painting, grouting, welding, cutting and spraying, wood-working, demolition and other operations which produce dust.

The substances whose use or import are prohibited under COSHH are those formerly prohibited under other regulations.

Substances hazardous to health and their Identification

The regulations apply to all substances which are hazardous to health, and to dust clouds of all kinds. They include substances categorised as TOXIC, VERY TOXIC, HARMFUL, CORROSIVE or IRRITANT under the Classification, Packaging and Labelling of Dangerous Substances Regulations (CPLR) 1984. Chapter 2 of this book gives details of many substances with health hazards which are used in construction, how they affect the human body and how to protect against them. They include cement, asbestos and other mineral fibres, many paint solvents, metal fumes, pigment dusts and vapours arising from welding, brazing, cutting and painting.

Identification of substances with health hazards which are used in construction depends in the first place on information provided by their suppliers. They are required to provide full information on any health hazards associated with their products, at the point of delivery. Construction firms and other organisations using these products must maintain a register of all such health and safety data and ensure that this is retained at site level and properly explained to all persons liable to be exposed.

1. **Assessment of Risk** (see HSE Guidance Note EH 42)

An assessment by a competent person must be made before work which may involve exposure of any employee to a substance hazardous to health is started. It requires consideration of the toxicity or harmfulness of the substance (e.g. as indicated by its published exposure limits), the process in which it is used and the level of operator exposure. This will depend on whether it is used in confined spaces or in the open, on the degree of ventilation and on the protective measures in force.

2. **Control of Exposure** (see HSE's EH 40 and 42)

Measures needed to eliminate or reduce exposure should be identified from the assessment and available on site. They should only rely on personal protective equipment as a last resort, in emergencies (e.g. as caused by plant failure) and in routine maintenance of plant and machinery handling or storing harmful substances. Special attention is required to prevent or control exposure to carcinogens. Control measures include the following:

a) Substitution with a less hazardous material,

b) Totally enclosing the process,

c) Partially enclosing the process,

d) Using exhaust ventilation,

e) Provision of safe means of storage and disposal,

f) Limiting working time with the hazardous substance,

g) Provision of adequate facilities for washing and changing, and storing and laundering contaminated clothing,

f) Reducing the number of persons exposed and exclusion of all non-essential access,

g) Prohibition of eating, drinking and smoking "on the job",

h) The use of personal protective equipment.

The nature of construction work often restricts use of control measures such as (b) which would be the most effective in a manufacturing process.

Substances to which maximum exposure limits (MELs) have been assigned require the strictest safeguards. Those to which occupational exposure standards (OESs) are assigned have next priority. Employers are required to set their own working control standards for harmful substances to which neither MELs nor OESs have been assigned, on the basis of all relevant information.

Respiratory protective equipment (RPE) must be properly selected for its purpose, of an HSE-approved type or to a HSE-approved standard, and matched to the job and the wearer. Eye protection should comply with the Protection of Eyes Regulations 1974 and BS 2092.

The types of protection available for the eyes, respiratory system and skin and criteria for their selection are discussed in Chapter 2.8 of this book.

Both employers and employees should ensure that the control measures provided are properly used or applied. Procedures should be established for their regular inspection and maintenance. Records of examinations, tests and repairs should be kept for at least five years.

3. Monitoring Exposure at the Workplace

This like assessment, should be related to the risk and is specially needed for very harmful substances such as asbestos. Monitoring the working atmosphere is more difficult and the results more uncertain in the ever changing conditions of a construction project compared with those of a factory. The methods and strategy recommended are discussed in Chapter 2.7 of this book and in HSE's EH 42.

Both record keeping and effective procedures for alerting managers and supervisors to dangerous conditions uncovered by monitoring are essential.

4. Health Surveillance of Workers

Where appropriate, this should be made to provide early detection of ill effects resulting from exposure to dangerous substances. Health records should then be kept for each worker involved. The level of surveillance needed depends on the nature and degree of the risk. Substances and processes considered "appropriate" are listed in Schedule 6 of COSHH. Supervisors and workers should be instructed on the early signs and symptoms of occupational diseases associated with harmful substances (e.g. asbestos and cement) which they encounter at work, and these should be reported promptly to the employment medical adviser or appointed doctor who is responsible for health surveillance. This may lead to the suspension of employees from work in which their health is adversely affected. Health surveillance aims to protect the health of employees, to assist in evaluating control measures, to collect and use data for detecting and evaluating health hazards and to assess the immunological status of employees in work involving micro-organisms. Procedures include biological monitoring, clinical examinations, enquiries about symptoms, inspection by a responsible person, review of records and occupational history. Employees are entitled to see their health records.

5. Informing, Instructing and Training persons at risk from harmful substances

These activities are included in the COSHH regulations and dealt with in Chapter 4.2 of this book.

Note on the responsibilities of subcontractors under COSHH

Many operations in construction are carried out by subcontractors who have the same responsibilities under COSHH as the main contractor. These should be spelt out in contracts and works orders between the main contractor and his sub-contractors.

ASTON UNIVERSITY

LIBRARY & INFORMATION SERVICES

Aston Triangle
Birmingham
B4 7ET   Tel: 0121 359 3611
England   Fax: 0121 359 7358